COMPOSITES IN CONSTRUCTION: A REALITY

PROCEEDINGS OF THE INTERNATIONAL WORKSHOP

July 20–21, 2001
Capri, Italy

EDITED BY
Edoardo Cosenza
Gaetano Manfredi
Antonio Nanni

1801 ALEXANDER BELL DRIVE
RESTON, VIRGINIA 20191–4400

Abstract: In the last two decades, fiber-reinforced polymer (FRP) composites for use in construction have generated considerable worldwide interest and expectation. FRP composites are expected to significantly improve the performance and durability of new or deteriorated constructed facilities. FRP composites can be used as stand-alone structural members, as reinforcement for prestressed and non-prestressed concrete, or in combination with other structural materials for new construction or repair/rehabilitation. The interest in FRP composites is fueled by innovative manufacturing technologies as well as significantly more stable, stronger constituent materials. These proceedings contain a collection of the papers presented during the workshop and a summary paper that outlines the outcomes of its discussions and deliberations.

Library of Congress Cataloging-in-Publication Data

Composites in construction : a reality : proceedings of the international workshop, July 20-21, 2001, Capri, Italy / edited by Edoardo Cosenza, Gaetano Manfredi, Antonio Nanni.
p. cm.
Proceedings of a workshop organized by the University of Naples.
Includes bibliographical references and index.
ISBN 0-7844-0596-4
1. Composite materials—Congresses. 2. Composite construction—Congresses. 3. Fibrous composites—Congresses. I. Cosenza, Edoardo. II. Manfredi, Gaetano, Dr. III. Nanni, Antonio. IV. Università di Napoli.

TA418.9.C6 C6322 2001
620.1'18—dc21 2001053992

Library of Congress Catalog Card No: 2001053992
ISBN 0-7844-0596-4
Manufactured in the United States of America.

Preface

In the last two decades, fiber-reinforced polymer (FRP) composites for use in construction have generated considerable worldwide interest and expectation. FRP composites are expected to significantly improve the performance and durability of new or deteriorated constructed facilities. FRP composites can be used as stand-alone structural members, as reinforcement for prestressed and non-prestressed concrete, or in combination with other structural materials for new construction or repair/rehabilitation. The interest in FRP composites is fueled by innovative manufacturing technologies as well as significantly more stable, stronger constituent materials.

The research programs and some of the most relevant field applications undertaken on FRP composites for construction have been presented at international conferences such as those in the series titled "Fiber Reinforced Plastics for Reinforced Concrete Structures (FRPRCS)." As a follow-up to the latest of these events held at the University of Cambridge, 16–18 July 2001, it was felt that a specialty workshop could provide an opportunity to re-think the state-of-the-art and possibly set the course for future initiatives. In this spirit, the University of Naples organized and hosted the two-day gathering titled: "Composites in Construction: A Reality." The meeting was held at Villa Orlandi, the International Center for Scientific Culture of Naples University "Federico II" (Anacapri, Island of Capri). The agenda items were related to the discussion of R&D progress on a worldwide basis. Particular emphasis was devoted to aspects relative to design philosophy, design codes, and public safety, having in mind that FRP materials are now being used in most parts of the world.

These proceedings contain a collection of the papers presented during the workshop and a summary paper that outlines the outcomes of its discussions and deliberations.

As shown in this archival volume, the Organizing Committee feels that this was an exceptional experience that should be repeated at periodic intervals. The success of the event is credited to its participants and the reviewers of the technical papers. Without the coordination and efforts provided by the workshop secretariat and staff, this idea would have not materialized. The Organizing Committee is particularly indebted to Andrea Prota and Roberto Realfonzo.

On behalf of all workshop participants, this volume is dedicated to its readers in the hope that it will further their knowledge and understanding of composite materials.

E. Cosenza, G. Manfredi, A. Nanni, and M. Pecce
Organizing Committee
Capri, July 21, 2001

Acknowledgments

Organizing Committee

Edoardo Cosenza	University of Naples Federico II, Naples, Italy
Gaetano Manfredi	University of Naples Federico II, Naples, Italy
Antonio Nanni	University of Missouri-Rolla, Rolla, Missouri, USA
Marisa Pecce	University of Sannio, Benevento, Italy

Scientific Secretaries

Andrea Prota	University of Missouri-Rolla, Rolla, Missouri, USA
Roberto Realfonzo	University of Naples Federico II, Naples, Italy

Administrative Secretaries

Carmine Citro	University of Naples Federico II, Naples, Italy
Carmen Ippolito	University of Naples Federico II, Naples, Italy
Ravonda McGauley	University of Missouri-Rolla, Rolla, Missouri, USA
Elena Sole	University of Naples Federico II, Naples, Italy
Gayle Spitzmiller	University of Missouri-Rolla, Rolla, Missouri, USA

Staff

Francesca Ceroni	University of Naples Federico II, Naples, Italy
Giovanni Fabbrocino	University of Naples Federico II, Naples, Italy
Maurizio Guadagnini	University of Sheffield, Sheffield, UK
Iunio Iervolino	University of Naples Federico II, Naples, Italy
Maria Polese	University of Naples Federico II, Naples, Italy
Arnaldo Stella	University of Naples Federico II, Naples, Italy
Gerardo Verderame	University of Naples Federico II, Naples, Italy

Sponsoring Institutions

The Organizing Committee would like to thank the following Institutions for the received financial support:

Regional Government of Campania

University of Naples Federico II – Polo delle Scienze e delle Tecnologie

University of Missouri-Rolla – University Transportation Center

Contents

Keynote Lecture

Codes and Standards

Manufacturing, Economics and Construction

Materials, Durability and Characterization

Analysis and Design

Outcomes

Indexes

Keynote Lecture

Research Supports and Policies in the Campania Region

Luigi Nicolais[1]

Abstract

Recent economic and social changes have affected the level of collaboration between research and industry. The role of Institutions and their decisions can be crucial in stimulating and promoting such relationships. This paper summarizes the presentation that the author made on the main programs and goals that the Regional Government of Campania is trying to achieve for local Universities and Research Centres. Its relevance for the academic world is that a new protocol in the relationship between University and Industry is proposed, which could also become a model to be implemented and improved upon in other countries. The author believes that the achievement of the fixed objectives and the development of similar political programs are key factors for the research of the next years and for the diffusion of composites in construction.

Introduction

In the opening ceremony of the Workshop "Composites in Construction: A Reality", the author communicated a greeting from the President of the Regional Government of Campania, A. Bassolino, to the participants. It was then pointed out that in the present social economic dynamics, improved scientific knowledge and more qualified human resources represent strategic factors, which are extremely important for adequate industrial and social growth. An efficient University-Industry relationship is a key on which to build a new *brain society* (i.e., knowledge-based society).

There are different ways to improve the University-Industry relationship. The three principal possibilities are:

- by education, both with institutional teaching and continuing education programs;
- by research, devoting time and energy to topics that may have industrial applications or that may provide useful industrial applications of present technology.

---

[1] Regional Minister of University, Research, Innovation and New Economy, Regional Government of Campania, via S.Lucia 81 - 80132 Naples, Italy, Phone +39 081 7962399 – 7962566, Fax +39 081 7962425 - 7962273, e-mail ass.nicolais@regione.campania.it

- by stronger collaborations, in order to increase industrial innovation and to develop tools that can help managers in the creation and growth of new highly technological companies.

The University, in which there is a significant amount of potentially useful technology and "know-how" that is not commercially utilised, can have a propulsive role in the development of small companies with high added value.

A structured relationship between these two worlds needs to be built that fosters a coordinated and continuous territorial growth.

According to the Organization for Security and Co-operation in Europe (OCSE) analysis, total expenditure for R&D in Italy is only 1.05% of the Gross Domestic Product (GDP), which is far from OCSE average value, but close to the Public expenditure (Figure 1).

98% of our productive system is constituted by Small and Medium Enterprises (SMEs), which usually do not invest enough in R&D.

The aims of European, national and regional programs are to integrate policies, support and investment activities regarding knowledge and technological transfer in order to allow Universities and Research Centres to contribute to local economic and social development.

Also, for the South of Italy, a policy focalised on research will foster the growth competitiveness of SMEs. Research is a key component in achieving these objectives and increasing employment.

The Campania model

To reach these goals, the Regional Government of Campania has instituted a Regional Minister for University, Research, Innovation and New Economy.

The first act realized under this minister has been the definition of a programme integrated with some guidelines and an agenda. As can be seen from this document, available on the regional web site (www.regione.campania.it/ricerca_scientifica), all actions are oriented to reaching the fixed objectives.

Another important aspect of this model is represented by an agreement signed on November 15, 2000, between the Ministry of University, Research and Scientific Development (MURST) and the southern Regions (Basilicata, Calabria, Campania, Puglia, Sardegna and Sicilia) belonging to the so called "Objective 1". This is a special program of funds and incentives that the Central Government issued for regions characterized by economic and employment problems. The agreement was based on the following objectives:

1. Strengthening centre-periphery connections for an integrated and coordinated management of PON (National Operative Program), PNR (National Research Program) and POR (Regional Operative Program) resources.
2. Redefinition of actions for a global development of potential of the South Italy Regions. The agenda and the program of activities include: identification of priority thematic areas, creation of thematic networks of

excellence centres and competence centres, specialization of thematic networks, and identification of coordinating nodes.

From November to the present, much progress has been made in research support, and Regione Campania has worked to realize its Regional Research System (Figure 2), which incorporates new management patterns, based on multidisciplinary cooperation and synergy within the regional, national and international scientific community.

In this system, reinforced support for basic research is forecast (Figure 3). There has been, for the year 2001, the introduction of innovative procedures for projects submitting, technical assistance, and an effective evaluation system based on curricula, international scientific context, relevance for regional economic and social development, and links to other national and international research programs.

This evaluation system, used also for research supported by EU funds, is made by an international peer review team adopting an internationally accepted evaluation method.

To optimise research with EU funds (Figure 4) we have drawn a network system based on competence centres (Figure 5). Centres of Competence are virtual structures that join intellectual, scientific and business resources for the coordination of the Regional Research Plan. They directly involve enterprises in the planning and implementation process and foster the establishment of knowledge-based companies. Each Centre integrates pre-competitive research activities focusing on development and provides high quality training activities in conjunction with other training institutions.

Their requirements are: high scientific competence in business sectors; integration of basic research, pre-competitive research and activities related to partnership research-enterprise; and management by a scientific board supported by an international advisory board.

The sectors in which the Centres are called to conduct strategic projects for the next years are:

- Analysis and monitoring of environmental risks;
- Advanced biology and its applications;
- Preservation, exploitation and improvement of cultural and environmental heritages;
- Agricultural products and food;
- New technologies for manufacturing sectors;
- Information and Communication Technology;
- Transport (by air, by sea, by road).

With the Centre of Competence we intend to reach the following objectives:

- Outcomes from pre-competitive research and patentable research
- Marketing of outcomes; self-financing
- Virtuous circle: Research-Development-Employment

In order to facilitate spin-off and to support a policy oriented to the knowledge transfer, the Region will realize:

- a collection of business concepts, agreement protocols and contracts with national agencies for the development;

- an integrated network of labs and services for enterprises with the creation and/or strengthening of liaison offices between Universities and Public Research Institutions;
- actions oriented to support and promote new technologies for products/processes and aimed at providing both scientific support for research projects and personnel training about new technologies. A Regional Research Observatory will be created to improve the activities of enterprises in high-tech fields.

Conclusions

These policies and activities will contribute to both the future of Campania and a new approach for research by PON, PNR and EU Programs. This South Italy Regions (Objective 1) experiencewill help generating a new model for R&D, based on integrated network systems.

The Regional Government of Campania hopes that this model can be assumedin the future, as an example for the international scientific community.

References

Official website of the Organization for Security and Co-operation in Europe, http://www.osce.org

	Total Expenditure on GDP	Fims'expenses on GDP	Public's expenses on GDP	Researchers per 10,000	% 1993-1999 Requests for Patents
Italy	1,05	0,57	0,48	32,00	25,00
Finland	3,11	2,15	0,96	94,00	446,90
France	2,18	1,35	0,83	60,00	36,10
Germany	2,29	1,55	0,74	60,00	35,90
Japan	3,06	2,18	0,88	96,00	9,60
The Netherlands	2,04	1,11	0,93	50,00	74,30
Spain	0,90	0,47	0,43	37,00	78,40
Sweden	3,70	2,77	0,93	86,00	85,70
United Kingdom	1,83	1,21	0,62	55,00	31,10
U.S.A.	2,84	2,16	0,68	74,00	21,60
OCSE -Total	2,23	1,54	0,69	-	67,40

Figure1. Research and Development (OSCE 2000)

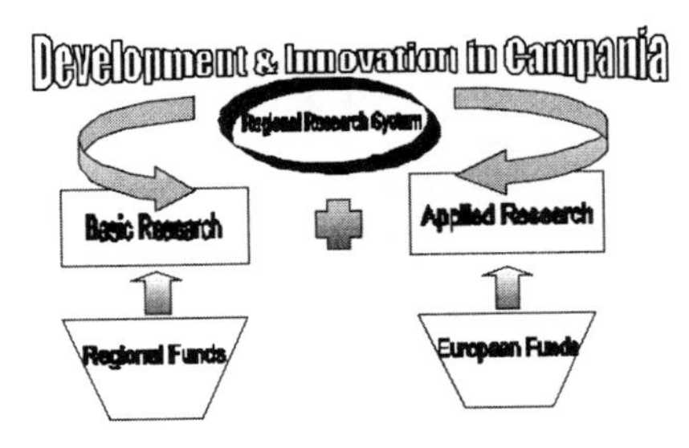

Figure 2. Regional Research System

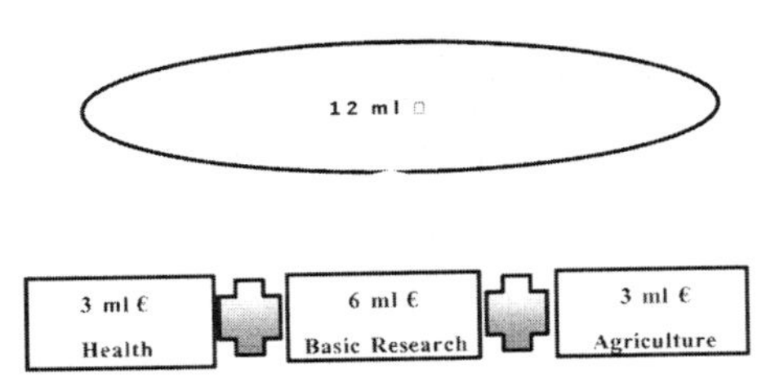

Figure 3. Regional Basic Research 2001 Funds

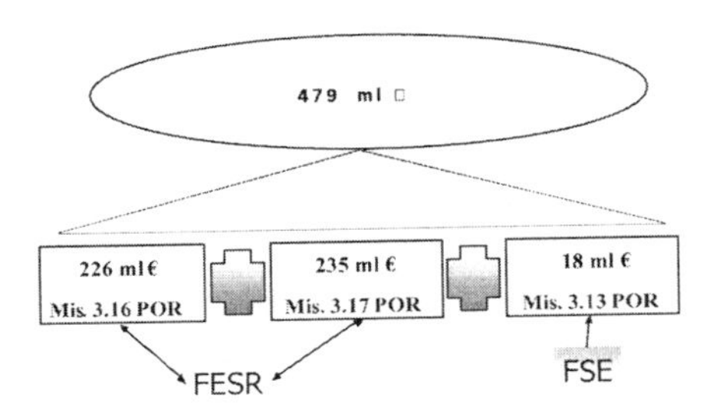

Figure 4. Regional Research Support, EU Funds

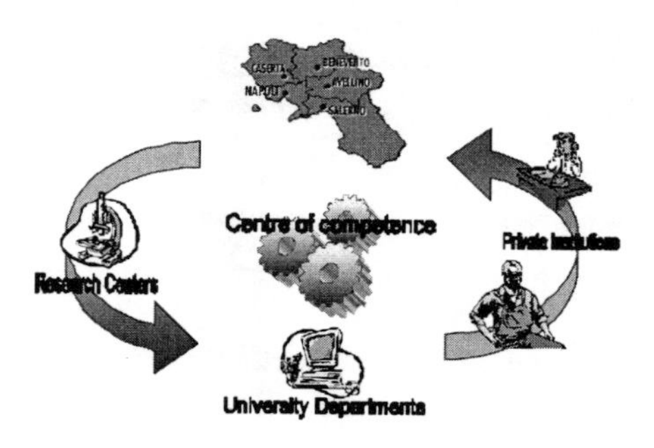

Figure 5. Centre of Competence. Connections

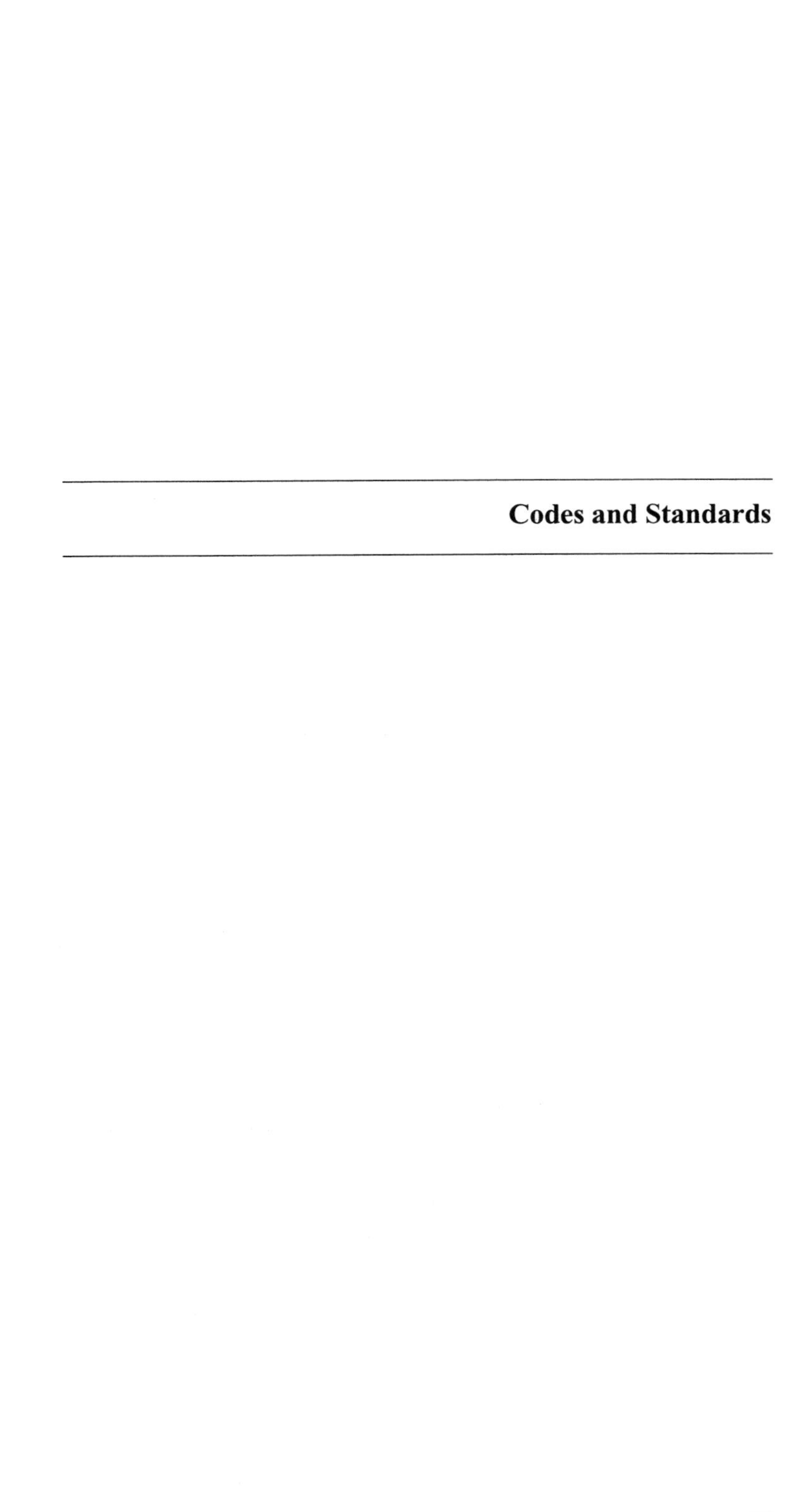

Codes and Standards

Guides and Specifications for the Use of Composites in Concrete and Masonry Construction in North America

Antonio Nanni[1]

Abstract

This paper covers the latest developments in the preparation of design guidelines, construction specifications, and inspection and quality control recommendations related to the use of composites in construction. The forms of FRP composites addressed are limited to bars and laminates for reinforcement of concrete and masonry structures (new construction and repair/rehabilitation). FRP bars are being used as the internal reinforcement in concrete members when the conventional steel bars may be undesirable for a host of reasons (e.g., corrosion), and principles for design and construction have been established and proposed to industry by the American Concrete Institute (ACI). Conversely, strengthening of concrete members with externally bonded FRP composites in the form of laminates or near surface mounted (NSM) bars can now be considered an "acceptable practice." Also in this case, the design and construction principles for use in practice are being finalized by ACI. The drivers for FRP strengthening technology are several, but perhaps the most relevant one is the ease of installation. On the wave of historical structures restoration projects conducted in Europe, there is an increasing interest in masonry-type applications even though no institution-sanctioned guidelines are available at present.

Introduction

In this paper, reference is made primarily to two technical documents produced by ACI Committee 440 under the new ACI series of emerging technology. The first one has been recently published (ACI Committee 440 2001) and provides recommendations for design and construction of FRP reinforced concrete (RC) structures. The second one is under development (ACI Committee 440 2001a) and provides guidance for the selection, design, and installation of FRP systems for externally strengthening concrete structures.

It should be noted that only notations critical to the understanding of the paper are defined herein and equations are expressed in US customary units.

----

[1] V&M Jones Professor of Civil Engineering, Center for Infrastructure Engineering Studies, 224 ERL, University of Missouri – Rolla, Rolla, MO 65409, USA; phone 573-341-4553; nanni@umr.edu

Design and Construction of Concrete Reinforced with FRP Bars

FRP materials are mostly anisotropic, do not exhibit yielding, and for design purpose, are considered elastic until failure. Design procedures should account for a lack of ductility in concrete reinforced with FRP bars. Both strength and working stress design approaches are considered by ACI. In particular, the guide makes reference to provisions as per ACI 318-95 Building Code Requirements for Structural Concrete and Commentary (ACI Committee 318 1995). A FRP RC member is designed based on its required strength and then checked for serviceability and ultimate state criteria (e.g., crack width, deflection, fatigue and creep rupture endurance). In many instances, serviceability criteria may control the design. This ACI document does not address prestressed concrete (PC) applications.

Design Values. The design tensile strength that should be used in all design equations is given in Eq. (1). The design rupture strain should be determined similarly, whereas the design modulus of elasticity is the same as the value reported by the manufacturer.

$$f_{fu} = C_E f^*_{fu} \quad (1)$$

where:

f_{fu} = design tensile strength of FRP, considering reductions for service environment

C_E = environmental reduction factor, given in Table 1 for various fiber types (column Int.) and exposure conditions

f^*_{fu} = guaranteed tensile strength of a FRP bar defined as the mean tensile strength of a sample of test specimens minus three times the standard deviation ($f^*_{fu} = f_{u,ave} - 3\sigma$)

Table 1: C_E factor for various fiber systems and exposure conditions (ACI 440 2001 and 2001a).

Exposure condition	Carbon		Glass		Aramid	
	Int.[a]	Ext.[b]	Int.[a]	Ext.[b]	Int.[a]	Ext.[b]
Interior exposure	1.0	0.95	0.8	0.75	0.9	0.85
Exterior exposure	0.9	0.85	0.7	0.65	0.8	0.75
Aggressive environment	n/s	0.85	n/s	0.50	n/s	0.70

[a] = New construction/internal; [b] = Strengthening/external; n/s = Not specified

Design parameters in compression are not addressed since the use of FRP bars in this instance is discouraged.

Flexure

Behavior and Failure Modes. If FRP reinforcement ruptures, failure of the member is sudden and catastrophic. However, there would be some limited warning of impending failure in the form of extensive cracking and large deflection due to the significant elongation that FRP reinforcement experiences before rupture. The concrete crushing failure mode is marginally more desirable for flexural members reinforced with FRP bars (Nanni 1993). In conclusion, both failure modes (i.e., FRP rupture and concrete crushing) are acceptable in governing the design of flexural members reinforced with FRP bars provided that strength and serviceability criteria are satisfied. To compensate for the lack of ductility, the member should possess a higher reserve of strength. The suggested margin of safety against failure is therefore higher than that used in traditional steel-RC design.

Figure 1 shows a comparison of the theoretical moment-curvature behavior of beam cross-sections designed for the same factored strength, ΦM_n, following the design approach of ACI 318 and ACI 440 (including the recommended strength reduction factors). Three cases are presented in addition to the steel reinforced cross section: two sections reinforced with GFRP bars and one reinforced with CFRP bars. For the section experiencing GFRP bar rupture, the concrete dimensions are larger than for the other beams in order to attain the same design capacity.

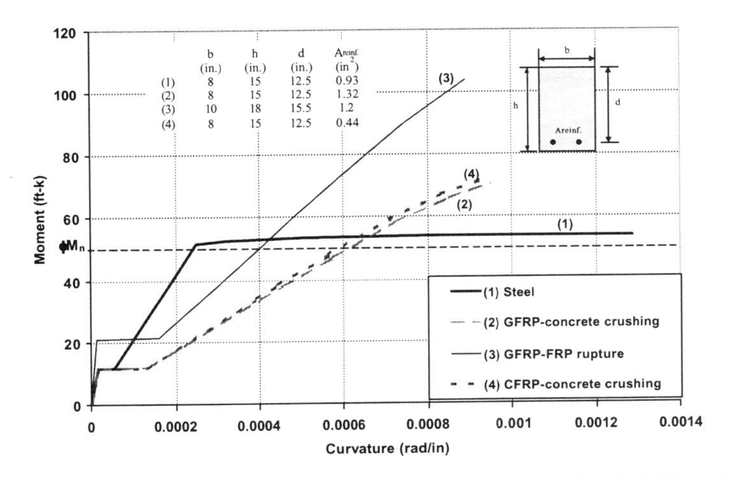

Figure 1. Moment-curvature relationships for RC sections using steel and FRP bars (ACI 440 2001).

Φ factor. Because FRP members do not exhibit ductile behavior, a conservative strength-reduction factor is adopted and set equal to 0.70 rather than 0.90 as per steel RC. Furthermore, because a member that experiences a FRP rupture exhibits less plasticity than one that experiences concrete crushing, a strength-reduction factor of 0.50 is recommended for FRP rupture-controlled failure. While a concrete crushing failure mode can be predicted based on calculations, the member as constructed may not fail accordingly. For example, if the concrete strength is higher than specified, the member can fail due to FRP rupture. For this reason and in order to establish a transition between the two values of Φ, a section controlled by concrete crushing is defined as a section in which the reinforcement ratio is larger or equal to 1.4 times the balanced reinforcement ratio ($\rho_f \geq 1.4\ \rho_{fb}$) and a section controlled by FRP rupture is defined as one in which $\rho_f < \rho_{fb}$.

Minimum reinforcement. If a member is designed to fail by FRP rupture, $\rho_f < \rho_{fb}$, a minimum amount of reinforcement, $A_{f,min}$, should be provided to prevent failure upon concrete cracking (that is, $\Phi M_n \geq M_{cr}$ where M_{cr} is the cracking moment). The minimum reinforcement area for FRP reinforced members is obtained by multiplying the existing ACI 318 limiting equation for steel by 1.8 (i.e., 1.8 = 0.90/0.50 which is the Φ ratio).

Crack Width. For FRP reinforced members, the crack width, w, can be calculated from the expression shown in ACI 318 with the addition of a corrective coefficient, k_b, for the bond quality. The k_b term is a coefficient which accounts for the degree of bond between the FRP bar and the surrounding concrete. For FRP bars having bond behavior similar to steel bars, k_b is assumed equal to one. Using the test results from Gao et al. (1998), the calculated values of k_b for three types of GFRP bars were found to be 0.71, 1.00, and 1.83. When k_b is not known, a value of 1.2 is suggested for deformed FRP bars.

Creep rupture and fatigue. Values for safe sustained and fatigue stress levels are given in Table 2. These values are based on experimental results with an imposed safety factor of 1/0.60.

Table 2. Creep rupture and fatigue stress limits in FRP reinforcement (ACI 440 2001 and 2001a).

Fiber type	Glass FRP	Aramid FRP	Carbon FRP
Creep rupture stress limit, $F_{f,s}$	$0.20\,f_{fu}$	$0.30\,f_{fu}$	$0.55\,f_{fu}$

Shear

Issues to be addressed when using FRP as shear reinforcement include: relatively low elastic modulus; high tensile strength; no yield point; tensile strength of the bent portion significantly lower than the straight one; and low dowel resistance.

According to ACI 318, the nominal shear strength of a steel RC cross section, V_n, is the sum of the shear resistance provided by concrete, V_c, and the steel shear reinforcement, V_s. Similarly, the concrete shear capacity $V_{c,f}$ of flexural members using FRP as main reinforcement can be derived from V_c multiplied by the ratio between the axial stiffness of the FRP reinforcement ($\tilde{n}_f E_f$) and that of steel reinforcement ($\tilde{n}_s E_s$). For practical design purposes, the value of ρ_s can be taken as $0.5\rho_{s,max}$ or $0.375\rho_b$. Considering a typical steel yield strength of 420 MPa (60 ksi) for flexural reinforcement, the equation for $V_{c,f}$ is that shown in Eq. (2) (noting $V_{c,f}$ cannot be larger than V_c).

$$V_{c,f} = \frac{\rho_f E_f}{90 \beta_1 f_c'} V_c \qquad (2)$$

The ACI 318 method used to calculate the shear contribution of steel stirrups, V_s, is applicable when using FRP as shear reinforcement, with the provision that the stress level in the FRP shear reinforcement, f_{fv}, should be limited to control shear crack widths, maintain shear integrity of the concrete, and avoid failure at the bent portion of the FRP stirrup, f_{fb}. Eq. (3) gives the stress level in the FRP shear reinforcement at ultimate for use in design. An expression for f_{fb} is given in ACI 440.1R-01.

$$f_{fv} = 0.002E_f \leq f_{fb} \qquad (3)$$

Development Length

The development length of FRP reinforcement can be expressed as shown in Eq. (4). This should be a conservative estimate of the development length of FRP bars controlled by pullout failure rather than concrete splitting.

$$\ell_{bf} = \frac{d_b f_{fu}}{2700} \qquad (4)$$

Manufacturers can furnish alternative values of the required development length based on substantiated tests conducted in accordance with available testing procedures. Reinforcement should be deformed or surface-treated to enhance bond characteristics with concrete.

Design and Construction of FRP Systems For Strengthening

Information on material properties, design, installation, quality control, and maintenance of FRP systems used as external reinforcement is presented in this ACI document (ACI Committee 440 2001a). This information can be used to select a FRP system for increasing the strength and stiffness of RC beams or the ductility of wrapped columns. This document does not address masonry walls.

It is recommended that the increase in load-carrying capacity of a RC or PC member strengthened with a FRP system be limited. The philosophy is that a loss of FRP reinforcement should leave a member with sufficient capacity to resist at least 1.2 times the design dead load and 0.85 times the design live load. Design recommendations are based on limit-states-design principles. This approach sets acceptable levels of safety against the occurrence of both serviceability and ultimate limit states (i.e., deflections, cracking, failure, stress rupture, fatigue). In determining the ultimate strength of a member, all possible failure modes and resulting strains and stresses in each material should be assessed. For evaluating the serviceability of an element, engineering principles, such as modular ratios and transformed sections, can be used.

FRP-strengthening systems should be designed in accordance with ACI 318 strength and serviceability requirements, using the outlined load and strength-reduction factors. Additional reduction factors applied to the contribution of the FRP reinforcement are recommended to reflect the limited body of knowledge. The environmental-reduction factor C_E to determine the FRP design strength and strain was given in Table 1 for different fiber types (see column Ext.). Similarity with values adopted for internal FRP reinforcement should be noted.

Flexure

Failure Modes. Guidance is given on the calculation of the flexural strengthening effect of adding longitudinal FRP reinforcement to the tension face of a rectangular RC member (concepts could be extended to cover T-sections and I-sections as well as PC). The nominal flexural capacity can be computed as per ACI 318. An additional reduction factor, $\psi_f = 0.85$, is applied to the flexural-strength contribution of the FRP reinforcement to account for uncertain reliability of the FRP reinforcement.

As FRP materials are linearly elastic until failure, the stress level in the FRP reinforcement is always proportional to strain. The effective strain level in the FRP reinforcement at the ultimate-limit state can be found from Eq. (5).

$$\varepsilon_{fe} = \varepsilon_{cu}\left(\frac{h-c}{c}\right) - \varepsilon_{bi} \leq \kappa_m \varepsilon_{fu} \tag{5}$$

with:

$$\kappa_m = \begin{cases} 1 - \dfrac{nE_f t_f}{2{,}400{,}000} & \text{for} \quad nE_f t_f \leq 1{,}200{,}000 \\ \dfrac{600{,}000}{nE_f t_f} & \text{for} \quad nE_f t_f > 1{,}200{,}000 \end{cases}$$

where: n is the number of plies; E_f is the elastic modulus of each ply; and t_f is the ply thickness. The term κ_m is a factor no greater than 0.90 that is meant to limit the strain in the FRP reinforcement to prevent debonding or delamination. This term recognizes that laminates with greater stiffness are more prone to delamination. The κ_m term is only based on a general recognized trend and on the experience of engineers practicing the design of bonded FRP systems.

Φ factor. The strength-reduction factor depends on the strain in the steel at ultimate, ε_s. Φ is set equal to 0.90 for ductile sections ($\varepsilon_s \geq 0.005$) and 0.70 for brittle sections where the steel does not yield. A linear transition for the reduction factor between these two extremes is then established.

Stress limits. To avoid plastic deformations, the existing steel reinforcement should be prevented from yielding at service load levels. The stress in the steel at service should be limited to 80% of the yield stress. Similarly, to avoid failure of a FRP-reinforced member due to creep-rupture of the FRP, stress limits for these conditions should be imposed on the FRP reinforcement. Limits on sustained and fatigue stresses are those listed in Table 2 and are identical to those for internal FRP reinforcement.

Shear

The nominal shear capacity of a FRP-strengthened concrete member can be determined by adding the contribution of the FRP reinforcing to the contributions from the reinforcing steel and the concrete. An additional reduction factor, ψ_f, is applied to the contribution of the FRP system. The additional reduction factor, ψ_f, should be selected based on the known characteristics of the application but should not exceed 0.85 for two- and three-sided wrapping schemes and 0.95 for completely wrapped elements.

The shear strength provided to a member by the FRP system should be based on the fiber orientation and an assumed crack pattern (Khalifa et al. 1998). It can be determined by calculating the force resulting from the tensile stress in the FRP along the assumed crack with an expression similar to that of steel reinforcement. To compute the tensile stress in the FRP shear reinforcement at ultimate, f_{fe}, it is necessary to calculate the effective strain in the FRP, ε_{fe}.

FRP systems that do not enclose the entire section (two- and three-sided wraps) have been observed to delaminate from the concrete before the loss of aggregate interlock of the section. For this reason, bond stresses should be analyzed to determine the usefulness of these systems and the effective strain level that can be achieved. The effective strain is calculated using a bond-reduction coefficient, κ_v, applicable to shear (see Eq. (6)).

$$\varepsilon_{fe} = \kappa_v \varepsilon_{fu} \leq 0.004 \text{ (for U-wraps or bonding to two sides)} \quad (6)$$

The bond-reduction coefficient is a function of the concrete strength, the type of wrapping scheme used, and the stiffness of the laminate (Khalifa et al. 1998). The methodology for determining κ_v has been validated for members in regions of high shear and low moment, such as simply supported beams. The methodology has not been confirmed for shear strengthening in areas subjected to simultaneous high shear and moment loads, such as continuous beams. In such situations, conservative values for κ_v are recommended.

Compression Members

The axial compressive strength of a non-slender member confined with a FRP jacket is calculated using the conventional expressions of ACI 318 substituting for f'_c the factored confined concrete strength $\psi_f f'_{cc}$. The additional reduction factor is set to $\psi_f = 0.95$.

Vertical displacement, section dilation, cracking, and strain limitations in the FRP jacket can also limit the amount of additional compression strength that can be achieved with a FRP jacket. The axial demand on a FRP-strengthened concrete member should be computed with the load factors required by ACI 318 and the axial compression strength should be calculated using the strength-reduction factors, ϕ, for spiral and tied elements required by ACI 318.

If the member is subjected to combined compression and shear, the effective strain in the FRP jacket should be limited based on the criteria given by $\varepsilon_{fe} = 0.004 \leq 0.75\varepsilon_{fu}$.

At load levels near ultimate, damage to the concrete in the form of significant cracking in the radial direction occurs. The FRP jacket contains the damage and maintains the structural integrity of the column. At service load levels, this type of damage should be avoided. In this way, the FRP jacket will only act during overloads that are temporary in nature. To ensure that radial cracking will not occur under service loads, the stress in the concrete is limited to $0.65f'_c$. In addition, the stress in the steel should remain below $0.60f_y$ to avoid plastic deformation under sustained or cyclic loads. By maintaining the specified stress in the concrete at service, the stress in the FRP jacket will be negligible.

Near-Surface Mounted (NSM) FRP Reinforcement

Although not directly addressed in ACI 440 (2001a), the use of NSM FRP bars is a promising technology for increasing flexural and shear strength of deficient RC and PC members (De Lorenzis and Nanni 2001). The advantages of NSM FRP bars compared with externally bonded FRP laminates are the possibility of anchoring the reinforcement into adjacent RC members, and minimal surface preparation work and installation time. The method used in applying the bars is as follows. A groove is cut in the desired direction into the concrete surface, the groove is then filled halfway with adhesive paste, and the FRP bar is placed in the groove and lightly pressed. This forces the paste to flow around the bar and fill completely between the bar and the sides of the groove. Finally, the groove is filled with more paste and the surface is leveled. As this technology emerges, the structural behavior of RC and PC elements strengthened with NSM FRP bars needs to be fully characterized.

Unreinforced Masonry (URM) Strengthening

At present, ACI provisions do not cover the use of composites for the strengthening of masonry structures even though some recommendations are available from other organizations (ICBO 1997). Also a subcommittee of the Masonry Standards Joint Committee (i.e., Strengthening, Repair, and Rehabilitation), which is a three-society coordinated effort, is considering the use of composites as a suitable technology for masonry repair/upgrade.

FRP materials in the form of laminates and bars have been used for the strengthening of URM elements similarly to concrete members. In the case of strengthening with NSM FRP bars, there is a preference in placing them into grooves cut into the bed joints (i.e., structural repointing). To be completely successful, masonry retrofit work should be carried out with the least possible irrevocable alteration to the building appearance, as many URM buildings are part of the cultural heritage.

For masonry walls strengthened with FRP laminates, research results have shown that debonding of the FRP laminate from the masonry substrate is the controlling mechanism of failure. This has been evident in masonry walls strengthened to resist either in-plane or out-of-plane loads. Therefore, there is a need to determine the effective strain of the laminate as a function of the amount of strengthening. For clay units, debonding may have a direct relationship with the porosity of the masonry itself. For this reason, walls built with different and representative types of masonry units should be investigated. To date, there is a tendency to use types and quantities of FRP composites similar to those used for the strengthening of RC elements.

Based on the premise of debonding as a controlling mode of failure, anchorage systems to avoid this failure mode need be developed. The use of steel angles may be effective when the wall is subjected to in-plane loads. However, when subjected to out-of-plane loads, the wall may be locally fractured in the anchorage regions due to the restraint of the wall movement (Tumialan et al. 2000). FRP bars have been successfully used for anchoring laminates in RC joists strengthened in shear. The anchoring technique consists of placing the fiber sheet under the bar embedded in a slot saw-cut in the base material.

Structural repointing is an alternative to strengthen masonry walls where aesthetics is important. Specifically for the case of walls subjected to in-plane loads, the effective strain developed in the bars needs to be estimated for different strengthening schemes. It has been observed that the failure in these walls is the result of the loss of bond between the adhesive and the masonry units (Tinazzi et al. 2000). The strengthening of only one side of a wall represents a frequently-encountered field situation where there is the presence of a veneer wall. Thus, the crack growth may be larger on the un-strengthened face of the wall. The crack between masonry units and the mortar in the un-strengthened face has been observed to travel through the wall thickness until debonding of the adhesive paste from the masonry units occurs. At this point the wall fails because the tensile stresses are no longer transferred to the FRP bars.

The interaction of strengthened walls with the surrounding structural elements (i.e. beams and columns) is of paramount importance since the effectiveness of the strengthening may be dangerously overestimated due to premature failures (e.g. crushing of masonry units at the boundary regions) in the masonry itself (Tumialan et al. 2000) or because the effects of the strengthening may produce undesirable changes in stiffness.

Conclusions

Even with some unresolved issues that should become a priority for future research, it can be concluded that the availability of design and construction guides developed by ACI for the use of FRP internal and external reinforcement for new and existing structures should allow the construction industry in North America to take full advantage of this emerging technology.

Acknowledgements

The author, who has drawn freely from the cited ACI documents wishes to acknowledge the direct and indirect contributions of the members of ACI 440 Committee and all individuals involved in the preparation of its documents.

References

ACI Committee 318 (1995). *Building Code Requirements for Structural Concrete and Commentary,* ACI 318-95/R-95, ACI, Farmington Hills, MI, USA.

ACI Committee 440 (2001). *Guide for the Design and Construction of Concrete Reinforced with FRP Bars*, ACI 440.1R-01, ACI, Farmington Hills, MI, USA.

ACI Committee 440 (2001a). *Guide for the Design and Construction of Externally Bonded FRP Systems for Strengthening of Concrete Structures*, ACI, Farmington Hills, MI, USA (in print).

De Lorenzis, L., and Nanni, A. (2001). "Characterization of FRP Rods for Near Surface Reinforcement," ASCE-Journal of Composites for Construction, 5(2), 114-121.

Gao, D., Benmokrane, B., and Tighiouart, B. (1998). "Bond Properties of FRP Rebars to Concrete," *Technical Report*, Department of Civil Engineering, University of Sherbrooke, Sherbrooke (Quebec), Canada, pp. 27.

International Conference of Building Officials (1997). *Acceptance Criteria for Concrete and Reinforced and Unreinforced Masonry Strengthening Using Fiber-Reinforced Composite Systems*, AC125-R1-1001, Whittier, CA.

Khalifa, A., Gold, W.J., Nanni, A., and Abdel Aziz, M.I. (1998). "Contribution of Externally Bonded FRP to Shear Capacity of Flexural Members" ASCE-Journal of Composites for Construction, 2(4), 195-203.

Nanni, A. (1993). "Flexural Behavior and Design of Reinforced Concrete Using FRP Rods," *Journal of Structural Engineering*, 119(11), 3344-3359.

Tinazzi, D., Arduini, M., Modena, C. and Nanni, A. (2000). "FRP-Structural Repointing of Masonry Assemblages," Proc., 3rd Inter. Conf. on Advanced Composite Materials in Bridges and Structures, J. Humar and A.G. Razaqpur, Editors, Ottawa, Canada, 585-592.

Tumialan G., Myers, J.J. and Nanni, A. (2000). "An In-Situ Evaluation of FRP Strengthened Unreinforced Masonry Walls Subjected to Out of Plane Loading," ASCE Structures Congress, Philadelphia, PA, CD version, #40492-046-004, 8 pp.

Survey of European Guidelines for the Use of FRP in Structural Applications

Luc Taerwe[1] and Stijn Matthys[2]

Abstract

Nowadays, there is a considerable interest in FRP (fibre reinforced polymer) reinforcement and structural shapes for building construction. Especially, the use of FRP for externally bonded reinforcement (EBR) to repair and strengthen existing structures is becoming more and more established world-wide. Although much research has been done and many applications already exist in various countries, the lack of codes of practice incorporating advanced materials is still seen as a barrier for its more extensive use. In 1996 the International Federation for Structural Concrete (*fib*) established Task Group 9.3, with the objective of developing guidelines for the design of concrete structures, reinforced, prestressed or strengthened with advanced composites. In December 1997 the EU TMR Network, ConFibreCrete, has been established and operates in support of the *fib* Task Group 9.3.

In this article an overview is given of the European guidelines proposed for the use of FRP in structural applications.

Introduction

The use of fibre reinforced polymers (FRP) as structural reinforcement gains more and more interest in construction practice world-wide. Continuing efforts in material development and research activities, with strong links to engineering practice, make FRP reinforcement where it stands today: a viable alternative to classical types of reinforcement, offering many potentials. Extensive research on

[1] Prof. dr. ir., Director, Magnel Laboratory for Concrete Research, Department of Stuctural Engineering, Ghent University, Belgium ; Luc.Taerwe@rug.ac.be

[2] Dr. ir., Post-Doc. Res. Assoc., Magnel Laboratory for Concrete Research, Department of Stuctural Engineering, Ghent University, Belgium ; Stijn.Matthys@rug.ac.be

the use of FRP in concrete structures started in Europe about 25 years ago and resulted, for example, in the world's first highway bridge using FRP post-tensioning cables in Germany in 1986. Clearly, a long way has been gone from these first applications and still much is to be done to stimulate the implementation and use of structural FRP reinforcement.

In the following the use of FRP reinforcement and structural shapes in Europe is discussed, as well as the European guidelines available or proposed for its use in structural applications.

Use of FRP in structural applications

General. Since many years there has been a great interest in FRP reinforcement in Europe (Taerwe and Matthys, 1999) and to some extent pioneering work in this field has been done in Europe. As these early developments were commercially less successful than hoped for, today the commercial application of FRP reinforcement in Europe is less exploited than in North-America and Japan. Also, compared to Europe, the market for FRP reinforcement is larger in Japan and North America. Nevertheless, in various European countries, a considerable amount of research and application projects are available, several research projects were established and developments in the use of advanced composites for construction have been taken place. Given these efforts and interest, FRP reinforcement and structural shapes are getting more generally known.

Externally bonded FRP reinforcement (FRP EBR). Although applications and demonstration projects have been taken place already in the late 1980's and early 1990's, commercial used of FRP EBR started mainly in Switzerland around 1993 and soon followed in other European countries. Whereas in the beginning it was expected that the use of FRP EBR would mainly apply to strengthening of bridges, it appears that a lot of the strengthening work with this technique is done in the building sector, among which the restoration of old buildings. The amount of FRP used per project varies from a few meters to some kilometres. Basically, the FRP EBR is used for concrete structures. To a lesser extent, yet various applications deal with FRP EBR on masonry, timber and metals.

Over the last years, the market for externally bonded FRP reinforcement is characterized by a clear dynamism. Different manufacturers and contractors are active in or enter this market segment, while researchers and commissions are trying to provide design guidelines on a short term basis. The type of applications range widely and different challenging projects have been executed. Furthermore, new types of products and techniques enter the market to extend the possibilities of the FRP EBR strengthening technique (e.g. mechanical anchorages, prestressed FRP EBR, etc.).

FRP reinforcement for RC and PC. The use of FRP reinforcement for reinforced (RC) and prestressed concrete (PC) is commercially much less established than that of externally bonded reinforcement. Nevertheless, FRP (prestressing) reinforcement is available in different European countries by local manufacturers or suppliers. Examples of recent initiatives are two prefabricated CFRP prestressed girders for a bridge over the high speed rail way line at Kortenberg (Belgium), CFRP PC poles in Switzerland, a cable stayed bridge at Herning (Denmark) with extensive use of CFRP reinforcement (both stay cables and reinforcement) and external post-tensioning with CFRP cables in the Dintelhaven bridge in Rotterdam (The Netherlands).

FRP structural shapes and profiles. Although the interest in advanced composites for civil engineering quit often relates to FRP reinforcement for concrete and to externally bonded FRP reinforcement, also the use of FRP as structural shapes and profiles is of interest. An important example of the use of structural FRP shapes is the Aberfeldy bridge in Scotland. This cable stayed pedestrian bridge, built in 1992, entirely consists of FRP materials, with aramid cables and glass reinforced polymer structural elements used in the deck and towers, the latter elements being bonded with epoxy adhesive. Currently, an all-composite bridge suitable for heavy traffic is also under study, to be build in Den Dungen in the Netherlands.

Unified Design Guidelines for FRP in Europe

A major obstacle in adopting FRPs widely in construction is the lack of accepted standards for design and quality control. As this aspect is generally recognised, several initiatives to develop design guidelines are ongoing on a national and international level. To stimulate the development of unified guidelines for Europe, a *fib* (Fédération International du Béton) Task Group has been established (*fib* TG9.3, 1998). The work started in a CEB (Comité Euro-International du Béton) Task Group (TG 3.10) in September 1996 which converted, with the merger of CEB and FIP in June 1998, into an *fib* Task Group (TG 9.3), under the supervision of Commission 9 "Reinforcing and Prestressing Materials and Systems". The title of TG 9.3 is "FRP Reinforcement for Concrete Structures". The task group consists of about 60 members, representing most European universities, research institutes and industrial companies working in the field of advanced composite reinforcement for concrete structures, as well as representing members from Canada, Japan and USA. Starting December 1997, the work of the *fib* TG 9.3 has been largely supported by the EU TMR (European Union Training and Mobility of Researchers) Network, ConFibreCrete that comprises of 11 teams from 9 European countries (TMR, 1998).

The development of design guidelines by these committees, as well as other European design guidance initiatives are discussed in the following.

FRP Reinforcement for RC and PC Structures

One of the early initiatives to develop guidelines followed from the work done by the EUROCRETE project, which has been included in interim guidelines published by the British Institution of Structural Engineers (1999) and SINTEF Norway (Thorenfeldt, 1998). These documents provide recommendations on the use of FRP reinforcement for RC, with respect to the British and Norwegian standards respectively.

EUROCRETE, a pan-European collaborative research programme aiming at developing FRP reinforcement for concrete started in December 1993 and ended in 1997. The project included one or more industrial partners from the United Kingdom, The Netherlands, Switzerland, France and Norway. The main areas which were investigated are : selection of suitable resins and fibres, development of appropriate manufacturing techniques, investigations to determine the durability of FRP rods exposed to aggressive environment either directly or embedded in concrete, determination of the structural behaviour through testing and analytical techniques, development of suitable design guidance, techno-economic and feasibility studies, development of case studies of trial structures and components.

Regarding durability, trials have been carried out on fibres, resins, laminates and FRP bars in a range of environments both in the laboratory and on site. FRP bars embedded in concrete have been on exposure sites in the UK, Norway and the Middle East for about 2 years. Much of the durability work has been with pull-out specimens as the failure of the surface layer will be the critical condition.

The structural testing has mainly been performed at Sheffield University. They have tested a large number of beams in bending and shear, some slabs in punching and done a lot of bond testing with a range of bars, using glass, carbon and aramid. They have also done testing in connection with the demonstration structures and the development trials outlined below.

Development trials on small precast units have been performed such as interlocking precast crib wall units, spreader plates for ground anchors and cladding panels. Some units have been succesfully redesigned, completely with FRP reinforcement and important savings in weight and cost have been achieved. The following demonstration structures have been realized :

- A footbridge using glass-fibre composite reinforcement was opened in May 1995 in Chalgrove, near Oxford (UK). In the bridge slab, measuring 5.0 m x 1.5 m x 0.30 m, bars with a diameter of 13.5 mm were used as mesh reinforcement with rods placed orthogonally at 150 mm spacing top and bottom.

- 60 m of precast concrete post and panel fence at Hemel Hempstead, constructed April 1996.
- A 10 m span footbridge in the Oppegård golf course near Oslo (Norway), completed in May 1997. Two types of Glass Fibre Reinforced Plastic (GFRP), developed in the Eurocrete project, were selected for use in the trial bridge. "Plytron" bars made of E-glass and polypropylene are used as shear links. The "Eurocrete reinforcement bar (E bar)", which is a composite with higher stiffness and strength, is adopted as main reinforcement. These bars are made of E-glass and vinylester matrix. For the post-tensioning of the bridge aramid basis tendons were utilized.
- A fender support beam (8.0 m x 1.5 m x 1.4 m) in Qatar, installed May 1996.

Unified design guidelines proposed by *fib* TG 9.3 are currently under development and will be published in a *fib* bulletin, called "FRP Reinforcement for Concrete Structures". This report deals with design aspects and state-of-the art knowledge on FRP materials, their properties and characterisation, as well as its use for RC and PC structures.

Externally Bonded FRP Reinforcement

Various national initiatives on design guidelines for the use of externally bonded FRP reinforcement (FRP EBR) have been taken or are ongoing, among which the following:

- In Sweden a report has been established on 'Strengthening of existing concrete structures with carbon fibre fabrics or laminates - Dimensioning, materials and execution', which has been excepted as a Swedish National Railroad and Road Code (Täljsten, 1999). A report by the Swedish Concrete Association, called 'Possibilities of non-metallic (FRP) reinforcement to day and in the future' is currently under development.
- In Germany, different so-called expert opinions, which are product related, have been issued. These documents are commissioned by the manufacturer and assess the suitability and design of the FRP product for use as adhesive-bonded reinforcement to strengthen concrete members.
- End 2000, a technical report called 'Design guidance for strengthening concrete structures using fibre composite materials' has been published by the Concrete Society in the UK(2000).
- In the Netherlands, a CUR working commission has reached its final stage in preparing recommendation on the use and design of externally bonded CFRP reinforcement for concrete members, to be published by the end of 2001.

- In Switzerland, a commission has been initiated which aims in the development of a SIA code on FRP EBR.
- In Greece, commissioned by the Seismic Planning and Protection Agency, a project has been started to develop design recommendations for seismic strengthening with FRP EBR.

A first bulletin by fib TG 9.3 Working Group EBR, called "Design and use of Externally Bonded FRP Reinforcement for Reinforced Concrete Structures", is in its final review stage and will be published in the second half of 2001. It gives detailed design guidelines on the use of FRP EBR, the practical execution and the quality control, based on the current expertise and state-of-the-art knowledge of the task group members (*fib* 2001).

A considerable effort has been made to present in the bulletin material which is state-of-the-art in the area of composites as strengthening material for concrete structures, as well as to offer detailed design guidelines on those aspects which form the majority of the design problems. Also, much consideration is given to the practical execution aspects and quality control. The report contains 9 chapters: Introduction, FRP Strengthening Materials and Techniques, Basis of Design and Safety Concept, Flexural Strengthening, Strengthening in Shear and Torsion, Confinement, Detailing Rules, Practical Execution and Quality Control, and Special Design Considerations and Environmental Effects.

Structural Shapes and Profiles

General. According to (J.T. Mottram, 1999), two European design handbooks exist (J.L. Clarke, 1996, Anon, 1995) to aid the design of structures erected from pultruded standard profiles. The EUROCOMP Design Code was published in 1996 in order to provide independent, general guidance to the construction industry. Unfortunately, many of its design expressions and procedures have not been verified by reliable research or practice. This design code is discussed in more detail in the next section.

Survey of the EUROCOMP Design Code. The book is in three parts. Part 1 is the EUROCOMP Design Code itself. The code follows closely the format of the European Eurocodes for the conventional structural materials of steel and concrete. It comprises eight chapters whose contents will be outlined below.

Chapter 1 gives the scope of the design code and introduces definitions and symbols. The basis of design is given in Chapter 2. Those practicing engineers familiar with design using conventional materials will here find a number of unusual and distinct features. With steel and concrete, the Eurocodes give a single partial safety factor based on well-founded experience. With fibre reinforced composite materials a range of material safety factors are proposed in Chapter 2, determined by the level of knowledge about the material being used and its method of manufacture. If properties of the composite are determined from constituent properties and micro-mechanical formulae, a high partial factor is to be used. If, however, material properties are determined by laboratory testing the factor is reduced.

Chapter 3 gives guidance on the constituent materials for fibre reinforced composites. Fibres and thermosetting resins are the main constituents the code deals with, but there are clauses for cores, gel coats, surface veils and additives.

In Chapter 4 the code deals with section and member design in terms of ultimate and serviceability limit states. The formulae given for calculations are generally those engineers will have used when designing structures of standard steel sections. However, these thin-walled formulae are appropriately modified to account for the differences in material strengths and stiffnesses. A number of these formulae for ultimate limit states have not, as yet, been fully assessed by experience, and to make allowance for this a high partial safety factor is needed. The long-term durability of the material is given prominence. The code has clauses to deal with the effects of long-term loading, such as fatigue, creep, creep-rupture and stress-corrosion. It also covers design for impact, for explosion or blast, for fire and for chemical attack.

Chapter 5 deals with the problem of designing efficient, safe and reliable connections. Polymeric composites differ from conventional materials in that adhesive bonding is used to form primary structural joints. In the first section of the chapter there is an introduction to different joint configurations, joining techniques and to the design approaches. A simplified design approach for plate-to-plate connections with loading in the plane is developed for mechanical joints and for bonded joints. All recommended design procedures have to consider the problem of predicting, to sufficient accuracy, the effect on joint resistance of local stress fields. These occur for example in a laminate in the region adjacent to the boundary between bolt and laminate. Such a problem is not as serious when the material is steel because yielding relieves such stress 'hot spots'. This single factor ensures that design for connections requires special attention and Chapter 5 was a difficult part of the code to prepare.

Issues related to construction and workmanship are to be found in Chapter 6. Special attention is given to the correct conditions in workshops and on sites, to enable manufacture and fabrication to be of a high quality. With the limited experience on the performance of structures in service, and of the continual introduction of new materials, sections, and joint details, laboratory testing at all levels will remain very important. The guidance on testing in Chapter 7 is therefore crucial to the future exploitation of polymeric composites in construction. Quality control, from material manufacture to erection of the structure, is given by clauses in Chapter 8.

Part 2 is the EUROCOMP handbook. It follows closely the format of Part 1 and, for each Chapter, provides detailed background information. Part 3 concentrates on research carried out by EUROCOMP partners. Results from five test programmes are used to assess and develop the information given in Parts 1 and 2.

Conclusions

The interest in FRP reinforcement in Europe is considerable and its use is getting more generally known, mainly with respect to externally bonded FRP reinforcement. A major part of the European activities (applications, research, development of guidelines) relate to the latter technique, in addition to FRP reinforcement for RC and PC, as well as to structural shapes and profiles.

In different European countries design guidance documents are available or under development. To develop unified European guidelines for the use of FRP reinforcement in concrete structures a *fib* (International Federation for Structural Concrete) Task Group 9.3 "FRP Reinforcement for Concrete Structures" has been established in 1996. On the research and dissemination of results level, *fib* TG9.3 is supported by a EU TMR Network "ConFibreCrete".

Two bulletins are being produced by *fib* TG9.3 on design guidelines for the use of FRP in concrete structures. A first bulletin, dealing with externally bonded FRP reinforcement, is in a final stage and will be published in the second half of 2001. A second bulletin on FRP materials and their use for RC and PC is under development.

Acknowledgements

The authors want to acknowledge the contribution of various individuals in sending useful information on the European activity regarding the use and design of advanced composites for structural applications, especially the information provided by J.T. Mottram is greatly appreciated. The authors wish to thank those members of *fib* TG 9.3 and ConFibreCrete who contributed to the development of guidelines.

References

Anon. (1995), "Fiberline design manual for stuctural profiles in composite materials", *Fiberline Composites A/S*, Kolding, Denmark.

Clarke, J.L. (1996), (Ed.), "Structural design of polymeric composites – *EUROCOMP design code and handbook"*, E & FN Spon, London, p. 751.

Concrete Society (2000), "Design guidance for strengthening concrete structures using fibre composite materials", *Technical report 55, The Concrete Society*, UK.

fib Task Group 9.3 (1998), "FRP Reinforcement for Concrete Structures", *International Federation for Structural Concrete, Commission 9, Task Group 9.3*, http://allserv.rug.ac.be/~smatthys/fibTG9.3 >(June 10, 2001).

fib (2001), "Design and Use of Externally Bonded FRP Reinforcement (FRP EBR) for Reinforced Concrete Structures", Final Draft, *Progress Report of fib EBR group*, International Federation for Structural Concrete.

Institution of Structural Engineers (1999), *Interim guidance on the design of reinforced concrete structures using fibre composite reinforcement*, Published by SETO Ltd., UK.

Mottram, J.T., (1999), "Design guidance with an emerging construction material", *Proc. COST C1 Inter. Conf., Control of the Semi-rigid Behaviour of Civil Engineering Structural Connections*, Liège 17-19 September, European Commission, Luxembourg, 539-546.

Taerwe, L., and Matthys, S. (1999), "FRP For Concrete Construction: Activities In Europe", *ACI - Concrete International*, 21 (10), 33-36.

Täljsten, B. (1999) "Strengthening of existing concrete structures with carbon fibre fabrics or laminates - Dimensioning, materials and execution", An extract from *Swedish National Railroad and Road Codes*, Lülea University, Sweden.

TMR (1998), "Development of Guidelines for the Design of Concrete Structures, Reinforced, Prestressed or Strengthened with Advanced Composites (ConFibreCrete)", *Training and Mobility of Researchers, Research Network TMR ConFibreCrete*, http://www.shef.ac.uk/~tmrnet >(June 10, 2001).

Thorenfeldt, E. (1998), "Modifications to NS3473 when using fibre reinforced plastic (FRP) reinforcement", *SINTEF Report STF22 A98741*, Trondheim, Norway.

Unified Code on Structural Use of FRP: Suggestions for Strengthening Design

Kiang Hwee Tan[1]

Abstract

Fiber-reinforced polymers (FRP) are increasingly used in Singapore and the surrounding region, in particular in repair and strengthening works. This paper presents the current status and potential application of FRP in the region, and discusses the code-related issues, in particular design concepts. Suggestions for a unified code on the structural use of FRP are made.

Introduction

The application of fiber-reinforced polymer (FRP) reinforcement in Asian countries outside Japan has witnessed a relatively slow start but it has since gained a footing in the construction industry, especially in the retrofitting sector, in the past half a decade or so. Several factors had hampered the widespread acceptance of FRP reinforcement. These include: (a) shortage of expertise in the industry; (b) high cost of FRP; (c) unavailability of appropriate test methods for material properties; and (d) unfamiliarity with the long-term behavior of the material.

Although several test methods, design codes and guidelines have been established in the American and European continents, as well as in Japan, they do not necessarily address the concerns of the South East Asian region, in particular those related to climatic conditions and construction practices (Tan 1997a). In addition, the diversities of codes in the various regions in terms of design philosophy make it complicated for them to be adopted in a different region.

This paper reviews the research work and explores code-related issues pertaining to FRP applications from a Singaporean perspective. Several suggestions regarding design concepts are made in relation to a proposal for a unified code on the structural use of FRP.

[1]Associate Professor, Department of Civil Engineering, National University of Singapore, Block E1A, #07-03, 1 Engineering Drive, Singapore 117576; phone 65-8742260; cvetankh@nus.edu.sg

Research and Application of FRP

Research activities related to FRP started in Singapore as early as in 1992 and have now become prevalent in the neighbouring countries. Most works however were focused on the basic member behavior and practical application of FRP in retrofitting works.

RC and PC. Use of FRP in new reinforced concrete (RC) or prestressed concrete (PC) structures is almost absent, due to the high cost and unfamiliarity associated with the material and absence of appropriate guidelines or codes. Some work on partial replacement of corroded steel reinforcing bars by FRP rods has however been carried out (Tan 1997b). Also, the fact that potential of FRP in terms of ultimate strength could not be fully realized without affecting the serviceability and the brittleness of the materials have somewhat restricted its use to PC structures. External post-tensioning using FRP tendons appears to be a promising area of application and considerable amount of research work has been carried (Tan, Farooq and Ng 2001, Tjandra and Tan 2001).

FRP Shapes and Systems. The potential in the application of FRP extruded shapes and systems has not been explored to any great extent. There were attempts to use FRP shells for column strengthening (Tan and Balendra 1997) and protection. However, the inflexibility of preformed sections has limited its widespread application. Nevertheless, the application of FRP in lightweight scaffold, communication towers and bridges are currently being pursued.

Repair and Strengthening. The bulk of FRP application is related to retrofitting works, in particular in the strengthening of slabs, beams and columns in buildings. There had also been unusual strengthening works carried out on dapped beams in overpass and masonry walls in buildings. Some studies on blast resistance of masonry walls retrofitted with continuous fibre sheets have also been carried out (Tan and Patoary 2001). There are also proposals to strengthen silos using continuous fibre sheets. Due to the lack of appropriate local codes, many applications are substantiated by laboratory tests.

Code-Related Issues

There is a need to co-ordinate research activities and work towards a unified code that would facilitate global construction activities. The following code-related issues, focusing on repair and strengthening using externally bonded FRP systems and external tendons, are discussed and suggestions are made for their consideration in a unified code.

Cost-Effectiveness. Although it has often been argued that FPR forms only a fraction of the total project, and the overall cost including maintenance would probably be competitive using FRP, there is room for further improvement for a cost-effective application. To achieve this, a good understanding of the intrinsic material properties and mechanics of the structure, coupled with a good knowledge of the construction practice and socio-economics of the region is essential (Tan 1999).

Test Methods. Much work (JSCE 1997; ACI 2001a) is still being carried out to establish appropriate test methods for the physical and mechanical behavior of FRP materials, especially in the case of FRP sheets. In particular, the long-term durability of FRP sheets is of great interests. The effect of large temperature fluctuations and humidity levels, rain and snow, and freeze-thaw had been studied in temperate countries, but for most Asian countries that lie in the tropics, high average temperatures, high humidity, thunderstorm and strong sun leading to larger dose of ultra-violet radiation, are of major concern.

Design Concepts. Some design guidelines on the use of FRP rods (JSCE 1997; ACI 2001b) and sheets (Concrete Society 2001; JSCE 2001; ACI 2001c) have recently been published or are being drafted. It is heartening to note that many of the clauses are based on international research activities, which indicates the possibility of the establishment of a unified code on the structural use of FRP. The author would like to provide some observations relating to the design clauses for strengthening works in view of the state-of-the art activities and issues described earlier.

Design for Axial Load. The axial load capacity of columns bonded with FRP systems arises from the effect of confining pressure provided by the FRP jackets at high loads. It has been argued that such a confinement effect is minimal with non-circular sections and that FRP jackets with fibers running longitudinally should not be relied upon to resist compression. Tests carried out by the author (Tan, 2001) indicated that it is possible to increase the strength of rectangular columns by as much as 20 percent, using a combination of transverse and longitudinal fibre sheets. It appeared that the longitudinal fibres could be relied on if they are confined by at least an equal amount of FRP sheets in the transverse direction.

Design for Flexure. It has been recognised that for an FRP strengthened section, the flexural failure modes are: (i) flexural compression due to concrete crushing; (ii) flexural rupture due to rupture of FRP sheets; (iii) flexural bond due to FRP debonding; and (iv) concrete cover delamination. Except for the last, the other three failure modes exhibit some degree of ductility, especially if the internal steel bars were to yield before failure, which is probably the case in most situations.

The author opined that cover delamination could be eliminated by the use of transverse sheets. By doing so, it is possible to derive the zones of flexural bond, flexural compression and flexural rupture for precracked beams (Tan 2001b). For a given internal tensile steel ratio ρ, the failure mode depends on the amount of externally bonded FRP sheets relative to the internal longitudinal reinforcement, quantified by the ratio of reinforcing indices ω_p/ω. As illustrated in Figure 1, the strengthened beam is likely to fail by flexural rupture if the internal tensile steel ratio is low and in flexural compression if it is high. For intermediate values of ρ, the failure mode could be any of the three, depending on the value of ω_p/ω. It is noted that the line BC depends on the primary crack spacing in the beam prior to bonding of FRP sheets.

Design for Shear. The shear capacity of members bonded with FRP sheets or strips, is generally evaluated by the conventional approach, that is, it consists of three components, carried by "concrete", steel reinforcement (stirrups, ties, spirals) and external FRP system. The contribution of the FRP system is evaluated using the conventional 45-degree truss method in the same manner as for the internal steel reinforcement, except that an effective strain in FRP and additional reduction factor are applied to account for the limited effectiveness of the FRP due to debonding or delamination.

For a general application of FRP sheets in shear strengthening of non-prismatic beams such as dapped beams (Tan 2001c), stepped beams, and beams with openings (Mansur, Tan and Wang, 1999) or recesses, a more refined approach using the strut-and-tie model is possible. Fig. 2(a) illustrates the strut-and-tie model for the design of the FRP reinforcement to strengthen a beam in which a recess has been created through the web. The forces in the strut and tie members for an increase in load carrying capacity are evaluated and additional FRP reinforcement as shown in Fig. 2(b) can be introduced with proper anchorages by assuming an effective strain in the FRP.

Anchorage and Detailing. One key issue that has to be further addressed is that of reinforcement detailing (Tan 2001d). The bond development length over cracked surface is of interest since strengthening works are likely to be carried out on cracked beams. The use of transverse FRP sheets and FRP grids as anchorages need to be further investigated. Anchoring FRP strips or sheets with mechanical or fibre bolts has been shown to improve strengthening effect in columns (Tan 2001a) and beams (Tan 2001c).

Figure 3 illustrates the case where the confinement effect of externally bonded transverse FRP sheets in rectangular columns could be enhanced by anchoring the sheets with a row of fibre anchors down the middle of the long faces of the column. It is seen that the fiber bolts act as anchor points and leads to a larger area of effectively confined concrete.

Durability and Fire Resistance. The long-term durability behaviour of FRP strengthened members has not been investigated to any great extent. Suitable design models to predict the service life under topical rainy climate in particular would be useful. Further research and recommendations regarding fire resistance are also desirable.

Concluding Remarks

Substantial work has been carried out on the structural use of FRP, especially in the North America, Europe and Japan, culminating in several design guidelines and recommendations. Extensive work is also currently being pursued in Singapore and other South East Asian countries. A concerted effort towards establishing a unified code that encompasses the needs and features of various regions is of significance as it would optimise research resources and promote international consensus on various code-related issues. Towards this end, some suggestions regarding design concepts are made.

Acknowledgment

The author acknowledges the provisions of research grants RP950683, RP981608 and RP992699 by the National University of Singapore for the studies reported.

References

ACI 440 (2001a). "Recommended Test Methods for FRP Rods and Sheets." *Draft February 2001*. American Concrete Institute, _ pp.

ACI 440 (2001b). "Guide for the Design and Construction of Concrete Reinforced with FRP Bars." American Concrete Institute, 41 pp.

ACI 440 (2001c). "Guide for the Design and Construction of Externally Bonded FRP Systems for Strengthening Concrete Structures." *Draft July 2000*. American Concrete Institute, _ pp.

Concrete Society (2001). "Design Guidance for Strengthening Concrete Structures using Fibre Composite Materials." *Technical Report 55*, 70 pp.

JSCE (1997). "Recommendation for Design and Construction of Concrete Structures using Continuous Fiber Reinforcing Materials." *Concrete Engineering Series 23*, Japan Society of Civil Engineers, 325 pp.

JSCE (2001). "Recommendationd for Upgrading of Concrete Structures with Use of Continuous Fiber Sheets." *Concrete Engineering Series 41*, Japan Society of Civil Engineers, __ pp.

Mansur, M.A., Tan, K.H. and Weng, W. "Effects of Creating an Opening in Existing Beams." *ACI Structural Journal*, 96(6), 899-905.

Tan, K. H. (1997a). "State-of-the-Art Report on Retrofitting and Strengthening by Continuous Fibers: Southeast Asian Perspective – Status, Prospects and Research Needs." Keynote Lecture, *Proceedings of the Third International Symposium on Non-metallic (FRP) Reinforcement for Concrete Structures (FRPRCS3)*, Japan Concrete Institute, Sapporo, Japan, October 14-16, 1997, Vol. 1, 13-23.

Tan, K. H. (1997b). "Behaviour of Hybrid FRP-Steel Reinforced Concrete Beams." *Proceedings of the Third International Symposium on Non-metallic (FRP) Reinforcement for Concrete Structures (FRPRCS3)*, Japan Concrete Institute, Sapporo, Japan, October 14-16, 1997, Vol. 2, 487-494.

Tan, K. H. and Balendra, T. (1999). "Strengthening of RC Columns using Preformed Glass FRP System." *Technical Report*, Prepared for Singapore Productivity and Standards Board.

Tan, K. H. (1999). "Towards a Cost-Effective Application of FRP Reinforcement in Structural Rehabilitation." Invited Paper, *Seventh East Asia-Pacific Conference on Structural Engineering & Construction (EASEC7)*, Kochi, Japan, August 27-29, 1999, Vol. 1., 65-73.

Tan, K. H. and Patoary, M. K. H. (2001). "Blast Resistance of Masonry Walls Externally Bonded with FRP Systems", *Interaction of the Effects of Munitions with Structures*, 10th International Symposium, California, USA.

Tan, K. H., Farooq, M.A. and Ng, CK. (2001) "Behaviour of Simple-Span Beams Locally Strengthened with External Tendons." *ACI Structural Journal*, 98(2), 174-183.

Tjandra, R.A. and Tan, K.H. (2001). "Strengthening of Reinforced Concrete Continuous Beams with External Carbon FRP Tendons." *Fifth International Symposium on Fiber Reinforced Polymer for Reinforced Concrete Structures (FRPRCS-5)*, Cambridge, UK, July 16-18, 2001.

Tan, K.H. (2001a). "Strength Enhancement of Rectangular RC Columns using FRP." *Journal of Composites in Construction*, Special Issue, American Society of Civil Engineers, USA (Submit for publication).

Tan, K. H. (2001b). "Effect of Interfacial Bond on Failure Mode of Precracked Members Retrofitted with CFS." *Concrete Engineering Series 42*, Japan Society of Civil Engineering, June 2001, II-59 - II-68.

Tan, K.H. (2001c). "Shear Strengthening of Dapped Beams using FRP Systems." *Fifth International Symposium on Fiber Reinforced Polymer for Reinforced Concrete Structures (FRPRCS-5)*, Cambridge, UK, July 16-18.

Tan, K. H. (2001d). "Detailing of FRP Reinforcement – An Overview." *International Conference on FRP Composites in Civil Engineering*, Hong Kong, December 12-14.

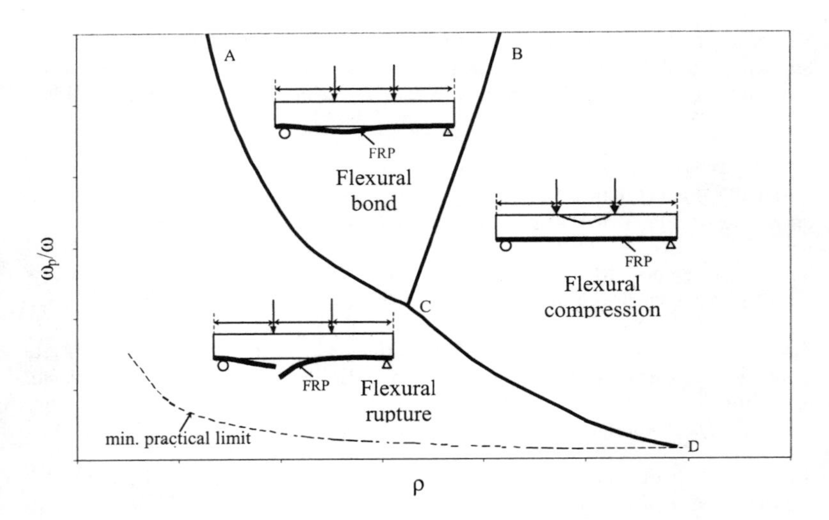

Figure 1. Flexural failure modes.

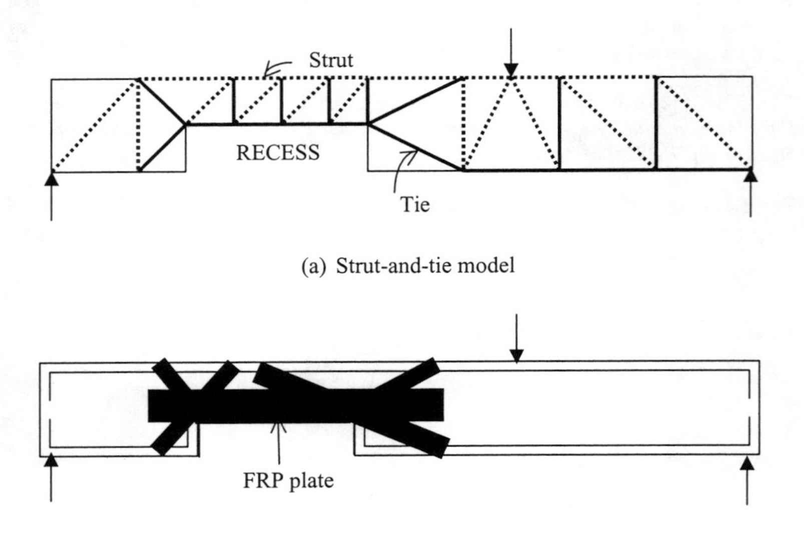

(a) Strut-and-tie model

(b) Strengthening using FRP plates

Figure 2. Strengthening using strut-and-tie method.

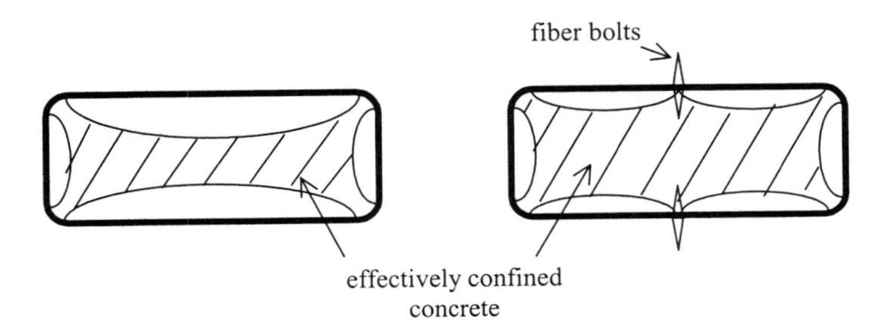

Figure 3. Use of fiber bolts in enhancement of confinement effect.

Codes and Standards for FRP as Structural Material in Japan

UEDA Tamon[1]

Abstract

This paper introduces briefly the Japanese codes and standards for FRP as reinforcement of new concrete structures and reinforcing material for upgrading existing structures. Some information on shaped FRP structural material is also provided.

Introduction

Unconventional materials may provide a better solution to construction under various environmental conditions or under space/time constraints. Introduction of the performance-based design, which only specifies the required performance of a structure, has made possible application of new materials that cannot be used under the specification-based design. As a result, fiber reinforced polymers (FRP) whose mechanical and chemical properties are quite different from the conventional ones have been gaining popularity steadily.

FRP as structural material for civil and building structures in Japan is mostly reinforcing material of concrete. FRP can be used as a substitute for steel reinforcement of new structures and as reinforcing material for upgrading (retrofitting and repairing) of existing structures. Usage of FRP itself as a material of the main structural component is still scarce. FRP used in Japan are carbon, aramid and glass with carbon fiber being the most commonly used for reinforcement of new and existing concrete structures.

This paper introduces briefly the codes and standards for FRP as reinforcement of new concrete structures and reinforcing material for upgrading the existing structures. At the end some information on shaped FRP structural material is provided. Due to space constraint, codes for construction are not introduced.

Standards for FRP Itself

The Japan Society of Civil Engineers (JSCE) and the Japan Concrete Institute (JCI) have proposed guidelines for testing method and standards for quality.

[1]Associate Professor, Division of Structural and Geotechnical Engineering, Hokkaido University, Sapporo, Japan 060-8628; phone +81-11-706-6218; ueda@eng.hokudai.ac.jp

Concrete Committee of JSCE issued the following standards for test methods of FRP (or called as "continuous fiber reinforcing material" in Japan) used in place of steel reinforcement or prestressing tendon in civil engineering structures (JSCE Second Research Committee on CFRM 1997):

- Test method for tensile properties of continuous fiber reinforcing material
- Test method for tensile properties at bend of continuous fiber reinforcing materials
- Test method for creep failure of continuous fiber reinforcing materials
- Test method for long-term relaxation of continuous fiber reinforcing materials
- Test method for tensile fatigue of continuous fiber reinforcing materials
- Test method for coefficient of thermal expansion of continuous fiber reinforcing materials by thermo-mechanical analysis
- Test method for performance of anchorages and couplers in prestressed concrete using continuous fiber reinforcing materials
- Test method for alkali resistance of continuous fiber reinforcing materials
- Test method for bond strength of continuous fiber reinforcing materials by pull-out testing
- Test method for shear properties of continuous fiber reinforcing materials by double plane shear

Later JCI proposed guidelines to improve and expand JSCE's standards for test methods to cover application to building structure (JCI TC952 1998). JCI methods do not include those for anchorage nor shear properties but the following new one:

- Test method for tensile properties at bent section of continuous fiber reinforcing materials

JSCE also presented quality specifications for FRP in place of steel reinforcement or prestressing tendon in civil engineering structures (JSCE Second Research Committee on CFRM 1997). The table of contents of the specifications is as follows: 1. Scope, 2. Presentation, 3. Category, Identification and Designation, 4. Quality of Fiber and Binding Material, 5. Mechanical Properties (volume ratio of axial fibers, guaranteed tensile strength, Young's modulus, elongation and relaxation rate), 6. Nominal Diameter and Maximum Size, 7. Test (sampling, tensile strength, creep failure strength and relaxation rate), 8. Calculation, and 9. Inspection. The types and configuration of FRP covered are: carbon, aramid, glass and vinylon; and rod, strand, braid, lattice and rectangle. Epoxy and vinyl ester are resins specified as binding material. Calculation methods are specified for volume ratio of axial fiber, nominal cross sectional area, nominal diameter, maximum size, nominal mass density, guaranteed capacity, tensile rigidity, elongation, creep failure capacity and relaxation rate. Inspection is conducted on mechanical properties and dimensions.

The standards for test methods of FRP sheet used as reinforcing material for upgrading existing structures were separately issued by JSCE (JSCE 292 Committee 2001). They include:

- Test method for tensile properties of continuous fiber sheets
- Test method for overlap splice strength of continuous fiber sheets
- Test method for bond properties of continuous fiber sheets to concrete

- Test method for bond strength of continuous fiber sheets to steel plate
- Test method for direct pull-off strength of continuous fiber sheets with concrete
- Test method for tensile fatigue strength of continuous fiber sheets
- Test method for accelerated artificial exposure of continuous fiber sheets
- Test method for freeze-thaw resistance of continuous fiber sheets
- Test method for water, acid and alkali resistance of continuous fiber sheets

Among the above standards the test methods for tensile and bond properties of continuous fiber sheet are based on the standards proposed by JCI (JCI TC952 1998). The test method for tensile properties (JCI TC952 1998) was prepared by modifying the national standard, JIS K 7073 "Test method for tensile properties of carbon fiber reinforced plastic" that was not for civil engineering structures.

Codes for New Concrete Structures with FRP Reinforcement

Recommendation for design and construction of concrete structures using continuous fiber reinforcing materials (CFRM) was issued by JSCE (JSCE Second Research Committee on CFRM 1997). The Recommendation was prepared to provide necessary information missing in the Standard Specification for Concrete Structures (JSCE Concrete Committee 1986). The design part of the Recommendation is based on the limit state design method in which ultimate, serviceability and fatigue limit states are examined. The main features of the design part are summarized as follows:

- Strength of bent portion of CFRM
- Tensile force-strain curves used for ultimate and serviceability limit states
- Flexural capacity
- Shear capacity of linear members (beams and columns)
- Punching shear capacity of planar members (slabs)
- Flexural crack width
- Basic development length
- Limiting stress of CFRM as tendon

Since CFRM has a significant lower strength at bend than its strength of straight part. The Recommendation provides Eq.1 for the design strength of bent portion of CFRM, f_{fbd}.

$$f_{fbd} = f_{fbk} / \gamma_{mfb} \tag{1}$$

with $$f_{fbk} = \left(0.05\frac{r}{h} + 0.3\right) f_{fuk} \tag{2}$$

where f_{fbk} is characteristic value of strength of bent potion, f_{fuk} is characteristic value of strength of straight portion, r is internal radius of bend, h is cross-sectional height of CFRM, and γ_{mfb} can generally be taken as 1.3.

Because of the fact that the tensile force-strain curve of CFRM shows some nonlinearity, different linear curves are specified for different limit states. The curve for ultimate limit states is defined as a linear line connecting the point of zero force and strain (or the origin) and the point of the tensile capacity and corresponding fracturing strain. On the other hand the curve for serviceability and fatigue limit states is the line connecting the points corresponding to 20% and 60% of the tensile capacity. Therefore, Young's moduli obtained from both curves are different and designated as E_{fu} and E_f respectively.

Steel shows yielding phenomenon and large fracturing strain. CFRM, however, does not show yielding until its brittle fracture in tension. Fracturing strains of CFRM is in the order of one tenth of steel fracturing strain. Therefore, steel reinforcement practically never reaches its fracturing point at flexural and shear capacity, while CFRM often fractures in tension. Another difference is the Young's modulus. The Young's modulus of steel is a constant of 200 GPa, while the Young's modulus of CFRM depends on the type and the volume content of fiber. Generally CFRM's Young's modulus is smaller than that of steel. Because of these differences, methods to calculate flexural and shear capacities are different. Design flexural capacity for members with CFRM as tension reinforcement should be calculated with not only the concrete strain limited to its ultimate strain in compression but also reinforcement strain limited to its fracturing strain. The limiting of reinforcement strain can usually be ignored for steel reinforcement.

The design equation for shear capacity of concrete members with steel reinforcement is generally experimental one. The equation is only applicable for members with reinforcement that yields but never fractures and whose Young's modulus is 200 GPa. The JSCE recommendation, therefore, introduced new equations for shear capacities. Equation 2 is for design shear capacity of linear members.

$$V_{yd} = V_{cd} + V_{sd} + V_{ped} \tag{2}$$

where V_{cd} is design shear strength carried by concrete that is equal to design shear strength of linear members without shear reinforcement in which γ_b=1.3, V_{sd} is design shear strength carried by shear reinforcement in which γ_b=1.15, and V_{ped} is component of effective tensile force of longitudinal tendon parallel to shear force in which γ_b=1.15.

$$V_{cd} = \beta_d \beta_p \beta_n f_{vcd} b_w d / \gamma_b \tag{3}$$

$$V_{sd} = \left[A_w E_w \varepsilon_{fud} (\sin\alpha_s + \cos\alpha_s)/s_s + A_{pw}\sigma_{pw}(\sin\alpha_{pw} + \cos\alpha_{pw})/s_p\right]z/\gamma_b \tag{4}$$

$$V_{ped} = P_{ed} \sin\alpha_p / \gamma_b \tag{5}$$

with f_{vcd}=0.20$f'_{cd}{}^{1/3}$, β_d=$d^{-1/4}$ (≤1.5), β_p=$(100p_w E_t/E_0)^{1/2}$ (≤1.5), β_n= 1+M_o/M_d (≤2.0) for $N'd$≥0 or = 1+2M_o/M_d (≥0) for N'_d<0, and

where N'_d is design axial force, M_d is design bending moment, M_o is bending moment to cancel stresses at tension end under the design bending moment, b_w is web width, d is effective width (in m for β_d), p_w=$A_f/(b_w d)$, E_t is Young's modulus of tensile reinforcement (=E_{fu}), E_0 is reference Young's modulus (=200 MPa), A_f is area of tensile reinforcement, f'_{cd} is design concrete compressive strength, A_w is total area of shear reinforcement within its spacing s_s, A_{pw} is total area of tendon as shear reinforcement within its spacing s_p, E_w is Young's modulus of shear reinforcement ($=E_{fu}$), ε_{fud} is design value of shear reinforcement strain at ultimate limit state, obtained from Eq.6, σ_{pw} is tensile stress in tendon as shear reinforcement at ultimate limit state ($=\sigma_{wpe}+E_{fpw}\varepsilon_{fud}\leq f_{fpud}$), σ_{wpe} is effective tensile stress in tendon as shear reinforcement, E_{fpw} is Young's modulus of tendon as shear reinforcement, f_{fpud} is design tensile strength of tendon as shear reinforcement, α_s is angle between shear reinforcement and member axis, α_{pw} is angle between tendon as shear reinforcement and member axis, z is distance between compressive resultant and the centroid of tension reinforcement, which may be $d/1.15$ in general, P_{ed} is effective tensile force in longitudinal tendon, α_p is angle between longitudinal tendon and member axis,

$$\varepsilon_{fwd} = \sqrt{f'_{mcd}\frac{p_w E_t}{p_{web} E_w}}\left[1+2\frac{\sigma'_N}{f'_{mcd}}\right]\times 10^{-4} \leq \frac{f_{fbd}}{E_{fu}} \tag{6}$$

where f'_{mcd} is design compressive strength of concrete considering size effect ($=(h/0.3)^{-0.1}f'_{cd}$ in MPa), $p_{web}=A_w/(b_w s_s)$, σ'_N is average axial compressive stress ($=(N'_d+P_{ed})/A_g\leq 0.4f'_{mcd}$), h is member height, A_g is gross sectional-area, and f_{fbd} is design strength of bent portion of CFRM (see Eq.1). From Equations 3, 4 and 6 the shear strength carried by concrete increases with an increase in Young's modulus of CFRM as tension reinforcement, while shear strength carried by shear reinforcement increases with an increase in Young's modulus of CFRM as both tension and shear reinforcement. This is because the ultimate strain in CFRM, ε_{fud} as shear reinforcement increases with their Young's modulus although it is limited due to its low strength of bent portion, f_{fb}.

Punching shear strength of planar members is also known to increase with stiffness of tension reinforcement similar to the shear strength of linear members without shear reinforcement. Therefore, JSCE equation for punching shear strength includes the same factor, β_p as in the equation for linear members (see Eq.3).

If bond properties of CFRM obtained by the standard test method are similar to those of steel reinforcement, flexural crack width can be evaluated by the same equation as that for members with steel reinforcement. CFRM available in Japan mostly show good bond properties like steel deformed bar. The development length of tensile reinforcement can also be calculated by the same equation for steel reinforcement as below:

$$l_d = \alpha_1 \frac{f_d}{4f_{bod}}\phi \tag{7}$$

where ϕ is diameter of tensile reinforcement, f_d is design tensile strength of reinforcement, f_{bod} is design bond strength of concrete ($=0.28\alpha_2 f'^{2/3}_{ck}/\gamma_c\leq 3.2$ MPa), α_2 is modification factor for bond strength of reinforcement (=1.0 where bond strength is equal to or greater than that of deformed steel bars, otherwise <1.0), f'_{ck} is characteristic compressive strength of concrete, α_1=1.0 for $k_c\leq 1.0$, 0.9 for $1.0< k_c \leq 1.5$, 0.8 for $1.5< k_c \leq 2.0$, 0.7 for $2.0< k_c \leq 2.5$, 0.6 for $2.5< k_c$, and

$$k_c = \frac{c}{\phi} + \frac{15A_{tr}}{s\phi}\frac{E_{tr}}{E_0} \tag{8}$$

where c is cover of reinforcement or half of the space between reinforcement, whichever is the smaller, A_{tr} is area of transverse reinforcement which is vertically arranged to the assumed surface of splitting failure caused by bond, s is distance between the centers of transverse reinforcement, E_{tr} is Young's modulus of transverse reinforcement, and E_0 is reference Young's modulus (=200 GPa). It can be seen from Eqs.7 and 8 that transverse reinforcement with a Young's modulus smaller than that of the steel reinforcement gives a longer development length than steel reinforcement.

Limiting stress for CFRM as tendon for serviceability condition should be equal to or less than the design value for creep failure strength obtainable by the standard test.

Code for Upgrading of Existing Structures with FRP Reinforcement

JSCE has issued two related codes lately: Guidelines for Retrofit of Concrete Structures – Draft – (JSCE Working Group, 2000; or JSCE 292 Committee 2001) and Recommendations for Upgrading of Concrete Structures with Use of Continuous Fiber Sheets (JSCE 292 Committee 2001). The former covers jacketing, external bonding and external cable with FRP reinforcement (sheet and cable), while the latter covers jacketing and external bonding with FRP sheet. Both adopt the performance-based concept and provide guidelines for both the design and construction. The main contents for design are as follows:

- Flexural capacity
- Shear capacity of linear members (beams and columns)
- Ultimate deformation of columns
- Flexural crack width
- Protection from chloride ion penetration

The equation for external bonding with continuous fiber sheet (FRP sheet) derived from the interfacial fracture energy between continuous fiber sheet and concrete, G_f as follows (JSCE 292 Committee 2001):

$$\sigma_f \le \sqrt{\frac{2G_f E_f}{n_f \cdot t_f}} \tag{9}$$

where σ_f is continuous fiber sheet stress at flexural crack caused by maximum moment, n_f is number of sheet layers, and t_f is thickness of sheet. If Eq.9 is satisfied, no delamination would occur. The ultimate flexural capacity can be calculated by the conventional bending theory with continuous fiber sheet stress limited to its fracturing stress. If Eq.9 is not satisfied, partial delamination would occur. In this case ultimate flexural capacity can be calculated as the lesser of the following two cases:

(i) Ultimate flexural capacity that can be calculated by the conventional bending theory with continuous fiber sheet stress limited to $0.9f_{fud}$. The reduction factor of 0.9 is to consider the fact that the hypothesis of linear strain distribution is no longer true after partial delamination.

(ii) Flexural moment when difference in continuous fiber sheet stress between location at flexural crack caused by maximum moment and location at the next flexural crack, $\Delta\sigma_f$ reaches the value given by Eq.10. This formula is to predict total delamination.

$$\Delta\sigma_f = \sqrt{\frac{2G_f E_f}{n_f \cdot t_f}} \tag{10}$$

Interfacial fracture energy G_f is obtained by pull-out test of continuous fiber sheet bonded to concrete, otherwise G_f=0.5 (N/mm) may be assumed.

In the case of FRP sheet jacketing the shear capacity of FRP sheet is quite different from that of steel shear reinforcement, since FRP sheet is an elastic material with rather small fracturing strain. The following formula that is applicable to both carbon and aramid fiber sheet was proposed (JSCE 292 Committee 2001):

$$V_{ud} = V_{cd} + V_{sd} + V_{fd} \tag{11}$$

with $$V_{fd} = K \cdot [A_f \cdot f_{fud}(\sin\alpha_f + \cos\alpha_f)/s_f] \cdot z/\gamma_b \tag{12}$$

where $K = 1.68 - 0.67R \qquad (0.4 \le K \le 0.8)$, $R = (\rho_f \cdot E_f)^{1/4}\left(\frac{f_{fud}}{E_f}\right)^{2/3}\left(\frac{1}{f'_{cd}}\right)^{1/3}$

$(0.5 \le R \le 2.0)$, and

$\rho_f = A_f/(b_w s_f)$, A_f is total cross-sectional area of continuous fiber sheets in space s_f, b_w is web width, f_{fud} is design tensile strength of continuous fiber sheet (N/mm^2), z is distance between compressive and tensile resultants in a cross-section, E_f is modulus of elasticity of continuous fiber sheet (kN/mm^2), α_f is angle formed by continuous fiber sheet about the member axis, f'_{cd} is design concrete strength, and γ_b is member factor (generally may be set to 1.25). The above formula was experimentally obtained, however it cannot distinguish failure modes with/without sheet fracture.

In the case of members that are upgraded with continuous fiber sheet externally bonded on the side of the members, ultimate shear capacity can be estimated by the method in which the stress distribution of the continuous fiber sheets is evaluated based on the bond constitutive law to determine the shear contribution of the sheet (JSCE 292 Committee 2001). This method uses numerical calculation to evaluate the stress distribution of the continuous fiber sheet in upgraded members to determine the shear contribution of the sheet.

As in the case of ultimate shear capacity the equation for ultimate deformation cannot be applied to the case of jacketing with continuous fiber sheet due to the difference in the mechanical properties between continuous fiber sheet and steel reinforcement. Therefore, JSCE Recommendations proposes the following formula (JSCE 292 Committee 2001):

and 2000 Part 1 [Design] and Part 2 [Construction] have been re-issued respectively in Japanese.)

JSCE Working Group on Retrofit Design of Concrete Structures in Specification Revision Committee (2000). "Guidelines for Retrofit of Concrete Structures – Draft –." *Concrete Library International*, JSCE, 36, 61-112. (Note: This is also included in another reference (JSCE 292 Committee, 2001) as appendix.)

Japan Association of Reinforced Plastics (1994). *Handbook for Structural Design of Reinforced Plastics.*

BEN Goichi (2001). "Design Criterion of FRP Structures and Their Applications." *Proceedings of The 1st Symposium on FRP Bridges, Structural Engineering Series*, JSCE, 21 19-26 (in Japanese).

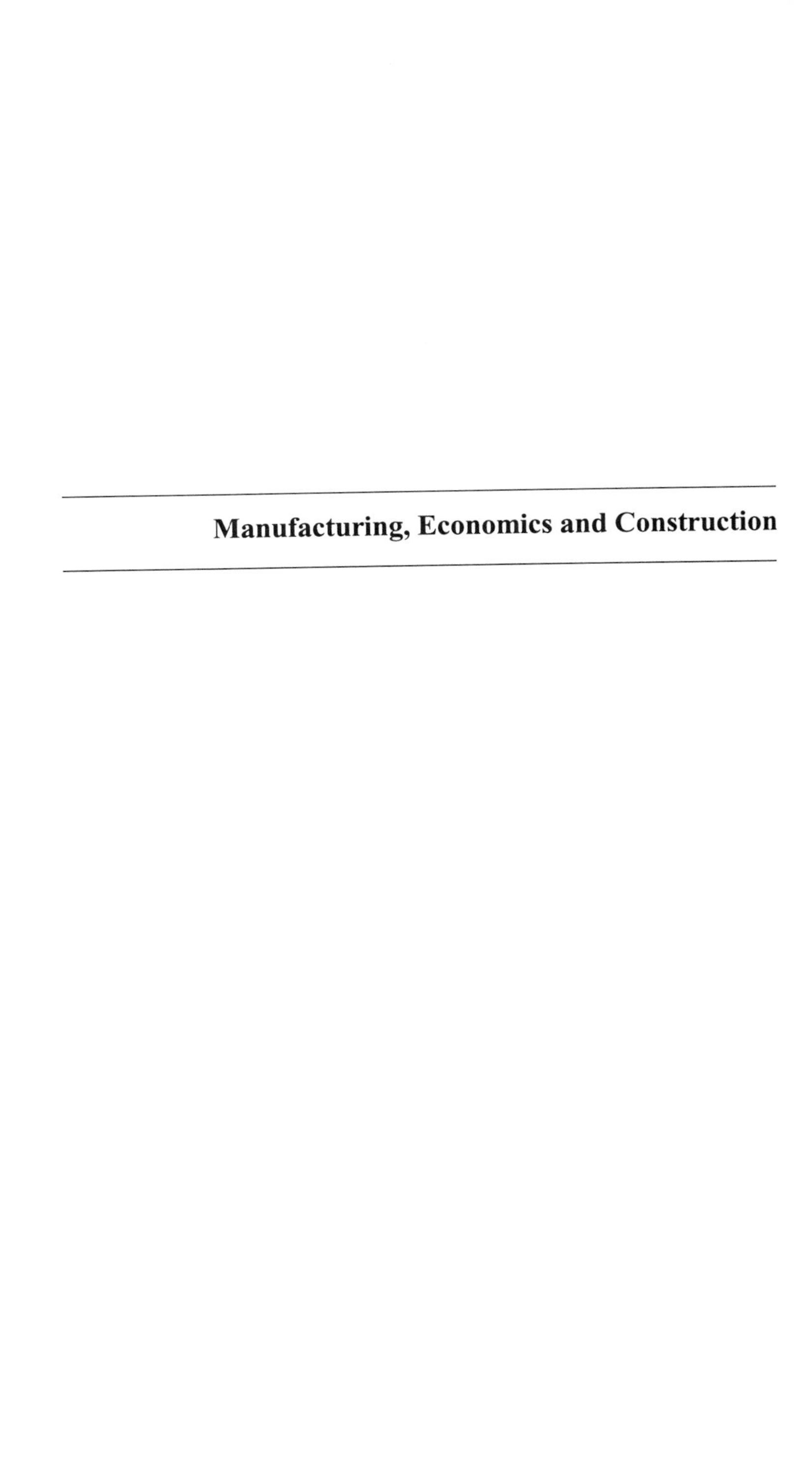

Manufacturing, Economics and Construction

FRP Rebar - A State of the Industry Report
Manufacturing, Construction, Economics and Marketing

Doug Gremel[1]

Abstract

An overview of the state of the FRP Rebar industry is given describing key milestones accomplished. Experiences of the author in terms of the construction, use and handling of GFRP bars are sited along with sales volumes in terms of lineal footage of GFRP rebar supplied. A discussion of the economics of GFRP rebar with specific examples from a bridge deck application noted as well as the successful implementation of GFRP rebar in the restoration of a dry dock at Pearl Harbor Hawaii. Future industry needs and barriers to further adoption of GFRP Rebar are discussed. Breakthroughs that would be beneficial to the composites industry in terms of infrastructure applications are expressed.

Introduction

It's an exciting time to be a producer of FRP materials. Our nascent industry is beginning to take form and enthusiasm is high among researchers, potential customers, institutional bodies and fellow competitors. Several very important milestones have been accomplished but we have more work ahead. A consensus seems to have emerged that FRP materials for reinforcing concrete are legitimate. Little concern is expressed these days in terms of the short -term performance of FRP reinforcing. Emphasis is on long-term durability, which will only be resolved with careful and conservative implementation. FRP reinforcing seems to be moving from the laboratory to the field, with a changing emphasis from research to implementation.

One of the chief accomplishments is the publication of ACI 440.1R-01 Design Guidance For The Use FRP Reinforcing. The growth of ACI committee 440 is a story in and of itself. A few years ago, the entire subcommittee could sit around the table in a small conference room. Our most recent 440 main meeting in Philadelphia required a ballroom size facility complete with sound system. Committee 440 is now one of ACI's largest and most active committees. 440 is an active group and has the buzz about it that says this is something exciting one should be a part of. In a recent copy of the ACI structural journal, Vol 98 No 3 May-June 2001, 7 out of 16 papers or approximately 40% of the journal dealt with topics in the field of FRP's. Under the Innovative Bridge Research program

[1] Director, Non-Metallic Reinforcing, Hughes Brothers Inc, Member ACI440, Chairman FRP Rebar Manufacturing Council, 210 N. 13th Street, Seward, NE 68434 Ph:402-646-6211, doug@hughesbros.com

sponsored by the Federal Highway Administration, approximately 60% of the projects funded involve FRP.

Publication of ACI 440.1R-01 is a distillation of all the significant research that has been performed around the world. It will allow the practicing engineer to use FRP rebar in a safe and conservative fashion. Publication of the design guidance will allow for expanded applications that aren't yet foreseen.

Another important milestone in our nascent industry is the formulation of the FRP Rebar Manufacturers Council. This group of five FRP bar manufacturers realize they need to be proactive in creating quality standards, product specifications and field application guidelines. The charter manufacturers are: Composite Rebar Technology, Inc, Hughes Brothers, Inc, Marshall Industries Composites, Inc. Pultrall, and Dow Chemical. It is hoped that other manufacturers will see the benefit of a common trade organization in the future. Initial efforts of the FRP Rebar Manufacturers Council include participation in HiTEC (Highway Innovative Technology Center) evaluation by various DOTS', FHWA and academia. Application for ICBO approval criteria for FRP bars, dissemination of information about ACI440 design guidelines and the participation in design guidelines for the use of FRP bars for strengthening of un-reinforced masonry. The FRP Rebar manufacturers Council is a working group of the Market Development Alliance (MDA) of the Composites Industry headed by Mr. John Busel.

The MDA has played a key role as a working trade organization in expediting the adoption of FRP materials in general. In addition to the FRP Rebar Manufacturers Council, working groups are focused on FRP applications for Bridges, FRP Dowel Bars for jointed concrete paving, FRP materials for strengthening of un-reinforced masonry, FRP reinforcing of engineered wood materials and the use of FRP materials in waterfront construction. The MDA plays a key role as an industry advocate separating the partisan interests of the individual companies and acting as a single voice or point of reference for unbiased information on topical areas.

For potential customers or specifiers of FRP bars, a key consideration is that there are several potential sources of FRP Bars. While each of the FRP Rebars may be somewhat unique in its surface treatment and specific physical mechanical properties, it is possible to design to a common denominator to achieve a suitable end use. The ACI 440 design guide is not based on any one proprietary product. Several large scale trial projects have been bid using common specifications to which several FRP Rebar suppliers have had to bid. For example, the Sierrita de la Cruz Creek bridge built by the Texas Dept of Transportation, the Morristown Vermont bridge built by the Vermont Transportation Agency and the rehabilitation of Dry Dock #4 at Pearl Harbor by the US Navy.

FRP Rebar suppliers owe a debt of gratitude to the many key academic researchers around the world who have performed research in the field of FRP's. Some of these organizations could be held as model examples of leveraging resources for maximum gain. International collaboration, coordination to reduce duplication of research, prodigious output of information, and working with both government and industry to produce practical results have been the hallmark of research organizations such as ISIS Canada, CIES of Missouri Rolla, and TMR ConfibreCrete of Europe just to name a few.

The field of FRP's is truly international in scope. Significant efforts have produced design guidelines in Japan, Norway, United Kingdom, Canada and the United States. The output of European researchers will likely leap frog the ACI 440.1R-01 document with the publication of FIB (Federation International du Béton) Task Group 9.3 efforts. The task group consists of about 60 members, representing many European universities, research institutes and industrial companies working in the field of advanced composite reinforcement for concrete structures, as well as representing members from Canada, Japan and USA.

Construction

While all this activity is taking place at the institutional and academic level, end users and specifiers continue to find applications in which to incorporate FRP Bars. It should be explained that the author has considerable experience in the field of GFRP rebar being in the employ of one of the leading suppliers. In calendar year 2000, our firm alone furnished approximately 10 million lineal feet of GFRP rebar. This volume has grown from nothing in 1995, our first full year of production.

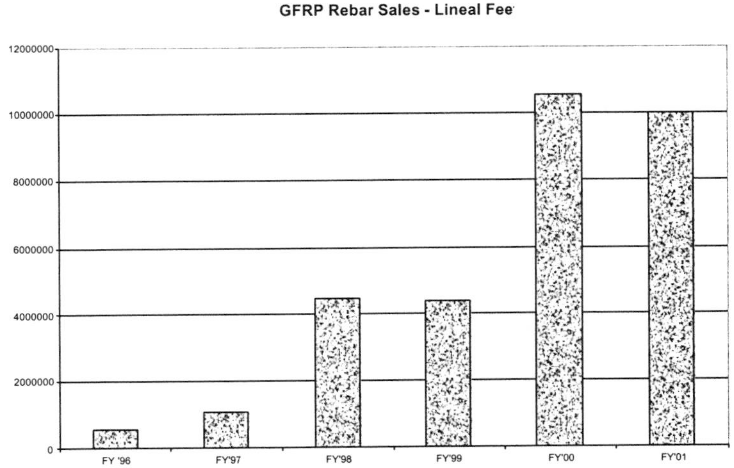

Figure 1. Hughes Brothers GFRP Rebar Sales - lineal feet

The majority of this GFRP rebar was used in "cosmetic concrete" or in constituent matrixes other than portland cement. Experience gained thus far has established a threshold for expanding the use of GFRP bars in more demanding applications such as top mat reinforcing for bridge decks, seismic upgrading of masonry, general waterfront and marina construction and uses of FRP bars in general civil engineering structures.

Another very common application has been the use of GFRP rebar in Magnetic Resonance Imaging areas in hospitals; used due to its magnetic transparency. A large number of other applications have used GFRP bars for their electrical neutrality in high voltage substations and reactive transformer pads, situations involving high voltage cabling in steel mills and foundries, magnetic and RF research laboratories and concrete near airports.

A large number of applications near marine salts have used FRP bars in seawalls, floating marine docks, general waterfront construction, dry dock rehabilitation and balcony restoration.

In terms of transportation applications, there have been several bridge decks in the US and Canada that have incorporated GFRP rebar in some portion of the bridge deck. West Virginia has used FRP rebar in the concrete decks of three box beam bridges, as well as the McKinleyville bridge deck. Texas has a highly instrumented application near Amarillo that will provide invaluable information on the performance of GFRP rebar in bridge deck application for the next several years. Bridge decks incorporating FRP Rebar are being built this summer in Iowa for the City of Bettendorf. Bids have been let for a bridge in Morristown Vermont also. The Salem Avenue bridge in Ohio and half a dozen bridge decks in Canada have used FRP bars. A full scale crash test has been performed at the Texas Transportation Institute of the Texas guardrail system incorporating GFRP bars in the top mat of a bridge deck.

Figure 2. TxDOT Sierrita de la Cruz Creek Bridge

As experience is gained in the construction and use of FRP bars, it's becoming evident that there are only very minor issues associated with implementation of the material over epoxy coated steel reinforcing. Common misperception is that GFRP bars are more delicate than epoxy coated bar. To the contrary, FRP bars seem to take the bumps associated with field installation better than the epoxy coating on steel bar. Field observations have been made that GFRP bars can become damaged in the field, but compared to the epoxy coating on steel bars, are much more durable and easily handled for the tying and placing of the rebar.

Precautionary measures taken thus far have included the tying of each rebar intersection. This appears to be overkill, but necessary for early applications due in part to the greater deflections encountered with construction foot traffic on bridge decks. As additional experience is gained, this restriction may be relaxed. The only other precautionary measure taken has been to use double the number of chair supports and to tie the mat down in several places for fear of having the mat float due to the lower specific gravity of the FRP bars. Experience has shown than in precast production facilities with large vibratory tables, the GFRP bars will float if not tied down. However, experience with placing concrete on bridge decks has shown that with the heavy foot traffic on the mat during placing and with the relatively less vibrating done in that application, by only tying the rebar mat down in a few places there have been no issues with the FRP mat floating.

Thorough design analysis has been performed on several bridge deck applications incorporating FRP bars. It should be emphasized that traditional design methods are still valid. While ACI 440.1R-01 does outline differences associated with FRP bar design, the basic principles of reinforced concrete design still apply. Intelligent use of the properties of FRP rebar dictates that the material be used in conjunction with traditional steel reinforcing in many applications placing the FRP bars in the areas of the structure most vulnerable to corrosion of the steel reinforcing.

Adopting FRP bars is a relatively minor step in the construction process. Traditional design methodology can be used taking into account the properties of the FRP bar and ACI Recommendations. The traditional bid letting and bidding process familiar to contractors, bar suppliers and rod busters and the use of multiple sources of supply of FRP bar and installation by the same methods make consideration of FRP rebar straight forward.

Economics

The economics of FRP bars are such that they will always be more expensive than black steel. The basic raw materials, glass fibers and polymer resin each sell for approximately \$1/lb, not including processing costs. While the strength to weight ratio is approximately $1/4^{th}$ that of steel, due to the low E-Modulus of the glass fibers and various safety factors, current utilization of the material is not very efficient. Pricing of epoxy coated rebar varies wildly by region (based on ENR Construction Economics), and its quite possible that compared to epoxy coated bar, GFRP bars may only represent a slight premium in some locals in the future.

While the light weight of the FRP Rebar does make the material easier to handle and install, its not clear that any saving found by the contractor will be passed along to the owner in the foreseeable future. This feature has been a tremendous benefit in terms of transportation costs. It's quite possible to ship large and small volumes of FRP bar around the world from a central factory.

We can begin to study some precise economic figures from actual installations. Mr. Timothy Bradberry, of the Texas Department of Transportation, performed a careful analysis for the Sierrita de la Cruz Creek Bridge. (TRB paper 01-3247: FRP Bar Reinforced Concrete Bridge Decks) For this project using GFRP Bars as top mat reinforcing in place of epoxy coated steel added $1.79/ft^2, which represents a 12.4% increase in the cost of the deck. The impact on the total cost of this bridge would be 6.4%.

Attempts to determine the economic impact of this 6.4% increase in cost become complicated by a number of assumptions. What is the anticipated service life of the structure using traditional steel reinforcing? What is the expected service life using FRP reinforcing? What is the future value of the additional money invested in the structure today? What are the present value expenses of rebuilding the structure at the end of its service life?

Figure 3. Dry Dock #4 Pearl Harbor - Steel Rebar degrading or gone

A recent installation at a government dry dock facility helps shed light on the economic reasons for implementing FRP bars. Approximately 10 years ago, rehabilitation was undertaken at a dry dock in Pearl Harbor that used steel reinforcing with 2" of cover to put a new façade on the interior face of a dry dock exposed to marine salts. After only 10 years of service, large portions of the face of the dry dock are deteriorated and spalling off the face of the dry dock.

Figure 4: Restoration made using GFRP Rebar

In many areas, the steel reinforcing is gone and all that remains is a stained area inside the concrete. A new cast in place façade is being patched over the spalled areas reinforced with FRP bars that have been doweled and epoxied into place in conjunction with a grid of FRP bars tied in place at the job site. In this very harsh service environment, the hurdle for FRP reinforcing to exceed the service life performance of steel is quite low.

Needs

With the ACI 440 design guidelines now published as an emerging technology series, the question becomes; What next? The most urgently needed institutional document is a set of standardized test methods for FRP Rebar. ACI 440-K subcommittee has reviewed and balloted drafts of proposed test methods. This effort has produced some excellent but incomplete methodologies for independent testing of FRP bars. This activity will evolve into an effort to publish ASTM test methods.

Another important requirement that will facilitate broad acceptance of FRP bars is the development of ICBO acceptance criteria. As an industry, the FRP Rebar manufacturers Council is just beginning an effort to develop these criteria. Individual manufacturers would then be required to submit supporting documentation to show that an individual FRP bar is acceptable.

It is hoped that the FRP Rebar Manufacturers Council can publish minimum performance criteria common to all FRP Rebar suppliers. This would aid the designer by reducing variations between the unique products on the market.

Other exciting developments are rapidly advancing in the area of using FRP bars for seismic strengthening of un-reinforced masonry structures. Along these lines a new task force is being set up to liaison between the ACI 440 committee and the masonry society.

To expedite further use by the design community, it would be helpful if a number of published applications notes describing actual implementation of FRP bars were available. They should be documented real life uses of the material to highlight the advantages of FRP rebar. The proliferation of this type of information will go a long way towards addressing the lack of familiarity with FRP by practicing engineers.

Barriers

Many of the barriers to further adoption of FRP rebar are being addressed. Lack of familiarity by practicing engineers will lessen in time due to the objectives of ACI440-D, professional education subcommittee. The goal of this effort is create an ACI seminar that will be given in several locations throughout the year. Participating engineers would receive continuing education credits for attendance.

One problem with published research papers is that they often lack detail as to the constitutive make-up of the composite materials being tested. Both researchers and industry share the blame for not being more open about constituent materials and processing information. Resin and fiber type, percentage of fiber, and any processing variables available should be dutifully reported in every paper. One would never undertake a metallurgical analysis without attempting to describe the type of steel being tested. The same situation exists with composites. As an industry, we need to realize that unless we are more open with our processes we will never advance. In a fast changing environment of research and development, our products continue to improve and without proper benchmarking we won't be able to know in hindsight what improvements if any were achieved. Decades old research on GFRP bar durability with undocumented material descriptions continues to hamper further advancement of the industry.

The historical development of fiber reinforced polymers stem from the military and aerospace industry that operate with completely different requirements than the construction industry. FRP fabricators who are used to dealing with customers in the same vein will quickly frustrate their potential customers and themselves. This is not to say that the requirements of the construction industry are less stringent. Contractors operate under a different set of deliverables than the mil-aero defense complex. Its not uncommon for a concrete contractor whose project has FRP bars specified to call needing product delivered to the job site in less than a weeks time. Prior to the phone call, the fabricator may have had no knowledge that their product was specified or of the project itself. Ideally, the FRP supplier would have a local representative working with engineering firms on specifying projects and a local stocking distributor to be able to service the local marketplace. Creating such a structured customer service network is a major undertaking.

Breakthroughs

Looking to the future, what kind of breakthroughs are needed to further advance the use of FRP reinforcing? Clearly carbon fibers are the material of choice but use is limited by their expense versus e-glass fibers. Any breakthrough in carbon fiber pricing will accelerate their use. While a long way from becoming commercially viable for use in pultrusion, one has to be excited about the possibilities of $1 to $2 per pound pitch based carbon fibers based on bi-products from the oil refining process.

In the area of GFRP bars, the cloud of durability still is on our horizon. The basic resin make-up, incorporating vinylester resin with greater strain capabilities has proven to be a superior matrix for encapsulating and protecting e-glass fibers. Perhaps a consensus will be achieved regarding the validity of accelerated durability testing for GFRP bars. Researchers should be cautioned against making broad brush conclusions for whole fiber types without detailing in their research the constitutive make up and processing of the materials they are testing. Even sizing chemistries vary between e-glass suppliers that affect performance with common resin systems.

Research using AR-glass fibers have proven frustrating in that similar or worse results to e-glass durability are achieved and yet the AR-glass fibers are used bare in concrete. It is hoped that advancements in sizing technology to enable better bond to the resin matrix will make the use of AR-glass fibers possible.

Advancements in resins will likely continue to improve the durability of e-glass GFRP Rebars. Thermoplastic resins, Urethanes and other resins with high toughness and greater elongation under strain have proven to have greater durability in preliminary testing. Additives and self-healing polymers to the resin may also show promise in protecting e-glass fibers from alkaline degradation.

Conclusions

The outlook for FRP Rebar is very promising. While the rate of adoption of this new material may seem slow to those of us whose careers are on the line, put in a historical perspective, the rate of adoption of FRP materials for reinforcing concrete has been very rapid.

The use of steel reinforced concrete design has had approximately 125 years to develop to its current state. Test methods, crack width equations and other basic parametric measures are still evolving for steel reinforced concrete design. While steel reinforced concrete is a tried and true methodology, there are some well understood limitations that FRP reinforcing can address to the overall benefit of concrete construction. Many solutions for mitigating the corrosion of steel reinforcing in concrete have been adopted by the industry, but altering the reinforcing gets to the root of the problem.

Inevitably, concrete structures will develop cracks which will ultimately lead to corrosion of the steel reinforcing. Careful use of this new technology by placing FRP bars in areas known to be susceptible to ultimate deterioration of the structure as a whole will improve the longevity of our civil infrastructure.

FRP in Construction: Applications, Advantages, Barriers and Perspectives

A. Balsamo[1], L. Coppola[2], P. Zaffaroni[2]

Abstract

The state-of-the-art for applications of fiber reinforced polymers (FRP) in civil engineering is illustrated, highlighting the advantages achievable in the strengthening and static improvement of structures by using these innovative materials compared to traditional systems. Possible future developments in their applications are then discussed. Limitations, barriers, and problems to be solved for the diffusion of FRP materials in construction are also outlined.

Introduction

The applications of FRP in civil structures are many. The interest of engineers and contractors in these applications is becoming stronger as the number of real jobs with FRP increases in many countries. Due to their high technical properties, composites show their best performance when used in the upgrade, restoration, repair or seismic strengthening of simple and complex structures. Typical examples of carbon fiber (CFRP) reinforcement (Figure 1) used in framed structures are:

- wrapping of columns using continuous sheets. The composite material is impregnated in situ and bonded to the columns. The wrap reduces the transverse expansion of the compressed element by confinement, therefore increasing its compressive strength.
- flexural and shear strengthening of beams realized by surface bonding of the impregnated material in situ or pultruded laminates; in some cases, the anchorage of sheets can be improved with the use of FRP rebars.

[1] Adjunct Lecturer, University of Naples "Federico II" – Dept. of Costruzioni e Metodi Matematici in Architettura, Via Monteoliveto 3 - 80139 Naples, Italy; phone +39-081-5523553; albalsam@unina.it
[2] MAPEI S.p.a. - Milan, Italy; phone +39-02-37673 ; seg_tec@mapei.it

(a)

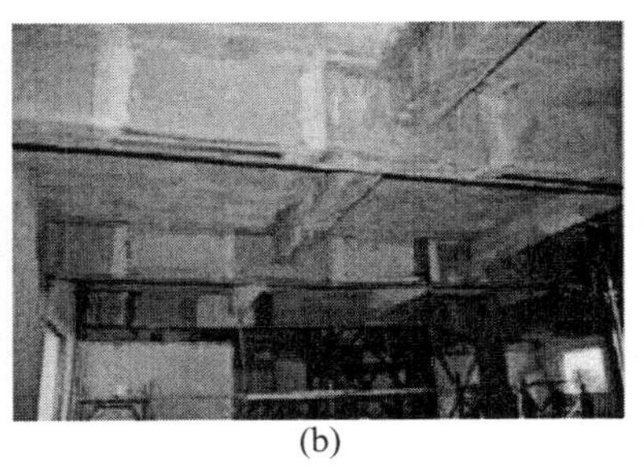

(b)

Figure 1. Strengthening of a Reinforced Concrete structure with CFRP – Municipal Slaughter-House (built in 1930) transformed into University Facility (Italy, 2001).

Advantages of FRP Compared to Traditional Materials

The real advantages of using FRP as an alternative to traditional methods should be evaluated from both technical and economic points of view.

Technically speaking, composite materials have significant advantages, most importantly their light weight along with high mechanical properties; resistance to aggressive chemical agents and impermeability to water. The use of steel, widely adopted in conventional construction, has several disadvantages, such as reduced durability caused by its vulnerability to chemical agents and corrosion. Moreover, the steel has little reversibility, whereas FRP is completely reversible since adhesive materials that transfer stresses can be removed.

Further important characteristics of composites include the following:

- unlimited capacity to be molded (fabric impregnated in situ) and perfect adaptability to the original shape of the structural element that needs to be reinforced, stiffened and/or integrated (e.g. frame knots, masonry vaults, etc);
- simple application methods with absence of complex preparation;
- non-invasive application. Their installation does not disrupt the original material of the structure;
- very brief time required to reach full mechanical performances of the applied reinforced composite.
- durability.

From an economic standpoint, the following are the main factors contributing to their competitiveness with respect to conventional materials:

- time saving;
- flexibility;
- low labor costs;
- low tooling and machinery costs on the construction site because of the light weight and manageability of tools and material used. The advantages of strengthening reinforced concrete with FRP compared to conventional methods (beton-plaquè) can be seen in the 65% drop of total costs of application because of reduced labor needed and ease of installation in situ;

- possibility of restoring a structure without interrupting its utilization by users. A typical example is the upgrade of piers and/or motorway bridge pillars damaged by unusual events (i.e., large/exceptional vehicle impact) with FRP. In this case, the work is generally carried out while the structure is still in use or with a limited and temporary closure of a traffic lane. This results in cost benefits and reduced effects on users and consumers.

Perspectives for the Application of FRP in Construction

Structural safety is always required, especially in seismic areas where social and economic concerns are very high. More resources are continually being devoted to the retrofit and upgrade of existing structures. The use of FRP is thus becoming more widespread as an optimal innovative system able to reduce the seismic vulnerability of reinforced concrete structures. The application of the FRP system does not increase the structural weight. It ensures an effective confinement of concrete. It can also perfectly adapt itself to complex shapes, increase the strength of critical regions of columns and can modify the strength hierarchy of the structure, inducing ductile controlled mechanisms of collapse.

The effectiveness of the FRP system for guiding and controlling the structural collapse can be also useful to optimize controlled "demolitions". This reduces demolition risks in regards to adjacent buildings, roads, etc.

More so than the conventional methods and materials, the use of the FRP is especially suitable for difficult and complex applications either on load bearing members (i.e., beams, columns; etc:) or on secondary elements (i.e., infill and partition walls) of structures having strategic interest and identified as *sensitive objectives* (in case of explosion risk, terrorist or attempted attacks, etc). In such situations, the adoption of FRP can help to limit damages to persons and structures.

The FRP technology is also very effective in cases of urgency: for safety and temporary preservation of structures damaged during special events.

Figure 2 refers to the immediate action taken in order to eliminate the hazardous conditions and consolidate the vaulting system of the St. Francis Basilica in Assisi (Italy). It was damaged by the 1997 earthquake and consequently strengthened using FRP wraps on the extrados. This technique allowed for the Giotto frescos to be saved even along the edges of the collapsed groins. This event highlights some important aspects of the use of FRP where a fast installation is a crucial factor. The work was carried out in extremely difficult conditions and continued even during following earthquake shakes. While suspended, the applicators installed the FRP avoiding the preliminary wetting of the substrate and an excessive penetration of the material that could damage the fresco underneath. Tests in situ proved the reversibility of such a job without endangering the frescos underneath. A monitoring system was installed to evaluate the effectiveness of the urgency actions taken. After the strong earthquake of April 1998, the maximum deformation of the vault was only 9 decimeters without damaging either the vault system or the frescos.

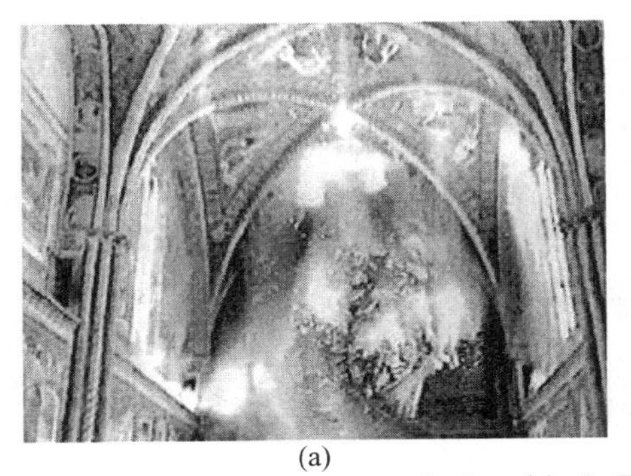

(a) (b)

Figure 2. High urgency strengthening of the St. Francis Upper Basilica in Assisi (Italy) damaged by the 1997 earthquake using FRP.

Another strategic field for the widespread use of the FRP system is in the restoration and seismic strengthening of historical masonry buildings. This is becoming of increasing importance in Europe where the lack of maintenance has seriously endangered the historical heritage. Finally, attention is being devoted to the preservation of the architectural heritage, especially in countries such as Italy and Greece where negligence and lack of preservation have had devastating effects after earthquakes. The effects worsen also due to the poor quality of the original construction materials used.

Figure 3 refers to the final consolidation and upgrade of the vault system of the Upper St. Francis Basilica in Assisi. FRP was used to build a series of very thin ribs on the extrados of the vaulting, following a typical Gothic design that would leave the original structure clearly visible. The ribs were fashioned in situ so they could be shaped to the deformed configuration of the vaulting. Moreover, while their thickness remained constant, their height could be adjusted depending on the deformation of the underlying vaulting. In this way, the extrados of the new ribs could follow a regular curve, substantially parallel to the original surface of the vaults before they experienced any deformation.

(a) (b)

Figure 3. Final upgrade of the St. Francis Upper Basilica in Assisi.

The FRP system can be advantageously used also in the strengthening of wooden structures. Figure 4 shows the partial reconstruction (2000) of the Flavian Amphitheatre arena floor in Rome (Italy). Wood (the original construction material) was used to rebuild the platform and runway. It was necessary to build a dense network of principal glulam beams that were reinforced with FRP to limit the height of the resisting beam sections in order to insert them correctly. The dimensions of the wooden beams where imposed by the original masonry of the structure.

(a)

(b)

Figure 4. Partial reconstruction of the Flavian Amphitheater Arena Floor in Rome.

Conclusions

The advantages of the FRP system foresee an incredible growth of the market not only applied to reinforced concrete structures, but also to other fields of civil construction. The market expansion must be stimulated further by the promotion of strategic synergies between academy and manufacturers. It is necessary to intensify a close interaction among interested people (i.e., industries, universities, research centers, contractors, engineers, etc.) aimed at reaching the following objectives:

- investigating the opportunity for an optimization of composites with reference to the different application fields (i.e., reinforced concrete, masonry, wooden structures);
- satisfying the needs of structural engineers. This means to provide a richer range in terms of type, orientation and weight of the fibers. A wider supply would allow a "tailor-made" *design* where the use of the fiber strength is optimized depending on calculations;
- developing of new matrixes;
- developing *packages* having a complete range of products that optimize the results of the use of FRP in repair and strengthening in the way that is more appropriate for the given substrate (reinforced concrete, masonry, wood, etc.) and observed level of damage;
- clarifying issues related to fire resistance;
- diffusing information about the opportunities offered by FRP and underlining their advantages in comparison to conventional systems;
- developing guidelines for the calculation and testing of jobs;
- proposing standards for quality and durability control of applications;

- carrying out a scientific validation of design and calculation procedure;
- promoting the diffusion of technical-scientific information of numerical procedures for designing the strengthening of typical reinforced elements (i.e., columns, beams, walls, etc.).

References

Balsamo A. (1999). " Static Reinforcement of the Masonry – Using Fiber-Reinforced Plastic (FRP) for Repair Work in Earthquake Zones ". *The Return of St. Francis*. Mapei S.p.a., Milan, Italy, November.

Balsamo A. and Battista U. (1999). " L'unghia di Giotto ". *The Return of St. Francis*. Mapei S.p.a., Milan, Italy, November.

Balsamo A., Battista U., Herzalla A., Viskovic A. (2000) " The Use of Aramidic Fibres to Improve the Structural Behaviour of Masonry Structures under Seismic Actions ", *International Council of Monuments and Sites ICOMOS Congress*, Bethlehem, Palestine, 16-19 October 2000.

Balsamo A., Cerone M., Viskovic A. (2001) " New Wooden Structures with Composite Material Reinforcements for Historical Buildings: the Case of the Arena Flooring in the Colosseum ". IABSE Conference in Lahiti.

Calandrino N., Zaffaroni P., Soffi R. (1999). " Through the Eye of St. Francis ". *The Return of St. Francis*. Mapei S.p.a., Milan, Italy, November.

Capponi G., D'Angelo C., Santamaria U. (ICR), Massa S., Omarini S., Filacchioni G. (1999) " I Materiali Compositi Fibrorinforzati per il Consolidamento delle Volte della Basilica Superiore di San Francesco. Verifica delle Caratteristiche Chimiche, Chimico-fisiche, Fisico-Meccaniche e della Durabilità ". *The Basilica of St. Francis in Assisi "* n. 8, November. Ministero Per i Beni e le Attività Culturali.

Carotenuto G., Giordano M., Nicolais L. (1999) " Materiali Compositi: Proprietà Elastiche e Resistenza al Danneggiamento ". ENCO Journal, year IV, n. 13.

Centroni C., Carluccio G. (1998). " I Controlli e le Indagini Strutturali nella Fase di Emergenza ". *The Basilica of St. Francis in Assisi "* n. 4, Ministero Per i Beni Culturali e Ambientali.

Centroni C., Croci G., Rocchi P., Carluccio G., Viskovic A. (1999). " Gli Interventi Strutturali ". *The Basilica of St. Francis in Assisi "* n. 8, November. Ministero Per i Beni e le Attività Culturali.

Croci G., Viskovic A. (1999) " The Use of Aramidic Fibers in the Restoration of the Basilica of St. Francis of Assisi ", Sixth International Conference on Structural

Studies, Repairs and Maintenance of Historical Buildings *STREMAH,* Dresden, Germany, 22-24 June 1999.

Massa S. (1999) " Valutazione della Probabilità di Condensazione nella Volta della Basilica ". *The Basilica of St. Francis in Assisi* " N° 8, November 1999. Ministero Per i Beni e le Attività Culturali.

Use of Composites by the Construction Industry – Issues and Challenges

Daniel W. Halpin[1] and Makarand Hastak[2]

Abstract

This paper discusses issues relating to the use of composite materials for the construction of infrastructure projects. Some of the practical construction problems encountered in the field are presented in the context of a demonstration project built in Dayton, Ohio. In addition to site adaptation problems, field support by the composite deck panel manufacturers proved to be a very important aspect of the project. From a management perspective, new project delivery formats such as Design-Build (DB) contracts offer a framework for innovation and the use of new materials and techniques.

Introduction

New civil engineering technologies are entering the market place at an astounding rate. During the past 10 years emerging technologies in the areas of subsurface construction, equipment automation and guidance, and composite materials have begun to reshape the way in which projects are constructed. This poses new challenges for the construction industry both from a technical and management perspective.

In a recent article in the Engineering News Record, James Roberts of the California Department of Transportation was quoted as follows:

----

[1] Professor and Head, Division of Construction Engineering and Management, Purdue University, West Lafayette, IN 47907-1294; phone 765-494-2244; halpin@ecn.purdue.edu

[2] Assistant Professor, Department of Civil Engineering, University of Cincinnati, Cincinnati, OH 45221; phone 513-556-3689; mhastak@uceng.uc.edu

"Quick-setting concrete, nighttime work, **composite materials for both decks and whole structures**, and large incentives for contractors will be tools for faster construction..." (ENR, June 11, 2001).

Composite materials are clearly having a major impact on how facilities are designed, constructed, and maintained. In fact, new and innovative project development methods as well as quantitative methods of cost evaluation will tend to help accelerate the acceptance of new materials by designers and contractors.

A number of demonstration projects have been undertaken to test the cost and reliability implications of using composite materials for construction and maintenance. This paper describes one of these test bed projects which uses composite deck sections. It addresses some of the issues which will influence the acceptance of composite materials as a means of constructing infrastructure projects.

A Demonstration Project

The Ohio Department of Transportation has made a major effort to evaluate the use of composite deck sections in conjunction with the renovation of the Salem Bridge located in Dayton, Ohio. Deck panels from four composite manufacturers were used in the deck reconstruction. A number of lessons relating to the installation/construction process were learned. As shown in figure 1 below, Hardcore Composites (HCI), Infrastructure Composites International (ICI), Composite Deck Solutions (DFI) and Creative Pultrusions (CPI) provided panels for various sections of the bridge. Originally, the bridge was divided into four equal sections representing 1.25 spans for each of the manufacturer's panels. In the event, ICI could only provide 7 panels rather than 21 as originally planned. This led to a larger section being completed with DFI panels.

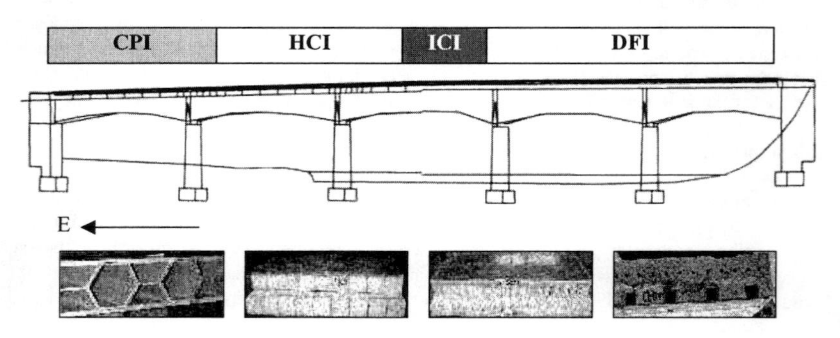

Figure 1. Approximate Locations of Various FRP Panels

Source: (Reising, 2000)

Description of Panels

The CPI panels were made of pultruded components bonded and interlocked to form the panel in the factory. The HCI panels were built up from fiberglass skin faces with multiple wrapped cells which provided stiffening webs in the longitudinal and transverse directions. These panels use cell core technology in conjunction with the Seeman Composite Resin Infusion Molding Process (SCRIMP). The ICI panels are fabricated using a corrugated sandwich system. The DFI system utilizes glass fiber reinforced polymer (GFRP) panels to support a concrete deck reinforced with tubular GFRP sections.

General Preparations

General preparation work for the bridge rehabilitation included partial reconstruction of the abutments, sandblasting of corroded areas and bearings, and installation of cross frame elements on the cross braces. Each of the panel systems required special support systems. Creative Pultrusions (CPI) provided four solutions for deck-to-girder connections depending upon the girder material and the required slope. HCI recommends the use of special connection clips for deck-to-girder connections and additional bolts at the abutment. The support of the ICI panels is based upon the use of haunches.

A common solution using haunches cast-in-place on top of the girders was developed by the Ohio Department of Transportation (ODOT) to support three of the panel systems (i.e. HCI, CPI, and ICI). The DFI system is, in effect, a stay in place form system. Therefore, a 16" wide haunch was cast for all but the DFI system support.

The haunches provided a leveling layer between the girders and the composite panels. Full contact was required between girders and panels. The haunches were also designed to allow transmittal of shear forces between the panels and the six steel girders of the bridge. This was accomplished using shear studs.

Forming the Haunches

Constructing the haunches turned out to be a significant source of problems. It proved difficult to achieve a uniform height of the haunch layer throughout the length of each girder. Therefore, the haunches were not perfectly level and not level transversely across the girders. Installation difficulties also arose due to the required stud holes for the panels. Since the contractor was not able to properly locate the stud holes, they had to be enlarged and adjusted leading to extra work. The location of stud holes in the support girders need to be carefully controlled in future projects. Due to the uneven levelness of the haunches as cast, the panels did not have even contact with each of the girders. Therefore, a filler level of grout and epoxy had to be applied to achieve reasonably even support of the panels through contact with each of the girders.

Panel Installation

Procedures for installation of the differing panel types varied. Production rates also varied from initial rates as low as 2 per day to a top rate of 7 to 8 panels per day. The panel requirement per span over the 5 bridge spans was approximately 17 panels per span (based on the initial allocation to each manufacturer of 21 panels for 1.25 spans per vendor).

Since ease of installation is one of the potential advantages that FRP deck panels would have over conventional cast-in-place methods, this is a key area of interest. In the case of DFI system, the procedure retains many of the steps involved in traditional cast-in-place deck construction. The most important aspects of installation relating to performance were (1) the method of support and load distribution to the supporting girders and (2) the method of joint connection between the panels as placed. In the cases of the full composite decks (e.g. CPI, HCI, and ICI), application of the wearing surface is also a critical issue. Table 1 shows special construction steps that were associated with each panel system.

Table 1. Special Installation Activities

ACTIVITY	CPI	HCI	ICI	DFI
Construction of haunch for full contact support	X	X	X	
Special joint preparation to remove excess epoxy resin			X	
Installation of special mechanism to inhibit uplift at joints		X		
Special welded grids for forming haunches used in panel support				X
Special procedure for installing shear studs	X			
Conventional cast-in-place concreting activities				X

Source: (Reising, 2000)

In the case of each of the full composite decks, special construction of a support layer – a haunch – on each of the six girders was required. As noted above, the difficulty in control of the level of these haunches both along the length of each girder as well as transversely across the girders required grouting and injection of epoxy to insure uniform load transfer to the supporting girders. In addition, special activities (see Table 1) relating to shear stud installation, joint connection preparation, and uplift at joints had to be addressed in the field by extra work. This impacted the productivity and required support from manufacturer's representatives at the site.

Joint Connection

The systems used by each of the manufacturers for joining the panels at their joints required differing actions from the construction contractor. The CPI system used a male-female joint with epoxy bonding along the joint. The integrity of the connection was achieved by the use of pressure applied after the glue was applied. Hydraulic presses provided by CPI could not be used effectively in the field. An expedient system using tension ropes attached to an excavator was used to press the panels together.

The HCI joint system is shown conceptually in Figure 2. It uses a butt contact area with FRP closure strips or tongues applied above and below the connection. This method does not require pressure to insure the bond. During construction, however, this system experienced uplifting due to lack of stiffness at the joint. This had to remedied by the use of a field expedient. Metal angle sections had to be installed to hold the panel edges in proper orientation and prevent the uplift phenomena. In fact, this joint connection system required more construction time and led to cracking in the wearing surface even prior to the opening of the bridge to service loads.

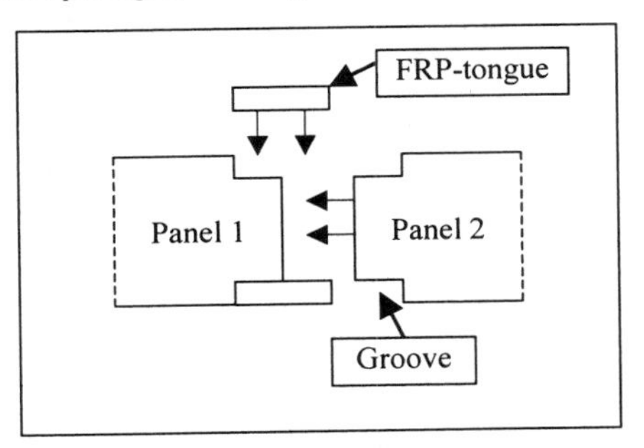

Figure 2. "Hardcore" Joint System

Problems were encountered with joining the ICI panels. Only 7 of the panels originally allocated ICI were supplied. The required deck panel thickness was 8 inches. The ICI panels showed a variation of +/-0.25 inches. This made the fit-up of the panel at the joint extremely difficult. The joints had to be specially prepared in the field to remove excess epoxy resin left from the manufacturing process. In many cases the lack of fit required epoxy filler since adequate closure at the joint using pressure could not be achieved.

Other construction challenges related to the installation of shear studs and the provision of openings in the panels for light post stanchions and sewer drains. The DFI system proved to be simplest in this respect since the field procedure is effectively the same as cast-in-place deck construction.

Lessons Learned

In all cases, field modifications had to be made to site adapt the 4 different systems to the project environment. Contractor personnel indicated that the system which was most easily installed was the CPI system. However, when asked about the labor required for installation, the consensus was that all systems would require a similar amount of labor. It would appear that once crews became familiar with the installation process, a learning curve effect would favor the CPI system and lead to the highest daily rates of panel installation.

Although the stay-in-place form (DFI) system was easiest to site modify, it retained many of the construction activities involved in conventional concrete deck construction. One advantage that obtained in all of the full panel systems was the reduction in dead load due to the lightness of the panel and the ability to move materials and vehicles on to the panels immediately following installation. This was not possible using the DFI system until the concrete decks had achieved acceptable strength.

Cracking in the road surface at the panel joints occurred with all of the full panel systems. This occurred even prior to the opening of the bridge to traffic and could be seen as separations in the wearing surface. This cracking might be due to inadequate support provided by the haunches. On the other hand, it may be due to poor selection of the wearing course material. The CPI and ICI panels showed breaks directly over the glue joint. The wearing course material could not deform to the same degree as the joints. In some cases the cracking was substantial and during strong rain showers water drained directly through the joint areas.

It was clear that careful detailing to achieve site adaptation was very important. The connection of the shear studs posed a problem in the field and required expedient solutions which could have been foreseen during the design phase. Strong manufacturer support was key to success in the panel installation. The level of support varied among the suppliers. The CPI system appears to have achieved the best success due to the fact that a strong onsite team helped solve problems and assisted the contractor in overcoming difficulties.

Barriers to Use of Composites

Strong support by the manufacturer is critical to insure acceptance of composite technology by the contracting community. In contrast to other sectors of the economy, the infrastructure and construction related sectors have traditionally been slow to accept innovation. The underlying technologies, which impact design and construction, tend to change very slowly. Innovation requires the support of a championing organization which is willing to accept the risk of implementing a new method, material or process improvement. Acceptance is also heavily dependent upon the culture of the professionals who are in key decision making positions and their perception of the utility of the innovation. Life cycle cost-benefit analysis (see Hastak and Halpin, 2000) can help establish the advantages associated with composite materials. Before new materials can gain wide acceptance, specifications defining the liability framework and potential commercial risks must be in place.

Demonstration projects such as the Salem Bridge are essential for showcasing and introducing new concepts. They also act as a test bed in which new materials and concepts can be tested under real world conditions. Just demonstration also assist in evaluating performance and associated liability issues. Such field installation projects are needed to work the kinks out of the construction process and to develop specifications for installation and performance.

Impact of Design Build Contracting

The Design Build (DB) method of contracting has become very popular during the past decade. The owner/client is typically given a stipulated-sum price for the project early in the project development process (at the completion of the conceptual design). The design-build team locks in on a specific price at this early time the life cycle of the project and guarantees that, barring major changes to the scope of the project, the owner can rely on the price quoted.

This project delivery approach provides an incentive to be innovative and use cost effective materials and processes to complete the project at the lowest cost while still retaining the functionality and life cycle performance features of the facility. If improved methods can be found to implement the design at a lower cost during construction, this is money in the pocket of the Design-Build team.

The DB approach tends to support new materials and processes if they lead to improved production while maintaining or improving performance. Composite elements, such as the FRP panels used in the demonstration project, are cost effective and lead to time savings. The modularity provided by composite elements should yield higher productivity at the construction site. Demonstration projects will hopefully confirm that savings both in the short term (e.g. during construction) and long term (e.g. during the project life cycle) can be obtained using composite materials.

Conclusion

The demonstration project indicated that improved rates of production can be achieved during construction. However, strong field support from the supplier of composite materials is essential to deal with field problems. It is essential that manufacturers provide training and problem solving assistance to assist construction contractors as composites and FRP elements become more widely accepted.

References

"Bridge Builders See Materials, Design Build as Strong Tools." (2001). *Engineering News Record*, McGraw-Hill Companies, New York, N.Y., 11 June 2001, p. 15.

Hastak, M. and Halpin, D. W. (2000). "A Model for Life-cycle Cost-Benefit Assessment of Composite Materials in Construction." *Journal of Composites for Construction*. Vol. 4, No. 3, August 2000, pp. 103-111.

R. Reising. (2000). "MOT 49 Installation Phase." *Project Report*, University of Cincinnati, Cincinnati, Ohio, Fall 2000.

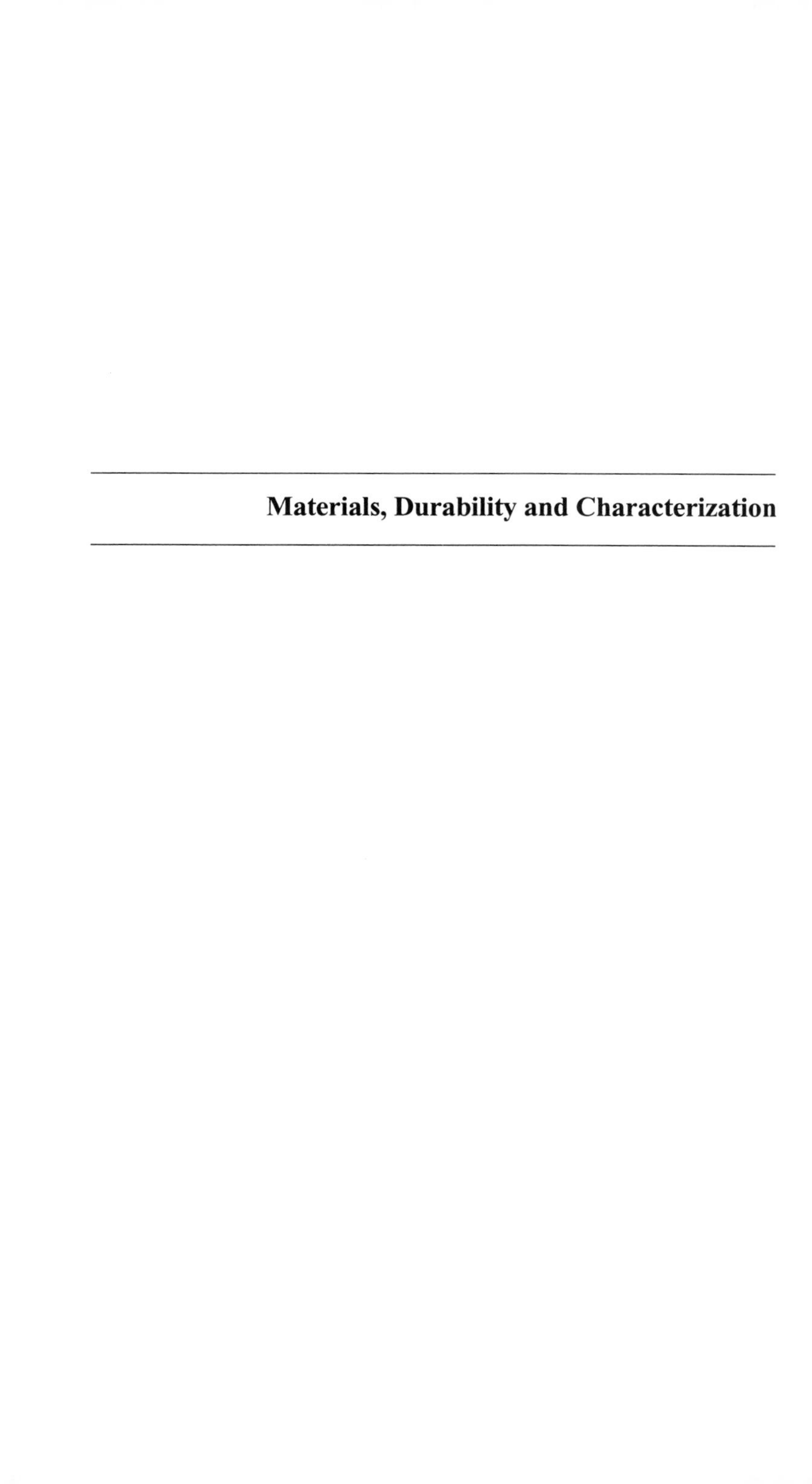

Materials, Durability and Characterization

Tests Methods to Determine Properties of FRP Rods for Concrete Structures

By B. Benmokrane[1], P. Wang[2], T. R..Gentry[3] and S. Faza[4]

Abstract

Fiber-reinforced polymer (FRP) materials have emerged as a practical alternative material for producing reinforcing rods and sheets for concrete structures. FRP reinforcing rods and sheets offer advantages over steel reinforcement in that FRP is non-corrosive and nonconductive. FRP rods include rods, grids and tendons for nonprestressed and prestressed concrete structures, respectively. FRP sheets are being used as external reinforcement for strengthening of existing concrete structures. Due to differences in the physical and mechanical behaviour of FRP materials compared to steel, unique test methods for FRP reinforcing rods and sheets are required. This paper provides test methods, which are being developed by ACI Sub-Committee 440K, and suitable test methods from ASTM standards, for determining physical and mechanical properties of FRP reinforcing rods for concrete structures.

Introduction

Recently, composite materials made of fibers embedded in a polymeric resin, also known as fiber reinforced polymers (FRP), have become an alternative to steel reinforcement for concrete structures (Saadatmenesh and Ehsani 1998; Benmokrane and Rahman 1998; Dolan et al. 1999). Because FRP materials are non-metallic and no corrosive, the problems of steel corrosion are avoided with FRP reinforcement. Additionally, FRP materials exhibit several properties, such as high tensile strength, that make them suitable for use as structural reinforcement. Currently, FRPs made with glass, carbon and aramid fibres are available with a wide range of mechanical properties (tensile strengths and bond strengths, elastic moduli, etc.). Generally, FRP concrete reinforcements do not meet national material standards as few such standards exist. Instead, manufacturers of FRP provide test data and recommend design values based on this test data. Unfortunately, in addition to the lack of material standards, few standard test methods exist for FRP concrete reinforcements. It is therefore difficult to compare test results between product manufacturers. In addition, much research has considered the durability of FRP concrete reinforcements in environments containing moisture, high and low temperatures, and alkaline environments. Test methods that allow for the comparison of

[1] ISIS-Canada, Department of Civil Engineering, University of Sherbrooke, Quebec, Canada J1K 2R1; Tel : (819) 821-7758; bbenmokrane@andrew.sca.usherb.ca

[2] Ph.D student, ISIS-Canada, Department of Civil Engineering, University of Sherbrooke, Quebec, Canada, J1K 2R1; Tel. (819) 821-8000 ext. 3181; wangpeng6_99@yahoo.com

[3] Professor, College of Architecture, Georgia Institute of Technology, Atlanta, Georgia, U.S.A Tel. :(404) 894-3845; russell.gentry@arch.gatech.edu

[4] Concrete Reinforcement Technologies Inc., Jacksonville, Florida, U.S.A.; Tel. :(904) 519-8803; salem.faza@yahoo.com

mechanical property retention in a wide range of standard environments are needed so that durable FRP-reinforced concrete structures can be assured.

The American Concrete Institute sub-committee 440K is developing a document titled "Recommended Test Methods for FRP rods and sheets". This document provides model test methods for determining the short-term and long-term mechanical properties of FRP reinforcing rods, including bars, grids, and tendons, for concrete, both prestressed and nonprestressed, and for FRP sheets as external reinforcement for concrete structures. The recommended test methods are based on the knowledge gained from research results and literature worldwide. Many of the proposed test methods for reinforcing rods are based on those found in "Recommendation for Design and Construction of Concrete Structures using Continuous Fiber Reinforcing Materials" published in 1997 by the Japan Society for Civil Engineers (JSCE). The JSCE test methods have been modified extensively to add details and to adapt the test methods to U.S. practice.

Additionally, several international organizations have developed standards for the testing of composite materials such as the most widely accepted ASTM standards. ASTM standards are developed through contributions of academics and industry and cover a range of topics that extend far beyond composite materials. The intention of this paper is to present some test methods proposed by ACI subcommittee 440K and suitable test methods from ASTM standards, for determining properties of FRP reinforcing rods for concrete structures.

Physical Properties

Since FRP reinforcing rods are made of fibre, resin matrix, and fillers, the short-term and long-term mechanical properties are significantly influenced by certain physical properties such as the types and ratio of the constituent components and processing conditions. These physical properties can serve as an indication of the mechanical properties of FRP rods. The described physical properties include fiber to resin ratio, void content, cross-sectional properties, and coefficient of thermal expansion.

Fiber to Resin Ratio

There are two test methods available for the determination of the fiber content of FRP: ASTM 3171 and ASTM D2584. ASTM D2584 is preferred to D3171 because this test method is easier and safer to perform (Tingley et al.1997). ASTM D2584 determines fiber content by using ignition loss of resin. This ignition loss can be considered to be the resin content only if the organic resin is completely decomposed under the test conditions and fibers do not decompose under these conditions. Aramid and carbon fibers will decompose under the test conditions in the presence of oxygen, therefore, the test must be performed in an inert atmosphere such as nitrogen gas when testing these fiber types.

Void Content

Void content refers to the total amount of space within a composite that is filled not by fiber, resin and fillers. High void content may result in lower strength, increased susceptibility to environmental effects, and discontinuous stress translation etc., therefore, it is necessary to determine void content and type to accurately assess mechanical test results when designing with FRP reinforcing rods.

ASTM D2734 is applicable for determining void content in composites, in which the effects of ignition of the resin, the reinforcement and the filler are known. Resin content is measured and a theoretical density is calculated. Densities of the reinforcement and the composite are measured separately. The measured density is compared to the theoretical density of composite. The difference in densities refers to void content. However, additional tests are required to measure the void type. Shearography is the method commonly used in the automobile and aerospace industries (Shang et al.1991). This method is effective in measuring the shape, size, and depth of the void. However, this method is more expensive than SEM (Scanning Electronic Microscopy) analysis. SEM analysis can give useful information related to content, distribution and type of voids.

Cross Sectional Properties

There are a variety of commercially available FRP rods for concrete reinforcement. They differ in terms of cross-sectional dimensions, due to varying forms including deformed, sand coated, ribbed, multi-strand cables, braided and hollow shapes. Therefore, a methodology is required to determine the nominal cross-sectional area, diameter and circumference of the various shapes. ACI 440K proposed "Test Method for Cross-Sectional Properties of FRP Rods". This method can be used to determine the cross-sectional area, the nominal diameter, and nominal circumference of an FRP rod. The principal test methodology is to measure length of specimen, determine the volume increase by immersing the specimen in the water/ethanol in a graduated cylinder. Based on the measured length and volume of specimen, the cross-sectional area, the nominal diameter, and nominal circumference of an FRP rod can be determined. FRP rod samples of 290-mm-long shall be used.

Longitudinal Coefficient of Thermal Expansion

Coefficients of linear thermal expansion are used for design purposes and to determine if failure by thermal stress may occur when a solid body composed of two different materials subjected to temperature variation. Thermal properties of fibres are substantially different in the longitudinal and transverse directions. FRP reinforcing rods manufactured from these fibres therefore have different thermal expansion in these two directions. ACI 440K proposed "Test Method for Longitudinal Coefficient of Thermal Expansion of FRP Rods". This method is used to measure the longitudinal coefficient of thermal expansion by thermal mechanical analysis (TMA) method as described in ASTM E 831-93. The principle of this method is to measure the changes in length of a test specimen caused by changes in temperature in order to calculate the coefficient of thermal expansion. The coefficient of thermal expansion of the test specimen within the measured temperature range could be determined.

Mechanical Properties

This section presents the proposed test methods related to tensile, bond, shear, flexural, fatigue, creep, relaxation and durability properties. The recommended anchor for testing FRP rods under monotonic, sustained and cyclic tension is also described.

Anchor

ACI sub-committee 440K specifies an anchor for performing tests for monotonic tension, creep, relaxation, fatigue, and pullout bond strength of FRP rods. The geometric dimensions of the anchor are shown in Figure 1. The cylinder wall thickness (t) shall be at least 3 mm and its inner diameter (D) shall be 10-14 mm greater than the diameter of the FRP rod (d). The length of the cylinder, L_g, shall be at least equal to $f_u A$ / 350 where f_u is the ultimate tensile strength in MPa and A is the cross-sectional area of specimens in mm^2, and shall not be less than 250 mm. The anchor may be adapted to fit into the grips of any testing machine or frame. The cylinder may be filled with either pure resin or a 1:1 mixture by weight of resin and clean sand. A resin compatible with the resin of which the test specimen is made shall be used. The anchor may be adapted to fit into the grips of any testing machine or frame.

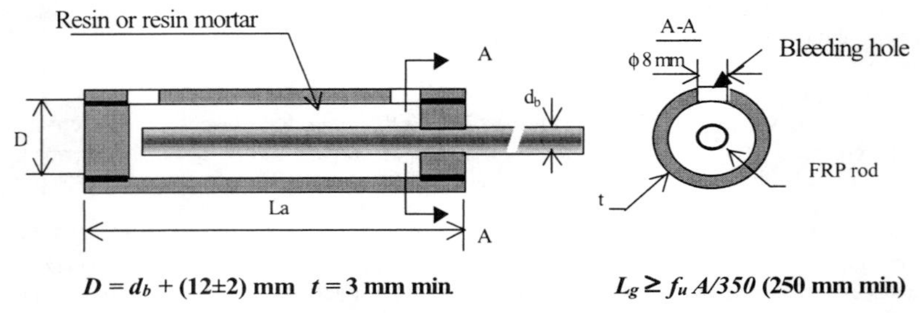

$D = d_b + (12\pm2)$ mm $t = 3$ mm min. $L_g \geq f_u A/350$ (250 mm min)

Figure 1. Anchor for Tensile Test

Longitudinal Tensile Properties

One of the most important and basic mechanical properties for concrete structural reinforcements is the tensile property of FRP rods. The tensile properties of FRP rods can vary significantly based on the type of fibers and resins, fiber volume fraction, fiber orientation, manufacturing process and quality control during manufacturing etc. ACI 440K proposed "Test Method for Tensile Properties of FRP Rods". This test method is used to measure tensile strength, modulus of elasticity and ultimate elongation of FRP rods, and the test method focuses on FRP rod itself, excluding the performance of the anchorage. Therefore, failure or pullout at an anchoring section shall be disregarded. The anchor specified in ACI 440 K "Anchor for testing FRP specimens under monotonic, sustained, and cyclic tension" shall be used. The length of the test section shall not be less than 100 mm, nor shall it be less than 40 times the nominal diameter of the FRP rod. The specified rate of loading shall be 100-500 MPa/min. The tensile strength and the ultimate strain can be calculated. The tensile modulus of elasticity shall be calculated from the difference between the load-strain curve values at 20% and 60% of tensile capacity.

Bond Strength by Pullout Test

Bond between concrete and FRP reinforcing rods affects development length, member deflection, crack spacing, and crack width. The bond properties of FRP reinforcing rods have been investigated extensively by numerous researchers through different types of tests, i.e. concentric pullout tests, splice tests, and cantilever beams etc. Each of these test methods has its merits and demerits; however, only pullout test method has been established as a recommended test method.

In this test method, two types of specimens are recommended. One containing one FRP rod embedded vertically to the direction of casting of the concrete, and the other containing two FRP rods embedded horizontally. For Specimens vertically embedded rod, specimens shall consist of concrete cubes, 150 mm (6 in.) on each edge, with a single FRP rod embedded vertically along a central axis in each specimen. Cover shall not less than $5d_b$ to avoid splitting of the concrete cover. For specimens horizontally embedded rod, these specimens shall consist of concrete prisms 150 by 150 by 300 mm (6 by 6 by 12 in.) with the longer axes in the vertical direction. Two rods shall be embedded in each specimen, perpendicular to the longer axis and parallel to and equidistant from the sides of the prism. The bonded length of the FRP rod shall be four times the diameter of the FRP rods.

The concrete shall be a standard mix, with coarse aggregates having a maximum dimension of 20 to 25mm. It shall be batched and mixed in accordance with the applicable portions of ASTM Practice C 192. The concrete shall have slump of 10±2 cm in accordance with ASTM C 143. Apply load to the RP rod at a load rate not greater than 22 KN/min, or at a testing machine head speed not greater than 1.27mm/min.The parameters that could be obtained include average bond strength, the curves for the pullout or bond stress versus slippage at both free-end and loaded-end displacement for each specimen, average bond stress causing slippage at the free end of 0.05 mm, 0.10 mm, and 0.25 mm, and the maximum bond stress at failure, nominal average bond stress causing slippage at the loaded end.

Transverse Shear Strength of FRP rods

ACI 440K proposed "Test Method for Shear Strength of FRP Rods". This test method specifies the test requirements for determining the transverse shear (dowel) properties by direct application of double shear. The shear testing apparatus shall be constructed so that a rod-shaped test specimen is sheared on two planes more or less simultaneously by two blades (edges) converging along faces perpendicular to the axis of the test specimen. The gap in the axis direction between the upper and lower blades (δ) shall be of the order of 0-0.5 mm, and shall be made as small as possible. The specification distance between shear planes (t) shall be 50 mm. The specified loading rate shall be such that the shearing stress increases at a rate of 30-60 MPa per minute. Failure, whether it is due to shear or not, shall be determined by visual inspection. If pullout of fibers etc. is obvious, the data shall be disregarded and additional tests shall be performed until the number of test specimens failing due to shear is not less than three. Shear strength could be calculated.

Flexural Properties of FRP Pultruded Rods

Flexural properties of FRP rods may vary with specimen depth, temperature, atmospheric conditions, and different strain rate. ASDTM D 4476-97 covers the determination of the flexural properties of FRP pultruded rods. The specimen is a rod with a semi-circular cross section, and is tested in flexure as a simple beam. The specimen rests on two supports and is loaded by means of a loading nose midway between the supports. The specimen is deflected until rupture occurs in the outer fibers, or until the maximum fiber strain of 5% is reached, whichever occurs first. The specimen length shall be 16 to 24 times its thickness, or depth, plus at least 20% of the support span to allow a minimum of 10% overhang at the supports. The following parameters could be determined: Maximum fiber stress, tangent modulus of elasticity, and maximum strain. Additional information can be found in ASTM D 4476-97.

Alkali Resistance of FRP Pultruded Rods

Alkaline reaction of FRP composite in concrete is a one of the major durability concerns for design engineers, since concrete is used in most of the civil infrastructure structures, and high alkalinity in concrete has a potential of severely degrade the FRP composite reinforcement. Test methods have to be developed to clearly evaluate this effect.

ACI 440K proposed "Test Method for Alkali Resistance of FRP Rods". This method is used to evaluate mass change and tensile capacity retention of the test specimen by immersing the specimens in aqueous alkaline solution through two types of procedures: procedure A and procedure B. In procedure A, specimens shall be immersed in the alkaline solution with no tensile load applied. The test control parameters are the pH value, temperature of the alkaline solution and immersion time. In procedure B, FRP specimens shall be immersed in the alkaline solution under sustained tensile load. The test parameters are the sustained load level, pH value, temperature of the alkaline solution and immersion time. The recommended composition of alkaline solution is 118.5g of $Ca(OH)_2$, 0.9g of NaOH and 4.2g of KOH in 1 liter of deionized water. The test can also be performed on specimens embedded in concrete. The concrete mix and the curing procedure shall be performed according to Specification C511 of the ASTM Standards. Specimens should be kept embedded in concrete for 28 days before testing.

Samples for Procedure A shall be immersed in the alkaline solution held at 60°C for exposure times of 35, 69, 104, and 138 days, designed to represent 25, 50, 75, and 100 years of service in a typical environment, unless longer exposure periods are specified. Samples for Procedure B shall be installed in anchoring devices at both ends in accordance with "Anchor for Testing FRP Rods Under Monotonic, Sustained and Cyclic Tension" the test section of the specimen shall be immersed in the alkaline solution at 60±3°C. The specimen shall be held in a conditioning fixture to subject to a constant tensile load of 1.1 time the design tensile strength for exposure time of 35, 69, 104 and 138 days, designed to represent 25, 50, 75, and 100 years of service in a typical environment, unless longer exposure periods are specified. The length of the test specimen shall be the sum of the length of the test section and the lengths of the anchoring sections. The length of the test section shall not be less than 40 times the nominal diameter of the FRP rod. Weight change test shall be performed in accordance with Procedure D of ASTM D 5229/D 5229M-92. For determining the tensile capacity retention, specimen shall be tested in tension to failure within 24 hours after removal from the conditioning environment. The tensile capacity retention rate could be calculated.

Tensile Fatigue of FRP Pultruded Rods

For FRP rods used as reinforcing rods, various versions of fatigue testing, such as tension-tension, tension-compression, compression-compression, etc., are possible. The tensile fatigue test method proposed by ACI 440K is considered to be the most basic for evaluating material characteristics. The principle of this tension fatigue test method is to measure the maximum repeated and relevant numbers of cycles for FRP rods to establish the S-N curve under a given set of controlled environmental conditions and corresponding load rate.

The total length of the specimen shall be 40 d_b+2L_a or greater where db is the nominal diameter of specimen in mm and La is the length of grip in mm. The load may be set in either of two ways: i.e. by fixing the average load and varying the load amplitude, or by fixing the minimum repeated load and varying the maximum repeated load. In either case, a minimum of three load levels shall be chosen such that the range of number of cycles to failure is between 10^3 to $4x10^6$. Typical S-N curves for FRP are used for maximum-minimum stress ratio R fixed at a certain value recommended as 0.1. In actual concrete structures subject to variable loads, permanent loads such as dead load weight can be considered as the minimum load, and the design load can be considered as the maximum load. Special procedure is also proposed where the maximum stress level for initial test is difficult to determine. The frequency shall be within the range of 1 - 10 Hz, preferably 4 Hz. The S-N curve shall be plotted in accordance with ASTM E739-91 with maximum repeated stress, stress range, or stress amplitude represented on a linear scale on the vertical axis, and the number of cycles to failure represented on a logarithmic scale on the horizontal axis. Where measurement points coincide, the number of coinciding points shall be noted. Right-facing arrows shall be added to indicate points from test results for test specimens that do not fail. The fatigue strength at $2x10^6$cycles shall be derived from the S-N curve.

Creep of FRP Pultruded Rods

Unlike steel reinforcing rods subjected to significant sustained stress for long time periods, creep failure of FRP rods may take place below the static tensile strength. Therefore, the creep strength must be evaluated when determining acceptable stress levels in FRP rods. Creep strength varies according to the type of FRP rods used. ACI 440K proposed "Test Method for Creep of FRP Rods". The proposed test method is intended to measures the load-induced, time-dependent tensile strain at selected ages for FRP rods, under an arbitrary set of controlled environmental conditions and the corresponding load rate.

The load shall be chosen between 0.2 and 0.8 of static tensile capacity. For each value of load ratio tested, failure shall not have occurred in at least five test specimens after 1,000 hours of loading. The load ratio-creep failure time curve shall be plotted on a semi-logarithmic graph, where the load ratio is represented on an arithmetic scale along the vertical axis and creep failure time in hours is represented on a logarithmic scale along the horizontal axis. A creep failure line chart shall be prepared by calculating an approximation line from the graph data by means of the least-square method. The load ratio at 1 million hours, as determined from the calculated approximation line shall be the creep failure load ratio. The load and stress corresponding to this creep failure load ratio shall be the million-hour creep failure capacity and the million-hour creep failure strength respectively. The million-hour creep failure strength shall be calculated.

Long-term Relaxation of FRP Rods

ACI sub-committee 440K proposed "Test Method for Long-Term Relaxation of FRP Rods". Test method is used to measure the load-induced, time-dependent tensile strain at selected ages for FRP rods under an arbitrary set of controlled environmental conditions and corresponding load rate. The relaxation value shall be calculated by dividing the load measured in the relaxation test by the initial load. The relaxation curve shall be plotted on a semi-logarithmic graph where the relaxation value (%) is represented on an arithmetic scale along the vertical axis, and test time in hours is represented on a logarithmic scale along the horizontal axis. An approximation line shall be derived from the graph data by means of the least-squares method. The relaxation rate after 1 million hours shall be evaluated from the approximation line; this value represents the million-hour relaxation rate. Where the service life of the structure in which the FRP rods are to be used is determined in advance, the relaxation rate for the number of years of service life ("service life relaxation rate") shall also be determined.

Summary

This paper discuss test methods, which are being developed by ACI Sub-Committee 440K, and suitable test methods from ASTM standards, for determining physical and mechanical properties of FRP reinforcing rods. It is anticipated that these model test methods will be considered, modified, and adopted, either in whole or in part, by a U.S. national standards granting agency such as ASTM or AASHTO. The publication of these test methods by ACI Committee 440 is an effort to aid in this adoption.

References

ACI 440K 2001 "Recommended test methods for FRP rods and sheets," draft, American Concrete Institute Sub Committee 440K, 89p.

Benmokrane, B., and Rahman, H. (1998) (eds.) "Durability of fiber reinforced polymer composites for construction," Proceedings of the CDCC98, Sherbrooke, Québec, 692p.

Dolan, C.W., Rizkalla, S., and Nanni, A. eds.1999 "Fiber reinforced polymer for reinforced concrete structures, " Proceedings of the Fourth International Symposium: Fiber Reinforced Polymer Reinforcement for Reinforced Concrete Structures, eds. Dolan, C. W., Rizkalla, S. H., and Nanni, A. ACI SP-188, 1182 p.

Japan Society of Civil Engineers (JSCE) 1997 "Recommendation for Design and Construction of Concrete Structures Using Continuous Fiber Reinforced Materials," Concrete Engineering Serises 23, ed. By A. Machida, Research Committee on Continuous Fiber Reinforcing Materials, Tokyo, Japan, 325 p.

Saadatmanesh, H., and Ehsani, M. R.. (eds.) 1998 "International Conference on Composites for Infrastructure," Proceedings of ICCI`98,Tuscson, Arizona, USA, (Vol.1) 723 p., (Vol.2) 783 p.

Shang, H. M., Toh, S. L., Chau, F. S., Shim, V. P. W., and Tay, C. J. (1991). "Locating and sizing disbonds in glass fiber-reinforced plastic plates using shearography," Journal of Energy Material and Technology, January, pp.99-103.

Tingley, D. A., Gai, Chunxu, and Giltner, E. E. (1997) "Testing methods to determine properties of fiber reinforced plastic panels used for reinforcing glulams," Journal of Composites for Construction, November, pp. 160-167.

Long-term behavior of FRP

György L. Balázs* and Adorján Borosnyói**

Abstract

Fiber reinforced polymer (FRP) reinforcements have some superior characteristics to steel reinforcements (especially they are non-corrosive), however, specific considerations may be needed for their behavior under long-term influences. Durability and time dependent mechanical characteristics of FRPs are essential parameters for design of concrete structures reinforced or prestressed with FRP. Based on an extensive literature review, present paper summarizes all aspects of FRP reinforcements considering durability (including effects of alkalis, chloride ions, UV radiation, water, elevated temperature and freeze/thaw cycles) as well as time dependent mechanical characteristics (such as creep, relaxation, fatigue) and their changes due to environmental effects.

Introduction

In the last decades considerable deterioration of concrete structures due to corrosion has turned the attention of researchers, designers, producers and owners on the application of non-metallic (FRP), therefore non-corrosive reinforcements. Short-term characteristics of FRPs can be more or less easily determined, however, long-term properties may require specific considerations. Lot of papers have been published so far dealing with durability and time dependent mechanical characteristics of embedded FRP reinforcements. Present paper intends to summarize the state-of-the-art (Table 1).

Durability of FRP reinforcements

FRP reinforcements are made of Fiber Reinforced Polymer composites. Composites have at least two different phases such as load carrying unidirectional fibers and a resin matrix.

Load carrying fibers of FRP reinforcements can be carbon, aramid or glass. Embedding matrix can be epoxy, polyester or vinyl ester resin. Long-term properties of both constituents are to be investigated.

---

* MSc (CE), PhD, Professor in Structural Engineering, Head of Department
** MSc (CE), PhD candidate
Budapest University of Technology and Economics, Hungary
Department of Construction Materials and Engineering Geology

Table 1 Long-term influences related to the behavior of FRP reinforcements.

Environmental influences		*Mechanical properties*
- alkalis	- thermal effects	- creep
- chloride ions	- fire	- relaxation
- water	- freeze/thaw	- fatigue
- UV radiation	- combined effects	

Effect of alkaline environment

Concrete is highly alkaline due to the high calcium hydroxide content of hardened cement stone (pH 12.5 to 14) that needs special attention for durability of FRP reinforcements (Fig. 1).

Carbon fibers cannot absorb liquids and are resistant to acid, alkali and organic solvents, therefore, do not show considerable deterioration in any kind of harsh environments (Machida 1993, Tokyo Rope 1993).

Deterioration of glass fibers in alkaline environment is well known. Therefore, the duty of resins is of great importance in protecting glass fibers. Experimental studies of GFRP reinforcement embedded in concrete or under accelerated aging tests in strong alkaline solutions have demonstrated that glass fibers show significant degradation due to alkali, independently of the of resin (Sen et al. 1993, Tannous and Saadatmanesh 1998, Uomoto and Nishimura 1999). Decrease in tensile capacity can be in the range of 30 to 100 percent according to saturation and acting time. Best protection is ensured with vinyl ester resin. Results of accelerated tests usually show more deterioration than tests with embedded reinforcement. Rate of deterioration of glass fibers in alkaline environment is highly dependent on the type of fibers.

Aramid fibers may also suffer deterioration in alkaline environment, however, to a less degree than glass fibers and may depend on the actual fiber product (Uomoto and Nishimura 1999). Decrease in tensile capacity can be 25-50 percent (Rostásy 1997).

Alkaline environment can deteriorate links between molecules of resins. Similarly to the resistance against water absorption the alkaline resistance of vinyl ester resins is the best while epoxy and polyester resins can give sufficient and poor resistance, respectively (Machida 1993).

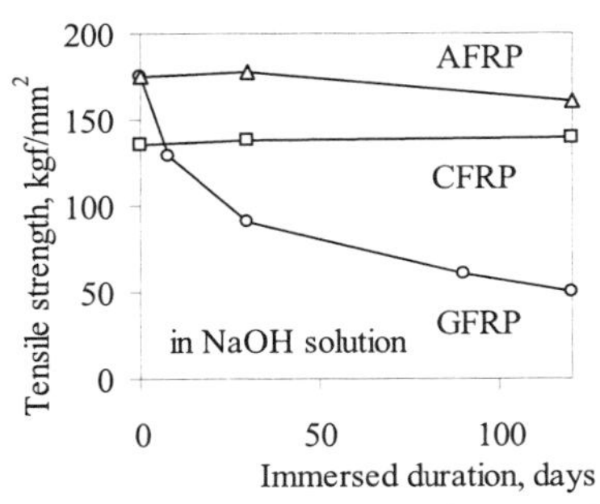

Fig. 1 Effect of alkali on tensile strength (after Uomoto-Ohga 1996)

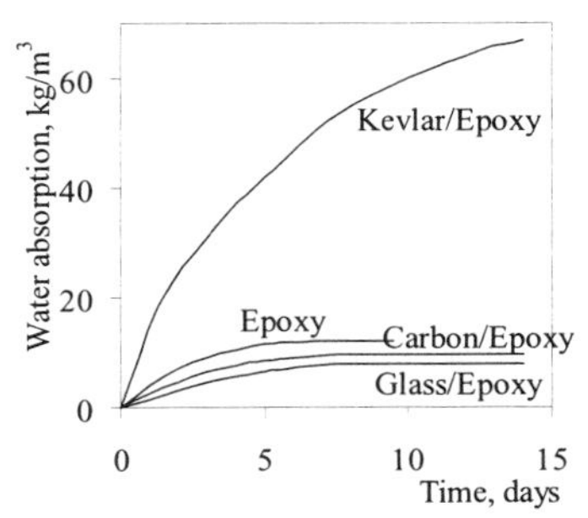

Fig. 2 Water absorption of FRPs (after Piggott 1980)

Effect of chloride ions

Chloride ions can penetrate into concrete in marine environment or by using of de-icing salts and can accelerate corrosion. In the presence of chloride ions the risk of corrosion of FRP reinforcement may also take place. CFRP and AFRP reinforcements are insensitive to chloride ions, however aramid fibers seem to be inapplicable in marine environment due to difficulties initiated by swelling (Sen et al. 1998a, Sen et al. 1998b). Experimental studies demonstrated that GFRP reinforcements can be seriously deteriorated in marine environment or in presence of de-icing salts led to corrosion induced failure (Saadatmanesh and Tannous 1997).

Effect of ultraviolet radiation

Polymeric materials can be considerably degraded by ultraviolet radiation (Piggott 1980). Embedded FRP reinforcements are protected from direct sunlight, however, when stored outdoors or applied as external reinforcement can be exposed to ultraviolet radiation. On one hand, deterioration of GFRP and CFRP materials are attributed to the degradation of resin matrix. After 2500 hours of exposure to direct sunlight decrease in tensile strength and in Young's modulus was less than 10 percent in the case of GFRP and negligible in the case of CFRP (Kato et al. 1997). Aramid fibers themselves may deteriorate due to UV radiation.

Effect of moisture and water

In fresh concrete contact of water and FRP reinforcement is evident. Material changes associated with water are usually of the resins. Water can absorb into polymer chains and can create weak chemical reactions causing considerable changes in characteristics (e.g. strength, Young's modulus, bond). These effects are mostly reversible, however, swelling of resin can cause micro-cracks in the matrix that can initiate fiber debonding and higher permeability. In general it can be stated that vinyl ester resins show the best resistance to water absorption, epoxy resins can provide sufficient resistance, while polyester resins usually have poor performance (Machida 1993).

For what concerning fibers, carbon and glass fibers cannot absorb water on the contrary to aramid fibers (Uomoto and Nishimura 1999). Water absorption of aramid fibers causes reversible decrease in tensile strength, Young's modulus or relaxation and irreversible decrease in fatigue strength (Piggott 1980). Decrease in characteristics of AFRP due to water absorption is about 15-25% (Gerritse 1993). According to swelling of AFRP reinforcement bond cracking can be induced by wet/dry cycles (e.g. in splash zones of marine structures) that cause deterioration (Sen et al. 1998a). Fig. 2 indicates water absorption capacity of various FRPs.

Combined and other effects

Effect of water and alkali absorption is clearly accelerated by elevated temperature. Rate of degradation of GFRP and AFRP can be doubled by the change of temperature from 20°C to 60°C (Rostásy 1997, Tannous and Saadatmanesh 1998).

Degradation of AFRP due to ultraviolet radiation can be accelerated by wet/dry cycles of marine splash zone (Uomoto and Ohga 1996).

As it is well known, carbonation of concrete has important role on the corrosion of ordinary reinforcement. Research work on the effect of carbonation on durability of FRP reinforcement is very limited. Available data show large scatter, however, carbonation seems to have no effect on the durability of FRP reinforcements (Sheard et al. 1997).

THERMAL ACTIONS ON FRP REINFORCEMENTS

General considerations

In the case of FRPs thermal actions can influence both mechanical characteristics and bond behavior. Coefficient of thermal expansion (CTE) of fibers, resins, FRPs and concrete are considerably different as indicated in Table 2. In the longitudinal direction FRPs have lower or nearly identical CTEs than that of concrete, however, in the transverse direction – governed mostly by the resin – reach 5 to 8-times higher values. In specific cases, when high temperature variation takes place the large difference between CTEs can lead to high radial pressure on the surface of the reinforcement that can cause longitudinal splitting of concrete cover. It reflects the importance of sufficient concrete cover especially if AFRP reinforcement is applied. Critical concrete cover (whenever splitting occurs) of AFRP tendons with sand coated surface was found to have 2.8×rebar diameter (Taerwe and Pallemans 1995). Authors found sufficient concrete cover of 2.5×rebar diameter to CFRP tendons with sand coated surface for pretensioned application with ten hours heat curing at a maximum temperature of 75°C (Balázs and Borosnyói 2001).

Table 2 Coefficients of thermal expansion of fibers, resins, FRPs and concrete

Material	Coefficient of thermal expansion, $\times 10^{-6}$ 1/K	
	longitudinal	transverse
carbon fiber	–0.9...+0.7	8...18
aramid fiber	–6.0...–2.0	55...60
glass fiber	5...15	5...15
resins	60...140	
CFRP	–0.5...1.0	20...40
AFRP	–2.0...–1.0	60...80
GFRP	7...12	9...20
concrete	6...13	

Thermal effects can also have influence on aging of resins, in this way the residual strength of FRP reinforcements. Experimental data on change of long-term residual strength of FRPs due to thermal cycles are not available.

Tensile strength and Young's modulus of FRPs may be influenced by the temperature. Under service temperature of concrete structures (from -20 to +60°C) the reduction in Young's modulus of CFRP is negligible, however, slight reduction of Young's moduli of AFRP and GFRP can be observed (Rostásy 1996). Change in tensile strength is attributed to only higher temperatures when deterioration of resin occurs.

Effect of elevated temperature

Polymeric materials are usually flammable or harming in the case of fire, therefore, basically resin determines the temperature/fire resistance of FRPs. Resins soften, melt or catch fire above 150-200°C. Fibers themselves are more or less able to resist to higher temperatures: aramid to 200°C, glass to 300-500°C while carbon in non-oxidizing environment up to 800-1000°C (Rostásy 1996).

Experiments demonstrated that – in addition to fiber type – the surface configuration of FRP reinforcement has a very important role on fire resistance and behavior under elevated temperature. Due to the temperature independence of carbon fibers themselves, CFRP shows the most favorable behavior. Decrease in tensile capacity of CFRP strands and braided tendons at 400°C was about 20 percent on the contrary to CFRP round rods that had no deterioration (Tanao et al. 1997). Decrease in tensile capacity of AFRP reinforcements at 400°C was about 60 percent independently of the configuration (Tanao et al. 1997). It has to be also noticed that CFRP specimens examined after cooling to room temperature from 300°C showed no decrease in tensile capacity (Tokyo Rope 1993).

Deflection of concrete beams reinforced with FRPs under high-temperature loading is also attributed to the surface configuration of reinforcement used (Sakashita 1997). Deflections of GFRP reinforced beams and beams reinforced with braided AFRP or braided CFRP tendons were greater than that of steel reinforced member. Good results were produced by beams reinforced with spirally wounded or straight CFRP rods with deflections of about one-fifth that of steel reinforced member.

Effect of freezing and thawing

In a lot of civil engineering applications reinforced concrete members are subjected to high number of freezing/thawing cycles (mostly combined with water and chloride ion penetration into the concrete). However, experimental data on the influence of such effects on the durability of FRP reinforcements is very limited. Due to freezing and thawing cycles (combined with water and chloride ion diffusion) degradation of fibers, resin and interfacial bond is possible. According to micro cracking of concrete under freezing and thawing cycles bond between concrete and FRP can be also influenced.

TIME DEPENDENT MECHANICAL CHARACTERISTICS OF FRP REINFORCEMENTS

Time dependent mechanical characteristics of FRP reinforcements can be different from that of ordinary steel reinforcements due to not only the different materials but also to the composite behavior: time dependent phenomena can take place within the material phases (fiber, matrix) and on the interface (bond, delamination).

Creep

Two different issues have to be distinguished related to creep: the creep strain under sustained load and the long-term tensile strength under sustained load (often called as stress rupture, residual strength or creep rupture strength).

CFRP shows excellent creep behavior in terms of creep strain: in general it can be stated that creep strain of CFRP at room temperature and humidity remains under 0.01% after 3000 hours at a tensile stress of even 80 percent of the tensile strength (Machida 1993, Saadatmanesh and Tannous 1999a, Tokyo Rope 1993). AFRP and GFRP give much higher creep strain than CFRP: 0.15-1.0% for AFRP and 0.3-1.0% for GFRP under conditions verified above (Gerritse 1993, Machida 1993, Piggott 1980).

Long-term tensile strengths of FRPs have a very important role in defining allowable sustained service stress in structural members. After 10^6 hours long-term tensile strength of CFRP, AFRP and GFRP can be estimated as 80-95%, 50-70% and 40-70% of the short-term tensile strength, respectively, considerably depending on the product examined (Fig. 3) (Ando et al. 1997, Machida 1997, Rostásy 1997, Uomoto et al. 1995, Wolff and Miesseler 1993, Yamaguchi et al. 1997).

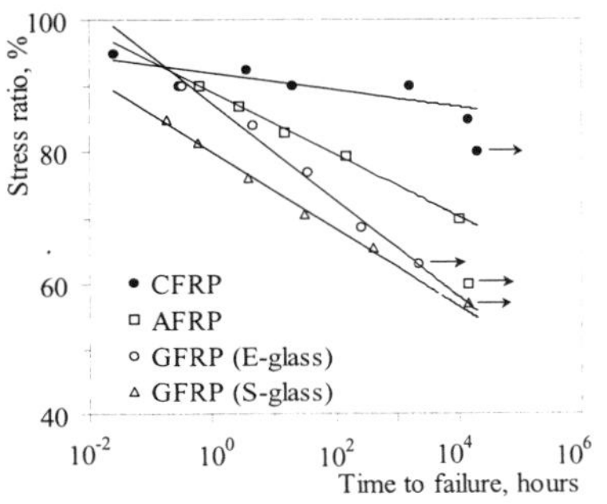

Fig. 3 Creep rupture of FRPs (after Machida 1997)

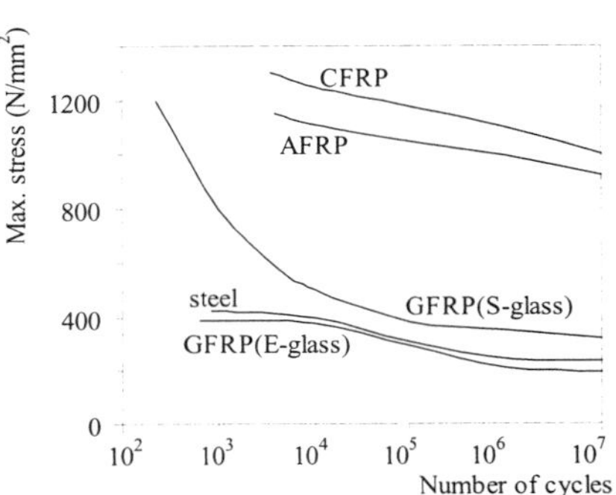

Fig. 4 $S-N$ curves of FRPs (after Machida 1997)

Relaxation

Relaxation of CFRP, AFRP and GFRP tendons after 50 years of loading can be estimated as 2-10%, 11-25% and 4-14 %, respectively, depending on the initial tensile stress (Ando et al. 1997, Machida 1997, Saadatmanesh and Tannous 1999a, 1999b, Tokyo Rope 1993, Wolff and Miesseler 1993). Relaxation after 1000 hours can be estimated as 0.5-1.0%, 5.0-8.0% and 1.8-2.0%, respectively.

Fatigue

Although limited number of fatigue tests of FRP reinforcements is available some obvious tendencies can be verified. Fatigue resistance of GFRP is usually less than that of prestressing steel (Wolff and Miesseler 1993, Machida 1993, 1997). AFRP tendons show similar or higher fatigue strength than prestressing steel with especially higher stress amplitudes (Gerritse 1993). Both GFRP and AFRP show similar dependency of stress level on fatigue strength like prestressing steel does (Uomoto et al. 1995). CFRP has excellent fatigue strength: 3 to 4 times higher than that of prestressing steel (Tokyo Rope 1993, Saadatmanesh and Tannous 1999a, 1999b). Fatigue of CFRP seems to be independent of stress level and amplitude (Uomoto et al. 1995). Fig. 4 indicates *S-N* curves of various FRPs.

Environmental effects on time dependent mechanical characteristics

Environmental effects can influence considerably time dependent mechanical characteristics of FRPs. Due to ingress of water, alkali or chloride ions into resins all types of FRPs are expected to deteriorate in their characteristics. Limited experimental data are available, however, basic assumptions can be drawn (Gerritse 1993, Rostásy 1997, Saadatmanesh and Tannous 1999a, 1999b). Relaxation of AFRP and CFRP is increased with 10-30 percent by absorption of water or solutions of chlorides or alkali.

Creep of AFRP and CFRP increases with 30-40 percent by alkaline environment. Long-term tensile strength of AFRP can be considerably decreased by alkali.

CONCLUSIONS

Based on an extensive literature review the following conclusions can be drawn considering long-term behavior of FRP reinforcements:

- In general, FRP reinforcements show better creep, relaxation and fatigue behavior than steel reinforcements. However, influence of environmental effects on the time dependent mechanical properties needs further considerations.
- Liquids (water, alkali and salt solutions) can diffuse into resins of FRPs that can cause deterioration of mechanical characteristics. Least diffusion can be found in vinyl ester resins. Glass fibers tend to deteriorate by alkali and chloride ions resulted in corrosion-induced failure. Aramid fibers can absorb water that cause reversible loss in strength, however, can cause bond failure by splitting due to swelling. Aramid fibers also may deteriorate in alkaline environment. Carbon fibers are resistant to any kind of harsh environment. UV radiation can be harmful on aramid fibers and resins.
- Coefficients of thermal expansion of FRPs in the longitudinal direction are smaller, in the transverse direction 5 to 8 times greater than that of concrete. This may have an influence on the minimum concrete cover against splitting. Behavior of FRPs under elevated temperature is governed by not only the resin and the fibers but also the surface configuration of tendons.

REFERENCES

Ando, N. – Matsukawa, H. – Hattori, A. – Mashima, M. (1997) "Experimental Studies on the Long-Term Tensile Properties of FRP Tendons", *Proc.* 3rd Int. Symp. FRPRCS-3, JCI, 1997, Vol. 2., pp. 203-210.

Balázs, G. L. – Borosnyói, A. (2001) "Prestressing with CFRP Tendons", *Proceedings* of the UEF International Conference on High Performance Materials in Bridges and Buildings, July 29 – August 3, 2001, Kona, Hawaii (in printing)

Gerritse, A. (1993) "Aramid-based prestressing tendons", *Alternative Materials for the Reinforcement and Prestressing of Concrete,* Ed. Clarke, Chapman & Hall, London, 1993, pp. 172-199.

Kato, Y. – Yamaguchi, T. – Nishimura, T. – Uomoto, T. (1997) "Computational Model for Deterioration of Aramid Fiber by Ultraviolet Rays", *Proc.* 3rd Int. Symp. FRPRCS-3, Japan Concrete Institute, 1997, Vol. 2., pp. 163-170.

Machida, A. (1993) "State-of-the-Art Report on Continuous Fiber Reinforcing Materials", JSCE, Tokyo, 1993.

Machida, A. (1997) "Recommendation for Design and Construction of Concrete Structures Using Continuous Fiber Reinforcing Materials", JSCE, Tokyo, 1997.

Piggott, M. R. (1980) "Load Bearing Fibre Composites", Pergamon Press, Oxford, 1980.

Rostásy, F. (1996) "State-of-the-Art Report on FRP Materials", *FIP Report, Draft*, 1996 (unpublished)

Rostásy, F. (1997) "On Durability of FRP in Aggressive Environment", *Proc.* 3rd Int. Symp. FRPRCS-3, JCI, 1997, Vol. 2., pp. 107-114.

Saadatmanesh, H. – Tannous, F. E. (1997) "Durability of FRP Rebars and Tendons", *Proc.* 3rd Int. Symp. FRPRCS-3, JCI, 1997, Vol. 2., pp. 147-154.

Saadatmanesh, H. – Tannous, F. E. (1999) "Relaxation, Creep, and Fatigue Behavior of Carbon Fiber Reinforced Plastic Tendons", *ACI Materials Journal*, V. 96, No. 2, March-April 1999, pp. 143-153.

Saadatmanesh, H. – Tannous, F. E. (1999) "Long-Term Behavior of Aramid Fiber Reinforced Plastic (AFRP) Tendons", *ACI Materials Journal*, V. 96, No. 3, May-June 1999, pp. 297-305.

Sakashita, M. – Masuda, Y. – Nakamura, K. – Tanao, H. – Nishida, I. – Hashimoto, T. (1997) "Deflection of Continuous Fiber Reinforced Concrete Beams Subjected to Loaded Heating", *Proc.* 3rd Int. Symp. FRPRCS-3, JCI, 1997, Vol. 2., pp. 51-58.

Sen, R. – Mariscal, D. – Shahawy, M. (1993) "Durability of Fiberglass Pretensioned Beams", *ACI Structural Journal*, V. 90, No. 5, September-October 1993, pp. 525-533.

Sen, R. – Shahawy, M. – Rosas, J. – Sukumar, S. (1998) "Durability of Aramid Pretensioned Elements in a Marine Environment", *ACI Structural Journal*, V. 95, No. 5, September-October 1998, pp. 578-587.

Sen, R. – Shahawy, M. – Rosas, J. – Sukumar, S. (1998) "Durability of Carbon Pretensioned Elements in a Marine Environment", *ACI Structural Journal*, V. 95, No. 6, November-December 1998, pp. 716-724.

Sheard, P. – Clarke, J. – Dill, M. – Hammersley, G. – Richardson, D. (1997) "EUROCRETE – Taking Account of Durability for Design of FRP Reinforced Concrete Structures", *Proc.* 3rd Int. Symp. FRPRCS-3, JCI, 1997, Vol. 2., pp. 75-82.

Taerwe, L. – Pallemans, I. (1995) "Force Transfer of AFRP Bars in Concrete Prisms", *Proc.* 2nd Int. Symp. FRPRCS-2,, E & FN Spon, London, 1995, pp. 154-163.

Tanao, H. – Masuda, Y. – Sakashita, M. – Oono, Y. – Nonomura, K. – Satake, K. (1997) "Tensile Properties at High Temperatures of Continuous Fiber Bars and Deflections of Continuous Fiber Reinforced Concrete Beams Under High-Temperature Loading", *Proc.* 3rd Int. Symp. FRPRCS-3, JCI, 1997, Vol. 2., pp. 43-50.

Tannous, F. E. – Saadatmanesh, H. (1998) "Environmental Effects on the Mechanical Properties of E-Glass FRP Rebars", *ACI Materials Journal*, V.95, No. 2, March-April 1998, pp. 87-100.

Tokyo Rope (1993), "Technical Data on CFCC®", Tokyo Rope Mfg. Co., Ltd. *Manual*, Tokyo, October 1993.

Uomoto, T. – Nishimura, T. – Ohga, H. (1995) "Static and Fatigue Strength of FRP Rods for Concrete Reinforcement", *Proc.* 2nd Int. Symp. FRPRCS-2, E & FN Spon, London, 1995, pp. 100-107.

Uomoto, T. – Ohga, H. (1996) "Performance of Fiber Reinforced Plastics for Concrete Reinforcement", *Proc.* 2nd ACMBS Conference, Montreal, CSCE, 1996, pp. 125-131.

Uomoto, T. – Nishimura, T. (1999) "Deterioration of Aramid, Glass and Carbon Fibers Due to Alkali, Acid and Water in Different Temperatures", *Proc.* 4th Int. Symp. FRPRCS-4, ACI SP-188, American Concrete Institute, 1999, pp. 515-522.

Wolff, R. – Miesseler, H.-J. (1993) "Glass-fibre prestressing system", *Alternative Materials for the Reinforcement and Prestressing of Concrete,* Ed. Clarke, Chapman & Hall, London, 1993, pp. 127-152.

Yamaguchi, T. – Kato, Y. – Nishimura, T. – Uomoto, T. (1997) "Creep Rupture of FRP Rods", *Proc.* 3rd Int. Symp. FRPRCS-3, JCI, 1997, Vol. 2., pp. 179-186.

Durability of FRP in Concrete
A State of the Art

Professor Peter Waldron[1], Dr Ewan A. Byars[1] and Valter Dejke[2]

Abstract

This paper discusses the concrete environment and its affect on fibre-reinforced polymers (FRP) in terms of internal and external aggressive conditions that may affect its durability. The specific conditions considered are the effects of moisture, chlorides, alkali, stress, temperature, UV actions, carbonation and acid attack.

The differences in performance between glass, aramid and carbon fibres and the polymers used to bind them have, where possible, been identified. However, higher variability in performance may be due to manufacturing techniques and details of some of the polymers and fibres tested are not disclosed in the literature.

Notwithstanding the above, the potential degradation mechanisms are discussed with reference to internationally published research and some very general recommendations are given at the end of each section in an attempt to give some guidance to engineers when selecting FRP for construction.

Introduction

FRP durability in concrete has predominantly been measured by accelerated test methods that expose specimens to environments harsher than they would normally encounter. This data is then used to extrapolate estimates of the likely long-term performance. It is important to bear in mind that accelerated exposure data and real-time performance are unlikely to follow a simple linear relationship and the relationships have yet to be confidently determined.

Mechanical changes in Young's modulus, tensile, interlaminar shear and bond strength are the best indicators of FRP deterioration. These may be complemented by studies of physical and microstructural properties using techniques including TGA (Thermogravimetric Analysis), Light and SE (Scanning Electron) Microscopy, DMA (Dynamic Mechanical Analysis), DSC (Differential Scanning Calorimetry), Potentiodynamic Polarisation Scans, Galvanic Coupling test and FTIR (Fourier Transform Infrared Spectroscopy).

Although some progress has been made towards understanding FRP deterioration in concrete, little data can be easily used by engineers. There is a lack of international agreement about FRP durability tests and variability in FRP production methods, fibre/polymer types and research approaches and lack of real-time performance data further complicates the issues.

[1]Centre for Cement and Concrete (CCC), Department Of Civil and Structural Engineering, University of Sheffield, Sheffield S1 3JD, UK. Tel +44 114 222 5061, Fax +44 114 222 5700 E-mail p.waldron@sheffield.ac.uk

[2]IFP Research Ab, The Swedish Institute for Fibre And Polymer Research

There is therefore a clear priority need for the international research community to identify and ratify a standard test procedure that could be confidently recommended to civil engineers as a basis upon which to select FRP materials for use as concrete reinforcement and pre-stressing bar. This paper reports on the work that has been done by researchers in this field and suggests some general guidelines that could be used when specifying FRP in concrete.

The Concrete Environment

Concrete contains calcium, sodium and potassium hydroxides creating pore water solutions of ~pH 13. This causes an oxide layer to form the steel surface, preventing movement of water and oxygen and inhibiting corrosion, which occurs after the decay of this oxide layer, either by pH-reducing carbonation or by chloride attack. Carbonation is the most common and occurs everywhere at a rate dependent on concrete W/C ratio, cement type, curing and CO_2 concentration. Chloride attack is limited to reinforced concrete structures with a supply of chlorides, such as sea structures, those susceptible to wind-borne chlorides or those sprayed by de-icing salts. Concrete structures may also experience heating/cooling, freezing/thawing and wetting/drying cycles, which also promote concrete decay and subsequent steel corrosion.

Durability of FRP

The important factors for FRP durability are different to those of steel. FRP does not appear to be affected by chlorides or carbonation that reduces alkalinity of the concrete, in fact the latter may extend its life, as might the use of pozzolans that reduce concrete alkalinity. The following sections give guidance on the use of FRPs in a variety of structural exposure conditions. It should be noted, however, that the available data is limited to short term testing and therefore a conservative approach should, and has, been taken.

Effects of Water. This has been studied in air at different %RH and temperature and by immersion at different temperatures and stresses (Bank and Gentry 1995; Saadatmanesh and Tannous 1997; Hayes et al 1998; Steckel et al 1998; Verghese et al 1998 and Djeke 2001). Common methods of measuring performance of FRP under these conditions are changes in tensile strength and Youngs Modulus.

Studies indicate that deterioration of polymer resins may occur when water molecules act as resin plasticizers and disrupt Van-der-Waals bonds in polymer chains (Bank and Gentry 1995). This has caused changes in modulus, strength, strain to failure, toughness, swelling stresses leading to polymer matrix cracking, hydrolysis and fibre-matrix de-bonding (Hayes et al 1998). The literature indirectly suggests that the last of these is fibre-dependent and this appears to be more serious at elevated temperatures (>60°C), in line with increased moisture absorption content at saturation, particularly for polyesters with higher water diffusivity. However Hayes et al (1998) found improved mechanical properties with some FRPs in water.

Effects of Chlorides. A potential application of FRP is in saline environments where steel is likely to corrode without additional protection. Researchers (Saadatmanesh and Tannous 1997; Sasaki et al 1997; Sen et al 1997; Gangarao and Vijay 1997; Chin et al 1997; Steckel et al 1998; Rahman et al 1998; Toutanji and El-Korchi 1998) have investigated glass, aramid and carbon fibre reinforced polymer (GFRP, AFRP and CFRP) products with different surface veil systems in different Cl^- concentrations up to 4%. Stressed and unstressed bars at ambient temperatures up to 70°C with varying RH have been used and in some cases effective weathering periods of >50 years have been claimed.

Results vary widely but differentiation between chloride attack and concomitant degradation due to moisture diffusion and/or alkali attack of the fibres is difficult. In broad terms, CFRP bars exposed to combined chloride/moisture attack in concrete show very little degradation with time, exposure or temperature. AFRP and GFRP elements may show up to 50% loss of strength and stiffness pre-stress relaxation of up to 30% and moment loss up to 20%.

It needs to be emphasised that deterioration may not be attributed to chloride attack but to alkali attack or resin plasticization by water uptake. However, there are some indications that saline solution is a slightly more severe environment than fresh water.

Effects of Alkali

Although concrete is traditionally protects reinforcement, alkalinity may affect glass fibres unless suitable polymer resins (Steckel et al 1998) protect them. Resistance is generally thought to be best with carbon, followed by Aramid and then glass fibres (Machida 1993).

Alkali attack is widely studied, but with no single durability test and many types of fibre/ polymer materials and production methods this leads to variable results and implications. In some investigations, FRP is embedded in concrete to study changes in the bond properties (Scheibe and Rostasy 1998) but the majority of research has used simulated concrete pore solutions containing $NaOH_2$, KOH and saturated $Ca(OH)_2$ with pH of 12-13.5. Temperature ranges used have been 20-80°C (Conrad et al 1998).

Mechanical Tests. Residual changes in tensile strength, Youngs Modulus and ultimate strain; physical analysis (TGA, DMTA, DSC, FTIR) (Bank et al 1998; Chin et al 1998) and diffusion tests (Alsayed and Alhozaimy 1998; Scheibe and Rostasy 1998) have also been used to correlate physical properties with mechanical behaviour. For concrete members, changes in moment capacity (Scheibe and Rostasy 1998; Gangarao and Vijay 1997) and pullout tests (Sheard et al 1997; Sen et al 1999) have been assessed. The following factors affecting the rate of alkali attack of FRP have been identified:

- The susceptibility of plain fibres to alkali attack
- The alkali-diffusivity of the resin and therefore the level of protection afforded to the fibre
- The quality of the fibre-resin bond, through which alkali can permeate and attack the fibre
- The temperature, which affects reaction rates and rates of diffusion
- The concentration of alkali (affected by cement type)
- Alkali ion mobility (affected by degree of saturation)

Due to commercial sensitivities, much of the research tends not to reveal technical details about fibres and polymers and this makes interpretation and robust conclusion difficult. In addition, no sound models yet exist for the conversion of accelerated results into reliable real-time data and the following discussion of alkali attack should be read with this in mind.

Exposure to alkaline solution - fibres. A study by Banks et al (1998) that immersed E glass/vinyl ester rods in ammonium hydroxide (NH_4OH) solution (30%) at 23°C (224 days) showed 12% tensile strength loss. TGA analysis showed deterioration in the matrix-fibre interface. Steckel et al (1998) immersed CFRP and GFRP systems in $CaCO_3$ solution (pH 9.5) at 23°C (125 days). The systems were unaffected except for 10% reduction in Young's modulus for one GFRP system and 30% reduction in short beam shear strength for another.

Tannous and Saadatmanesh (1998) immersed one type of AFRP and two types of CFRP tendons in saturated $Ca(OH)_2$ solution (pH 12) at 25°C and 60°C. The AFRP specimens showed reductions of 4.3 and 6.4% in tensile strength after a year but the CFRP tendons were unaffected. Another investigation (Porter et al 1997) exposed GFRP and CFRP to an environment equivalent to 50 years real exposure (3 months at pH 12.5-13 and 60°C). Three

GFRP systems lost 55-73% tensile strength but CFRP tendons were unaffected. Uomoto and Nishimura (1997) immersed GFRP, AFRP and AGFRP (aramid-glass hybrid) to Na $(OH)_2$ solution at 40°C for 120 days. GFRP lost 70% tensile strength but AFRP and AGFRP specimens were unaffected. Using EPMA (Electron Probe Micro-Analysis) the Na intrusion was shown to be deeper in GFRP. Combined freeze-thaw/alkali-exposure testing by Gangarao and Vijay (1997) generated 7-49% tensile strength loss and 3-31% drop in Youngs Modulus for E-glass GFRP systems (with vinyl ester or polyester). Saadatmanesh and Tannous (1997) immersed CFRP, AFRP and GFRP specimens in saturated $Ca(OH)_2$ solution at 25 and 60°C and showed that Fick's Law could predict FRP tensile strength losses.

Exposure to alkaline solution - resins. Chin et al (1998) immersed polymeric resins in alkali at ambient and elevated temperatures and tested the specimens for tensile strength and using DMTA, DSC, TGA and FTIR. The results showed that vinyl ester polymers had higher resistance than iso-polyester (80% and 40% tensile strength remaining respectively). Bakis et al (1998) tested three different GFRP rods by 28-days immersion in a saturated solution of $Ca(OH)_2$ at 80°C. The 100% vinyl ester rods were less affected than vinyl ester/polyester blended matrixes. Alsayed and Alhozaimy (1998) examined two types of GFRP bars (40% unsaturated polyester/60% urethane modified vinyl ester and an unspecified second type) coated with cement paste (w/c=0.5) and immersed in water. The tensile strength of the GFRP bar decreased by 20% in 4 months. When the same specimens were immersed in an alkaline solution (20 gm/l NaOH) the corresponding value was 30%. For the other GFRP type (resin type not declared), the corresponding reductions were practically zero, suggesting that resin type and manufacturing process may be the key factors for alkaline durability of GFRP bars.

Alkali exposure under mechanical stress. In real concrete structures, most reinforcement is stressed and the influence of stress and alkali has been studied. Rahman et al (1998) subjected GFRP and CFRP made with vinyl ester resin to NaOH (58g/l) solution at 70°C for 370 days with tensile loadings of 0.3 and 0.5 ultimate tensile strength for the GFRP and CFRP specimens respectively. The GFRP specimen failed after 45 days probably because of rapid diffusion of hydrated hydroxyl ions (OH^-), but FTIR analysis showed most of the resin to be unaffected. Gangarao and Vijay (1997) found strength reductions (1-76%) for stressed GFRP bars in alkaline solution of pH 13 for 201 days. Best resistance was with vinyl ester. Sheard et al (1997) reported reduced interlaminar shear strength for some GFRP and CFRP systems in pH 11.5-13.5 solutions but others were almost unaffected. In a related study, Clarke and Sheard (1998) showed that CFRP specimens performed less well than GFRP after 6 months exposure in accelerated conditions (pH 12.5, 5% ultimate bending stress at 38°C). Benmokrane et al (1998) also found reduced strengths in stressed alkali exposure when evaluating the influence of resin type and manufacturing processes and concluded that vinyl ester is the most suitable polymer for GFRP bars. Arockiasamy et al (1998) found 0% strength reduction on CFRP cables after exposure under tension (0.65 ultimate tensile strength) in an alkaline environment (pH 13-14) after 9 months.

Alkali attack at elevated temperature. Scheibe and Rostasy (1997) conducted stress rupture tests for AFRP (Arapree with epoxy resin) bars under different conditions. Embedded bars were loaded to 0.75 ultimate tensile strength in air (65% RH, at 20°C) or immersed in 0.4m KOH solution saturated with $Ca(OH)_2$ at temperatures up to 60°C. The lifetime in concrete exposed to 20°C alkali solution and 20°C air was 714 and 3308 hours respectively.

Porter et al (1997) immersed embedded E-glass/vinyl ester rods to 60°C water and 60°C alkali (pH 12). In pullout tests the rods were unaffected, possibly because thick concrete cover protected the bars from full exposure. Pantuso et al (1998) embedded GFRP bars in concrete and subjected them to wetting/drying cycles in water for 60 days. Tensile strength decreased by up to 21% compared to 7% for naked rods immersed in a water bath. To simulate a tidal zone, CFRP specimens in concrete were subjected to wetting/drying cycles for 18 months at 20-60°C by Sen et al (1998). The bond strength increased due to swelling of the FRP bars, but flexural tests on reinforced beam specimens did not show similar improvement. Sheard et al (1997) also reported no mechanical or physical deterioration in GFRP or CFRP after 12 months in various alkaline solutions at 20-38°C. Porter et al (1997) studied prestressed beams (0.4 ultimate tensile strength) immersed in highly alkali solutions and reported that GFRP/polyester resin tendons lost their pre-stressing force whilst CFRP, also made with polyester resin appeared unaffected. Adimi et al (1998) studied tension-tension fatigue of GFRP and CFRP reinforcement in varying alkalinity and reported only negligible effects.

An AFRP durability study by Scheibe and Rostasy (1998) tested prestressed (ultimate tensile strength 0.7-0.85) slabs, pre-cracked and stored for 2 years. The moment capacity was unchanged. However Gangarao and Vijay (1997) immersed GFRP-reinforced beams in salt water for 240 days and showed a reduced moment capacity of 18%, attributed to alkali-induced bond deterioration. Tomosawa and Nakatsuji (1997) exposed reinforced beams on the Japanese coast for 2 years and found no flexural strength reduction for AFRP, CFRP or GFRP bars but a small reduction for prestressed beams with AFRP and CFRP tendons. Field exposure tests using GFRP and CFRP pullout specimens by Sheard et al (1997) found a slight increase in pullout strength after 12 months, attributed to increased concrete strength. Similar results by Sen et al (1999) for CFRP epoxy rod specimens in an outdoor environment for 18 months. In this case the increase was ascribed to swelling of the CFRP material.

Table 1 gives a summary of results from Gothenburg University for tensile strength reductions obtained for GFRP bars in alkaline solutions, concrete and water at 60°C and 20°C.

EXPOSURE CONDITION	TEMP (°C)	% ORIGINAL TENSILE STRENGTH Age at Test (days)				
		28	90	180	365	545
Alkali	60	82	55	37	32	31
Concrete	60	91	80	57	51	45
Water	60	93	84	75	73	72
All (average)	20	95	92	90	88	80

TABLE 1
Effect Of Temperature on Alkali, Concrete and Water Attack of GFRP

This shows that the most aggressive environments, in descending order are alkali 60°C, concrete 60°C, then water 60°C. The results obtained for the alkali, concrete and water at 20°C were close and an average of these conditions is shown. At 20°C, GFRP tensile strength deteriorates by around 20% after 18 months. The hotter environments demonstrate the significant effects of temperature on GFRP degradation and demonstrate that extra, carefully evaluated safety factors should be used when bars are subjected to elevated temperatures. A new approach is proposed in another paper (Byars et al 2001) and has factors that take into account temperature, moisture and time and therefore allows environment-specific durability design of FRP to be applied to the structural design process.

Utra Violet Rays

Ultraviolet rays affect polymeric materials (Bank and Gentry 1995). Although FRP reinforcing bars are not exposed after use, UV rays may cause degradation during storage or if FRP is used as external reinforcement. Exposure tests have been performed in the laboratory (Kato et al 1997) and under field conditions (Tomosawa and Nakatsuji 1997), who measured and compared the tensile strength of aged and virgin samples to evaluate degradation. Kato et al (1997) examined the effect of the UV rays on AFRP, CFRP and GFRP rods exposed in a high-UV intensity laboratory environment for 250, 750 and 1250 wetting/drying cycles, with UV-intensity of 0.2MJ/m^2/hour and temperature of 26°C. In addition, fibres were also tested after UV exposure of up to 1,000 hours. AFRP rods show around 13% reduction in tensile strength after 2500 hours exposure, GFRP rods 8% after 500 hours (no reduction thereafter) and CFRP rods showed no reduction. The pattern was almost identical for the fibre testing.

Thermal Actions

Deterioration by thermal action may occur in FRP when constituents have different coefficients of thermal expansion and transverse thermal expansion particularly important for good bond. Sen and Shahawy (1999) studied diurnal/seasonal temperature change on durability of 12 pre-cracked, pretensioned CFRP pre-tensioned piles, designed to fail by rupture of the pre-stressing rods. These were stored in tanks and subjected to wetting/drying and temperature cycles (20-60°C). The durability was assessed over three years by periodic flexural tests. The results of the tests indicated that the performance of the piles was largely unaffected, but both bond degradation and reductions in ultimate load capacity were observed for some specimens in which the pre-exposure pre-cracking damage was greatest.

Carbonation

A small amount of research work on the effect of carbonating concrete on FRP was carried as part of the EUROCRETE project (Sheard et al 1997), which studied wide range of FRP durability aspects. The data obtained was more variable than that for other accelerated conditions, however no deterioration due to carbonation was observed.

Acid Attack

There is little published data on the effects of acid attack on FRP. Indeed it is likely that in acid conditions, deterioration of concrete would be of greater concern. There is clearly a need to investigate this issue and produce some guidance for circumstances when acid resistant cement, such as high-alumina cement, was used in conjunction with FRP reinforcement.

Conclusions

1) In moist conditions above 40°C, the use of FRP as reinforcement bar should be backed by laboratory data about the performance of the specific fibre/polymer combination selected in more aggressive conditions than the proposed exposure environment.
2) Chloride attack data are insufficient to draw conclusions and the use of FRP as reinforcement should be based on knowledge of the performance of the bar in a Cl^- environment and the effects of moisture and alkali attack on the selected system.
3) Performance of FRP reinforcement and pre-stressed tendons in alkali varies with the materials (fibres and resins) used and manufacturing processes. Literature suggests that FRP deteriorates much faster in alkali solution than in concrete, probably due to the relative mobility of OH^+ ions. Specific observations are given overleaf:

a. Extensive degradation has been evidenced in GFRP rods after exposure to alkaline solutions at high temperature. Bars embedded in concrete at various temperatures and with good fibre-resin combinations show only limited degradation, but this increases with temperature and stress level.

b. Alkali affects AFRP bars and tendons less than GFRP, but a combination of alkali solution and high tensile stress (in the order of 0.75 ultimate tensile strength) may damage AFRP bars significantly.

c. There is no significant alkali attack problem for **CFRP** with a proper fibre-resin system.

d. Vinyl esters have much better alkali resistance than polyesters resins.

4) For embedded FRP reinforcement UV-attack poses no problems but rods and tendons should be protected direct sunlight whilst in storage. Plate bonded or wrapped sheets of FRP should be protected from sunlight using a proprietary system.

5) Literature suggests that temperatures over 60°C may present significant problems for FRP, but further research is needed to make robust conclusions and recommendations.

6) It is unlikely that carbonation promotes deterioration of FRP bars in concrete. The associated reduction in pH is likely to improve the durability of concrete since it reduces the concrete pore-water alkali that attacks some fibres and polymers.

General Remarks

This paper has discussed durability-related aspects of FRP embedded in concrete and the approaches taken by various researchers. It is clear that whilst broad conclusions can be drawn about the relative performances of FRP materials, these cannot be applied strictly in all cases due to variations in the materials and manufacturing processes used to make FRP.

It is also clear that differences in design approach to FRP durability will make it difficult for the international construction community to have confidence about predictions of FRP service life in aggressive environments. The biggest problem is the perception that GFRP is sensitive to alkali attack and that the concrete environment is therefore intrinsically highly aggressive. Research has shown that the concrete environment is not as aggressive as alkaline solution and that alkali resistance can be significantly improved by the selection of appropriately treated glass fibres, suitable resins and better production techniques.

This underpins the need for further research and a move towards an internationally accepted performance-based durability test methodology that is calibrated against defined performance specifications that are useable with a concrete design context. This is the next stage of the durability work being carried out by Task Group 9.3 of the *fib*.

References

Alsayed S and Alhozaimy A (1998). Durability of FRP Reinforcements under Tension-Tension Axial Loading Cyclical Loading. Proc. 1st Intl. Conf. On Dur. of Fiber Rein. Polymer (FRP) Composites for Construction, Sherbrooke, pp. 635-647.

Bakis CE, Freimanis AJ, Gremel D and Nanni A (1998). Effect of resin material on bond and tensile properties of unconditioned and conditioned FRP reinforced rods. Proc. 1st Intl. Conf.

on durability of fiber reinforced polymer (FRP) composites for construction, Sherbrooke, August, pp. 525-535.

Bank LC, Gentry T R (1995). Accelerated Test Methods to Determine the Long-Term Behaviour of FRP Composite Structures: Environmental Effects, Journal of Reinforced Plastic and Composites, Vol. 14, pp. 558-587.

Bank LC, Gentry TR, Barkatt A, Prian L, Wang F and Mangala SR (1998). Accelerated ageing of pultruded glass/vinyl ester rods. Proc. 2nd. Intl. Conf. On Fibre Composites in Infrastructure, Vol. 2, pp. 423-437.

Byars, EA, Dejke V and Demis, S (1999-2001). Development drafts of a model European durability specification for FRP concrete. Lisbon, Delft, Riga, Venice and Cambridge meetings of fib Task Group 9.3.

Byars, EA, Waldron P, Dejke V and Demis S. Durability of FRP in Concrete – Current Specifications and a New Approach (in draft).

Chin JW, Nguyen T and Aouadi K (1997). Effects of Environmental Exposure on Fiber-Reinforced Plastic (FRP) Materials Used in Construction, Journal of Composites Technology and Research, Vol 19, No 4, pp. 205-213.

Clarke, JL and Sheard P (1998). Designing Durable FRP Reinforced Concrete Structures, Durability of Fibre Reinforced Polymer (FRP) Composites for Construction. Proc. 1st Intl. Conf. (CDCC' 98), Sherbrooke, pp 3-24.

Dejke, V (2001). Durability of FRP Reinforcement in Concrete. Literature and Experiments. Dept of Building Materials, Chalmers University of Technology, Goteborg, Sweden, 210pp.

Gangarao HVS and Vijay PV (1997). Aging of Structural Composites Under Varying Environmental Conditions, Non-Metallic (FRP) Reinforcement for Concrete Structures, Proc. 3rd Intl. Symp., Vol 2, pp 91-98.

Hayes MD, Garcia K, Verghese N and Lesko JJ (1998). The Effects of Moisture on the Fatigue Behavior of a Glass/Vinyl Ester Composite, Proc. 2nd Intl. Conf. on Fibre Composites in Infrastructure ICCI'98, Vol. 1, pp. 1-13.

Kato Y, Yamaguchi T, Nishimura T, and Uomoto T (1997). Computational model for deterioration of aramid fibre by ultraviolet rays, Non-Metallic (FRP) Reinforcement for Concrete Structures, Proc. 3rd Intl. Symp., Vol. 2, pp 163-170.

Pantuso A, Spadea G and Swamy RN (1998). An experimental study on the durability of GFRP bars. Proc. 2nd Intl. Conf. on fiber composites in infrastructure, ICCI'98, Vol. 2, Tuscon, pp 476-487.

Porter ML, Mehus J, Young KA, O'Neil EF and Barnes BA (1997). Aging for fibre reinforcement in concrete. Non-Metallic (FRP) Reinforcement for Concrete Structures, Proc. 3rd Intl. Symp., Vol 2, pp 59, 66.

Porter ML, Mehus J, Young K, Barnes B, and O'Neil EF Eds (1997). Aging Degradation of fibre composite reinforcement for concrete structures, Advanced Comp. Mats.in Bridges and Structures, 2nd Int. Conference, Canada, pp 641-648.

Rahman AH, Kingsley C, Richard J and Crimi J (1998). Experimental Investigation of the Mechanism of Deterioration of FRP Reinforcement for Concrete. Proc. 2nd Intl. Conf. on Fibre Comp.in Infrastructure ICCI'98, Vol. 2, Tucson, pp. 501-511.

Saadatmanesh H, and Tannous F (1997). Durability of FRP rebars and tendons, Non-Metallic (FRP) Reinf. For Conc. Structures, Proc. 3rd Intl. Symp., Vol 2, pp 147-154.

Sasaki I, Nishizaki I, Sakamoto H, Katawaki K and Kawamoto Y (1997). Durability Evaluation of FRP Cables by Exposure Tests, Non-Metallic (FRP) Reinforcement for Concrete Structures, Proc. 3rd Intl. Symp., Vol. 2, pp 131-137.

Scheibe M and Rostasy FS (1997). Stress-rupture of AFRP subjected to alkaline and elevated temperatures. Non-Metallic (FRP) Reinforcement for Concrete Structures, Proc 2nd Intl. Symp. on fiber composites in infrastructure, ICCI'98, Vol. 2.

Sen R, Shahawy M, Rosas J. and Sukumar S (1997). Durability of AFRP & CFRP pretensioned piles in a marine environment, Non-Metallic (FRP) Reinforcement for Concrete Structures, Proc. 3rd Intl. Symp. Vol 2, pp 123-130, October.

Sen R, Shahawy M, Sukumar S and Rosas J (1999). Durability of Carbon Fibre Reinforced Polymer (AFRP) pretensioned elements under Tidal/Thermal Cycles, ACI Structural Journal, may/June, Vol. 96, No. 3, pp 450-457.

Sen R, Shahawy M, Sukumar S and Rosas J (1998). Effects of Tidal Exposure on Bond of CFRP Rods. Proc. 2nd Intl. Conf. on Fibre Composites in Infrastructure ICCI'98, Vol. 2, Tucson, pp 512-523.

Sheard P, Clarke JL, Dill M, Hammersley G and Richardson D (1997). EUROCRETE – Taking Account of Durability for Design of FRP Reinforced Concrete Structures. Non-Metallic (FRP) Reinforcement for Concrete Structures, Proc. 3rd Intl. Symp., Vol 2, pp 75-82.

Steckel GL, Hawkins GF and Bauer JL (1998). Environmental Durability of Composites for Seismic Retrofit of Bridge Columns. Proc. 2nd Intl. Conf. on Fibre Composites in Infrastructure ICCI'98, Vol. 2, Tucson, pp. 460-475.

Tannous FE and Saadatmanesh H (1998). Durability and long-term behavior of carbon and aramid FRP tendons. Proc. 2nd Intl. Conf. on fiber composites in infrastructure, ICCI'98, Vol. 2, Tuscon, pp 524-538.

Tomosawa F and Nakatsuji T (1997). Evaluation of ACM Reinforcement Durability By Exposure Test. Non-Metallic (FRP) Reinforcement for Concrete Structures: Proc. 3rd Intl. Symp., Vol 2, Sapporo, pp. 139-146.

Toutanji H and El-Korchi T (1998). Tensile Durability Performance of Cementitious Composites Externally Wrapped with FRP Sheets, Proc. 2nd Intl. Conf. on Fibre Composites in Infrastructure ICCI'98, Vol. 2, pp. 410-421.

Uomoto T and Nishimura T (1997). Development of new alkali resistant hybrid AGFRP rod, Non-Metallic (FRP) Reinforcement for Concrete Structures. Proc. 3rd Intl. Symp., Vol 2, pp 67-74.

Verghese NE, Hayes M, Garcia K, Carrier C, Wood J, and Lesko JJ (1998). Temperature Sequencing During Hygrothermal Aging of Polymers and Polymer Matrix Composites: The Reverse Thermal Effect. Proc. 2nd Intl. Conf. on Fibre Composites in Infrastructure ICCI'98, Vol. 2, Tucson, pp. 720-739.

FRPs for Strengthening and Rehabilitation: Durability Issues

K.W. Neale[1], P. Labossière[2] and M. Thériault[3]

Abstract

Issues regarding the long-term durability of fibre reinforced polymer (FRP) strengthening and rehabilitation technologies are addressed. A brief overview of some recent research activities in this field is first presented. In particular, we discuss work related to performance in corrosive environments, applications in cold regions, and fatigue behaviour. On the basis of this review, some important needs for future investigation are identified. The importance of developing standardized accelerated laboratory test conditions, as well as of establishing appropriate field sites for calibrating results from accelerated tests, is emphasized.

Introduction

The needs for civil engineering infrastructure rehabilitation are obvious. First, aggressive environments have resulted in the serious deterioration of large numbers of existing structures. In addition, as many structures no longer comply with the load and design criteria specified by current codes, numerous strategies for structural upgrading are being explored. Included in these potential solutions are methods based on the use of fibre reinforced polymers (FRPs) for structural rehabilitation and strengthening.

A particular concern related to the use of FRPs for structural rehabilitation is the long-term performance of this technology in harsh and corrosive environments. As a result, this is a topic that is receiving considerable attention. Indeed, conference proceedings over the past few years (e.g., Meier and Betti 1997, Benmokrane and Rahman 1998, Dolan et al. 1999) contain numerous contributions related to durability aspects of FRP repairs. This is a topic that is currently

[1] Professor, Department of Civil Engineering, University of Sherbrooke, Sherbrooke, Qc, Canada, J1K 2Rl; phone 819.821.7752; kenneth.neale@courrier.usherb.ca
[2] Professor, Department of Civil Engineering, University of Sherbrooke, Sherbrooke, Qc, Canada, J1K 2R1; phone 819.821.8000(x 2119), pierre.labossiere@courrier.usherb.ca
[3] Assistant Professor, Department of Civil Engineering, University of Sherbrooke, Sherbrooke, Qc, Canada, J1K 2R1; phone 819.821.8061, michele.theriault@courrier.usherb.ca

receiving significant research support in the U.S. (Chong 1998). An example of the importance of assessing the environmental durability of FRP retrofitting systems is a recent study which showed that unacceptable reductions in mechanical properties can occur if resins with inadequate moisture absorption characteristics are employed (Hawkins et al. 1998).

In Canada, the major research effort in the field of FRPs for infrastructure rehabilitation is being conducted through the Network of Centres of Excellence *ISIS Canada* ("Intelligent Sensing for Innovative Structures"). An important focus in ISIS Canada is the theme on *FRPs for Structural Rehabilitation*. This theme, which is centred at the University of Sherbrooke (Sherbrooke, Quebec), has as its primary objective the development and field implementation of FRP technologies for the rehabilitation and strengthening of civil engineering structures. Despite the obvious advantages of FRPs, rehabilitation techniques using these high-performance materials are far from being accepted on an everyday basis. Reservations exist in part because of open questions concerning the performance, cost-effectiveness, and long-term durability of FRP retrofit techniques in severe climatic conditions such as those encountered in Canada. The ISIS network is therefore focusing on those aspects of performance and durability that are of particular relevance to Canada. The main objective is to develop and implement FRP rehabilitation methodologies for the maximum service-life extension of corrosion damaged or structurally deficient structures, with a minimum of cost, time and complexity.

In this paper, various issues regarding the long-term durability of FRP rehabilitation techniques are addressed. A brief overview of research related to the performance of FRP repair and strengthening technologies in corrosive environments and cold regions is first given. Fatigue and creep behaviour are also discussed. The needs for future research in this field are identified, and possible opportunities for appropriate fieldwork are suggested.

Overview of the Durability of FRP Rehabilitation Technologies

An exhaustive review of the entire body of work on the long-term durability and performance of FRP strengthening and rehabilitation techniques is beyond the scope of this paper. In this section, a representative overview of the field is given to illustrate the typical areas of interest and concern. The intent is to highlight key areas where progress has been significant, and to identify important unresolved issues.

FRP Repairs in Wet-Dry and Corrosive Environments. A particular concern related to the use of FRPs for structural rehabilitation is the long-term performance of such repairs in wet-dry and severe corrosive environments. In Canada, extensive research on this topic is being conducted at the University of Toronto (Lee et al. 2000, Pantazopoulou et al. 2001), the University of Waterloo (Soudki and Sherwood 2000), and the University of Sherbrooke (Beaudoin et al. 1998, Raîche et al. 1999, Lacasse et al. 2001). These investigations have been limited to ambient temperature environments.

The University of Toronto group has focused on the use of FRP wraps to repair reinforced concrete columns damaged by steel reinforcement corrosion (Lee et al. 2000, Pantazopoulou et al. 2001). In this work, improved techniques for the laboratory simulation of field corrosion in reinforced concrete columns have been developed. Large-scale column specimens with various steel reinforcement configurations were prepared and subjected to accelerated corrosion regimes in the laboratory, leading to extensive cracking and partial delamination of the concrete cover. Various FRP repair procedures were considered, consisting primarily of jacketing the damaged columns with either glass or carbon FRP wraps. The columns were then tested to structural failure and/or subjected to post-repair accelerated corrosion, monitoring, and testing. The results showed that the FRP repairs can greatly improve the strength of the repaired columns, and can also retard the rate of post-repair corrosion. Moreover, subjecting the FRP-repaired columns to extensive post-repair corrosion results in no loss of strength or stiffness, and only slight reductions in ductility.

The University of Waterloo studies have focused on the viability of using carbon FRP laminates for the strengthening of corrosion damaged reinforced concrete beams (Soudki and Sherwood 2000). Beam specimens with variable chloride levels were constructed, and some of these specimens were strengthened using externally-bonded carbon FRP laminates. The specimens were subjected to accelerated corrosion, and subsequently tested in flexure. The test results revealed that the FRP laminates can successfully confine the corrosion cracking, and that the strengthening scheme is able to restore the capacities of the corrosion damaged beams.

In addition to the durability aspects associated with steel reinforcement corrosion in concrete elements, research has also recently been conducted on the effects of wet-dry environmental conditions on the performance of both FRP materials used for repair (Raîche et al. 1999), as well as on FRP repaired beams (Toutanji and Gomez 1997, Beaudoin et al. 1998). These studies indicate that FRP-strengthened concrete beams are not damaged significantly by exposure to wet-dry environments. Additional studies at the University of Sherbrooke have been conducted on the FRP rehabilitation of concrete beams suffering from alkali aggregate reaction damage (Lacasse et al. 2001). In this investigation, FRPs have been used as external reinforcement, and preliminary results indicate that this technique can be quite effective at reducing the expansions due to alkali aggregate reactions.

FRP Repairs in Cold Regions. The performance of FRP rehabilitation methods in cold regions is another issue of particular concern, and much research has recently been conducted on the behaviour of concrete structures strengthened with FRP sheets and laminates subjected to cold climate conditions. Tests on carbon FRP sheets subjected to natural and accelerated exposure have shown that these materials have adequate weatherproofing properties with regard to tensile strength and bond to concrete, as well as being quite durable under freeze-thaw cycling (Yagi et al. 1997). However, tests on FRP-wrapped concrete cylinders exposed to freeze-thaw action have revealed that freeze-thaw cycling can significantly reduce the strength and ductility of FRP-wrapped concrete in comparison to specimens kept at room temperature (Soudki and Green 1997, Toutanji and Balaguru 1998). Moreover, Soudki and Green (1997) report that FRP-wrapped specimens subjected to freeze-thaw cycling fail in a more brittle manner than similar specimens kept at room temperature. The FRP wraps of their freeze-thawed specimens ruptured suddenly in a series of hoops, in contrast to the more continuous failure line observed for the wrapped cylinders kept at room temperature. Nevertheless, as wrapped cylinders exposed to freeze-thaw action show a significant increase in strength over unwrapped cylinders exposed to freeze-thaw, adequate FRP wrapping is a possible remedy for restoring the strength of a freeze-thawed cylinder to that of an unwrapped specimen kept at room temperature. The temperature ranges in this study were from –18C to +18C.

Research on the effects of freeze-thaw action, for similar temperature ranges, has also recently been reported for FRP-strengthened concrete beams. A preliminary indication is that freeze-thaw cycling does not induce significant deterioration of the bond durability between FRP plate reinforcements and concrete (Green et al. 2000). Other research has been conducted on the behaviour of concrete structures strengthened with FRP sheets subjected to cold climate conditions (Baumert et al. 1996). In these studies beam specimens were strengthened with carbon fibre sheets. One half of the specimens were subjected to short-term low-temperature exposure, of –27C, while the others were kept at room temperature. Unstrengthened control beams were also exposed to either room temperature or low-temperature conditions. The failure mode in these tests was by shear peeling of the sheets, and was unaffected by the low-temperature conditions. Furthermore, the low-temperature specimens generally failed at higher loads than the room-temperature specimens because the concrete exhibited an increase in strength at the lower temperatures. The authors conclude that the carbon fibre sheets were unaffected by the short-term exposure to low temperatures.

Fatigue and Creep Behaviour. Another topic related to the long-term behaviour of FRP strengthened structures is their behaviour under conditions of sustained loads or fatigue loading. The extended use of a structure at loads often greater than the original design loads may compromise the safety of the structure if no provisions are made with regard to creep and fatigue performance. Limited research has been carried out on the behaviour of FRP-strengthened reinforced concrete beams under fatigue-type loading; however, particularly noteworthy contributions are the investigations by Heffernan (1997) and Shahawy and Beitelman (1998). These investigations have shown that the fatigue performance of concrete beams can be enhanced significantly by means of FRP strengthening.

Test results have been presented for both rectangular and T-section concrete beams loaded monotonically and cyclically to failure. Up to six million cycles of loading were applied in these experiments. Analytical methods for simulating fatigue behaviour have also been proposed and compared to the experimental results (Heffernan 1997). The ability of the post-strengthened beams to carry the stresses through repeated cycles has been assessed, and a design model has been proposed for the use of FRP sheets to forestall fatigue failure. This analytical model provides good, yet conservative, results for the fatigue life prediction of both conventional reinforced concrete beams as well as for reinforced concrete beams strengthened with carbon FRP sheets. This study demonstrates the interesting result that equivalent strength beams, conventionally reinforced and strengthened with FRPs, have equivalent fatigue lives.

Field Studies. The use of FRP strengthening and rehabilitation technologies for practical field applications has grown tremendously in recent years. Initially, many of these applications were demonstration projects, in that their main objective was to convince the user sector of the merits of these new and relatively unproven technologies. In Canada, field applications date from 1995 (Neale and Labossière 1998, Rizkalla and Labossière 1999). There is thus very little field data here related to long-term behaviour; nevertheless, to date the performance of the FRP field applications have proven to be very successful. In many of the field projects, provisions have been made for the long-term monitoring of the FRP repairs using fibre optic sensors. This type of monitoring is very beneficial in providing assurance to the user sector.

Research Needs and Opportunities

The above overview shows that, although studies regarding the long-term durability of FRP strengthening and rehabilitation techniques have begun only quite recently, progress has been impressive and quite rapid. A considerable amount of research has been carried out to date, and research productivity continues to escalate. This research is being complemented by a rapidly growing number of important field applications that are providing valuable information on the long-term performance of FRP repairs. The results from such field projects are essential to convince the user sector of the long-term reliability and durability of FRP strengthening and rehabilitation technologies. However, as discussed below, the needs for further research are significant, and there are excellent opportunities for concerted efforts in this field.

Needs for Research. Research into durability-related aspects of FRP strengthening and repair is a relatively new field, and the information currently available is consequently rather limited. There is thus an obvious need for more data on the effects of various environments (wet-dry, marine, freeze-thaw, UV, etc.), temperatures, and loading conditions (fatigue, creep, etc.) on both the properties of FRP material systems used in infrastructure rehabilitation, as well as on the performance of FRP strengthening and repair schemes. Better understanding of differences in behaviour due to changes in resin and fibre types is required. The fire resistance of FRP-repaired structures is also a major concern. In addition, as the effects of sustained loading in combination with environmental aging have generally been neglected, more research is required on this subject. Another topic to be further investigated is the effect of the state of damage in a structure, at the time of an FRP repair, on the long-term performance of the retrofit. In addition to the above needs for data on durability, adequate analytical and numerical models are needed to simulate the various phenomena and to develop appropriate guidelines for design.

Among the outstanding issues that must be addressed are the correlations of accelerated laboratory tests to actual field conditions, as well as the validity of extrapolating results from small-scale specimens to full-scale structures. Furthermore, the literature review indicates that simulated laboratory environments vary considerably from one research establishment to another, with the result that it is extremely difficult to fully synthesize the existing test data. As such, it is virtually impossible at present to arrive at definite conclusions regarding the durability of a given FRP retrofit scheme in a particular environment.

Opportunities for Concerted Activities. The research needs identified above suggest that concerted efforts are required to fully address all issues and, moreover, that there exist many interesting opportunities for collaboration. As the various parameters involved in durability investigations are numerous, strategic planning among researchers in the field is essential to ensure that all critical topics are adequately covered and that unnecessary duplication is avoided. Collaborative endeavours to standardize accelerated laboratory test conditions would undoubtedly contribute greatly towards improving our basic understanding of the durability characteristics of FRP strengthening technologies. Finally, coordinated activities with regard to establishing controlled exposure test sites for FRP repair materials and FRP-repaired elements would be of great value. As part of a joint research program between the University of Sherbrooke and the Public Works Research Institute of Japan, three identical exposure sites (one in Sherbrooke and two in Japan) have been set up to assess long-term durability in three very different climates. A comprehensive international program to construct similar sites elsewhere would undoubtedly represent a significant contribution towards establishing correlations between accelerated laboratory simulations and real environments. This would provide as well invaluable information on the durability of FRP rehabilitation technologies.

Acknowledgements

The authors gratefully acknowledge the financial support of the Natural Sciences and Engineering Research Council of Canada (NSERC) and the Canadian Network of Centres of Excellence on Intelligent Sensing for Innovative Structures (ISIS Canada).

References

Baumert, M.E., Green, M.F. and Erki, M.A. (1996). "Low temperature behaviour of concrete beams strengthened with FRP sheets." *CSCE Annual Conference*, Edmonton, Alberta, 29 May – 1 June, **IIa**, 179–190.

Beaudoin, Y., Labossière, P. and Neale, K.W. (1998). "Wet-dry action on the bond between composite materials and reinforced concrete beams." *Durability of Fibre*

Reinforced Polymer (FRP) Composites for Construction, B. Benmokrane and H. Rahman, Eds., Dept. of Civil Engineering, Université de Sherbrooke, 537–546.

Benmokrane, B. and Rahman, H. (Eds.) (1998). *Durability of Fibre Reinforced Polymer (FRP) Composites for Construction*, Dept. of Civil Engineering, Université de Sherbrooke.

Chong, K.P. (1998). "Durability of composite materials and structures." *Durability of Fibre Reinforced Polymer (FRP) Composites for Construction*, B. Benmokrane and H. Rahman, Eds., Dept. of Civil Engineering, Université de Sherbrooke, 1–12.

Dolan, C.W., Rizkalla, S.H. and Nanni, A. (1999). *Fourth International Symposium on Fiber Reinforced Polymer Reinforcement for Reinforced Concrete Structures*, SP-188, American Concrete Institute.

Green, M.F., Bisby, L.A., Beaudoin, Y. and Labossière, P. (2000). "Effect of freeze-thaw cycles on the bond durability between fibre reinforced polymer plate reinforcement and concrete." *Canadian Journal of Civil Engineering,* **27**, 949–959.

Hawkins, G.F., Steckel, G.L, Bauer, Jr., J.L. and Sultan M. (1998). "Qualification of composites for seismic retrofit of bridge columns." *Durability of Fibre Reinforced Polymer (FRP) Composites for Construction.* B. Benmokrane and H. Rahman, Eds., Dept. of Civil Engineering, Université de Sherbrooke, 25–36.

Heffernan, P.J. (1997). "Fatigue behaviour of reinforced concrete beams strengthened with CFRP laminates." *Ph.D. Thesis*, Royal Military College of Canada, Kingston, Ontario.

Lacasse, C., Labossière, P. and Neale, K.W (2001). "FRPs for the rehabilitation of concrete beams exhibiting alkali-aggregate reactions." *Fifth International Symposium on Fiber Reinforced Polymer Reinforcement for Reinforced Concrete Structures FRPRCS*-5, Cambridge, U.K., 16–18 July, (in press).

Lee, C., Bonacci, J.F., Thomas, M.D.A., Maalej, M., Khajehpour, S., Hearn, N., Pantazopoulou, S. and Sheikh, S. (2000). "Accelerated corrosion and repair of reinforced concrete columns using carbon fibre reinforced polymer sheets." *Canadian Journal of Civil Engineering*, **27**, 941–948.

Meier, U. and Betti, R. (Eds.) (1997). *Recent Advances in Bridge Engineering – Advanced Rehabilitation, Durable Materials, Nondestructive Evaluation and Management*, EMPA, Switzerland.

Neale, K.W. and Labossière, P. (1998). "Fiber composite sheets in cold climate rehab." *Concrete International,* **20**(6), 22–24.

Pantazopoulou, S.J., Bonacci, J.F., Sheikh, S., Thomas, M.D.A. and Hearn, N. (2001). "Repair of corrosion-damaged columns with FRP wraps." *Journal of Composites for Construction, ASCE*, **5**, 3–11.

Raîche, A., Beaudoin, Y. and Labossière, P. (1999). "Durability of composite materials used as external reinforcement for RC beams." *CSCE Annual Conference*, Regina, Saskatchewan, 2–5 June, **1**, 155–164.

Rizkalla, S. and Labossière, P. (1999). "Structural engineering with FRP – in Canada." *Concrete International*, **21**(10), 25–28.

Shahawy, M. and Beitelman, T.E. (1998). "Fatigue performance of RC beams strengthened with CFRP laminates." *Durability of Fibre Reinforced Polymer (FRP) Composites for Construction*, B. Benmokrane and H. Rahman, Eds., Dept. of Civil Engineering, Université de Sherbrooke, 169–178.

Soudki, K.A. and Green, M.F. (1997). "Freeze-thaw response of CFRP wrapped concrete." *Concrete International*, **19**(8), 64–67.

Soudki, K.A. and Sherwood, T.G. (2000). "Behaviour of reinforced concrete beams strengthened with carbon fibre reinforced laminates subjected to corrosion damage." *Canadian Journal of Civil Engineering*, **27**, 1005–1010.

Toutanji, H. and Balaguru P. (1998). "Durability characteristics of concrete columns wrapped with FRP tow sheets." *Journal of Materials in Civil Engineering, ASCE*, **10**, 52–57.

Toutanji, H. and Gomez, W. (1997). "Durability characteristics of concrete beams externally bonded with FRP composite sheets." *Cement and Concrete Composites*, **19**, 351–358.

Yagi, K., Tanaka, T., Sakai, H. and Otaguro H. (1997). "Durability of carbon fiber sheet for repair and retrofitting." *Non-Metallic (FRP) Reinforcement for Concrete Structures*, Japan Concrete Institute, **2**, 259–266.

Characterization and Durability of FRP Structural Shapes and Materials

John J. Lesko[1], Michael D. Hayes[2], Timothy J. Schniepp[2], and Scott W. Case[3]

Abstract

Fiber-reinforced polymer (FRP) composite structural shapes and systems represent a significant opportunity to enhance national infrastructures through improved durability and by enabling rapid replacement techniques. We present an overview of the materials, component fabrication techniques, methods of analysis for stiffness and strength, component durability issues, and design codes and guidelines for these shapes and systems. The state-of-the-art is presented and suggestions for paths forward are outlined.

Introduction

Fiber-reinforced polymer (FRP) composites are receiving attention as a potential replacement for traditional materials in construction, as the community seeks reductions in maintenance and losses due to corrosion, weight, and installation time. With each of these potential benefits comes a challenge for design, questions about performance and concerns about how to handle new forms and materials in the context of present practices. Yet many experimental programs and FRP installations are proving the efficacy of these new materials and forms in buildings, bridges, offshore oil and gas rigs, cooling towers, power transmission and other commodity members. Many of these structural shapes and systems are commercialized and available on the market for general use. Others have been designed and manufactured for specific applications with the hopes of reaching a commodity market.

Thus, in this discussion structural shapes and systems are defined here as those elements used as beams and/or columns for load bearing applications. Structural applications of structural shapes include girders, stringers and frames. In addition, adhesive bonding or mechanical fastening of structural elements can be accomplished to develop other sections or systems to serve a structural function.

---

[1]Associate Professor, [2]Graduate Research Assistants, [3]Assistant Professor, Materials Response Group, Department of Engineering Science & Mechanics (0219), Virginia Tech, Blacksburg, VA 24061; phone (540) 231-5259; jlesko@vt.edu

Bridge decks constructed from constant cross section members and custom components are typical of this approach (Bakis 2001).

We present in this manuscript an overview of the materials, component fabrication techniques, methods of analysis for stiffness and strength (design), component durability, and design guidelines employed for FRP shapes and systems.

Discussion

Materials and Fabrication.

Resins. Overwhelmingly these structural components are produced from glass fiber and free radically cured styrene containing polymers such as vinyl ester and polyester thermosetting resins. These resins are comparatively inexpensive (US$1.00 to 4.00/kg), rapidly cured (under 2 minutes) and possess low saturation moisture contents and good resistance to caustics. Although epoxy resins possess good mechanical properties, long cycle times and high temperature curing requirements make them less attractive in the construction markets (as compared to traditional materials) due to their high initial material costs and detailed manufacturing requirements. Additives for fire and UV resistance are typically combined with thickening (viscosity control) and release agents for processing. Clay fillers act as both a thixatrop and a means to reduce shrinkage. Excessive shrinkage leads to the development of high residual stress, and cracking within the laminates of styrenated resins (typically 6-8% shrinkage upon curing) (Li 1997).

Although these materials are functional in their present form, they were not originally intended or designed for civil infrastructure applications. This is particularly true of polyester resins, which lack moisture stability. Vinyl esters are known to be very brittle and present toughening schemes can change the cost and processing conditions drastically. Thus the materials suppliers should seek to develop resins that are compatible with the high volume low cost production methods (e.g. pultrusion and VARTM), with low shrinkage and improved toughness. Moreover, these styrenated resins are very sensitive to UV and should be modified to possess long-term resistance to this radiation.

Fibers. When considering fiber, the use of carbon or aramid is limited to those applications where the performance benefits outweigh cost concerns. It is very difficult to establish designs in these industries where the added costs are outweighed by the desired performance as are observed in aerospace and marine applications (weight critical structures). Yet, engineered fabrics (multi-directional) are employed for tailoring stiffness and strength of the structural elements in place of the commodity shapes traditionally formed from glass continuous strand mat (CSM) and roving. Although the cost of engineered fabrics is higher, and require added attention to inventory as compared to CSM and roving, optimized stiffness and strength can be achieved. Moreover, consolidation of fiber architectures into generalized laminate configurations (including hybridization) can reduce fabrication costs.

Although headway is being made to reduce the costs of carbon fiber, glass fiber will continue to be used even though it lacks chemical/moisture resistance (even with it's low stiffness). Future efforts to improve the environmental stability of glass fiber could be made through radical changes in the way glass fiber is made. Presently water is used to cool and size (along with polymer compounds); eliminating this practice by for instance sizing with a thermoplastic powder could greatly improve initial fiber strength and moisture environmental resistance.

The Fiber Matrix Interphase. One of the "constituents" of the composite that is often overlooked is the interphase formed between the fiber and matrix during processing of the composite (Madhukar 1991). This three-dimensional volume of affected polymeric material around the fiber can also be distinguished from the interface. The interface is the two-dimensional boundary that joins the fiber to the polymer. Both play a role in the micro-mechanics of the composite function and possess a profound effect on the strength and durability of the resulting composite (Reifsnider 1994). Oxidative surface treatments for carbon fiber and the addition of polymer coatings to the fiber surface (sizing) for both carbon and glass fibers are used to influence the formation and properties of this interphase region. When attempting to incorporate carbon fiber into vinyl ester resins, it has been recognized that traditional sizing materials for aerospace grade carbon fiber are not adequate (Broyles 2001) where traditional glass sizings have shown adequate and commercially available. Verghese and co-workers (2001) have identified thermoplastic sizing agents applied from aqueous solutions, and their mechanics, that improve static composite properties, fatigue performance and resistance to moisture. The formation of this interface/phase are strongly influenced by the resin viscosity, the diffusivity of the sizing and the time the two components are in contact prior to curing.

Again further work on sizing technologies including the use of thermoplastic materials is recommended for both glass and carbon fiber. Acceptance by the industry is critical where sizing technologies are sometimes a proprietary and value added means for distinguishing fiber products.

Composite Shape and System Fabrication. Structural shape fabrication is accomplished through one of several means (listed in order of increasing volume of production): hand lay-up (and co-cured structures), vacuum assisted resin transfer molding (VARTM) and pultrusion. While hand lay-up and VARTM practices are simple and easily adapted to custom laminates and shapes, it is costly from a labor and cycle time perspective. Pultrusion represents a very efficient means to produce structural shapes however with limitations of constant cross section production and high initial tooling costs (Myer 1985). However, given the standardization of shapes and cross sections, the process can be fitted to produce high quality parts with various resins and laminate configurations.

Although the pultrusion and VARTM communities have made great strides in producing standardized and quality controlled parts, the industry as a whole lacks standards that can assist the designers and owners in ensuring the construction of reliable systems. Developing such standards between corporate entities presents a formidable challenge. However, the Market Development Alliance (MDA) has organized such an effort with the goal of building guidelines for product implementation. The need for such guidelines was made evident by the recent installation of Salem Avenue bridge deck replacement in Dayton, Ohio (Henderson 2000). Attention to details of production, assembly and installation were shown critical to the field performance of the structural systems.

Characterizing the Stiffness and Strength of FRP Shape. The characterization of FRP shapes is well documented using Timoshenko beam theory (Bank 1989), which becomes increasingly important for FRP materials with low shear stiffness. Designing FRP shapes and components is made easy through closed-form, mechanics-based beam lamination theory methods detailed by Barbero et al. (1993). One can start with fiber and matrix properties and content, laminate lay-up and shape dimensions and predict with reasonable accuracy the stiffness of the structure. The performance of assemblies of such shapes and components can also be combined to produce reasonable mechanistic descriptions of stiffness and structural deflection. For example, mechanics of materials concepts were used to estimate the equivalent orthotropic plate properties of FRP decks by Qiao et al. (2000). Sandwich structures are also generalized with similar smearing techniques, where the effective structural properties for a corrugated core are derived through an analysis by Davalos et al. (2001). With such techniques available, shapes and components (and likewise their assemblies, e.g. decks) can be adjusted to derive structurally efficient and easy to manufacture sections. While systematic methods of optimizing pultruded shapes have been developed (Davalos et al. 1996), FRP shape (and structure, e.g. deck/plate) designs are largely derived by trial and error (McGhee et al. 1991) due to manufacturing constraints.

Predicting Section Capacity. While these methods are adequate for predicting and designing for stiffness and deflection, they can bear some improvements in the strength prediction. Typically Tsai-Wu or Tsai-Hill failure criteria are applied to laminate stresses to predict both mode and capacity. These theories do not always allow for tracking progressive damage and accurately assessing ultimate capacity, as composites and their shapes are efficient at redistributing stresses around damage. The presence of the shape only complicates matters. Finite element schemes have been used successfully to describe the damage progress and large-scale deflections.

In most cases the delamination (matrix dominated) or shear failure mode controls the residual stiffness and strength of the component or structural system used in the civil infrastructure. These shear (in and out of plane) failures are typical of thick section composites and structures loaded out of plane and/or in the presence of concentrated loads. Hybridized laminates within these shapes only exacerbate this problem with local changes in stiffness leading to higher in and out of plane shear and peel stresses.

In an attempt to deal with such issues our laboratories have begun to investigate the capacity of pultruded shapes and the controlling mechanisms. Our work suggests that capacity is controlled by shear at points of concentrated load (Hayes 2001). These observed phenomena suggest that new approaches to assessing capacity and local stress states are necessary for design and prediction of long-term performance.

Sandwich Theories. Our group is presently investigating the use of sandwich theories to describe FRP shape damage modes and how they may influence the design guidelines for civil structures. A sandwich structure is typically composed of two thin face sheets (flanges) bonded to a thick central core (web). The face sheets are typically metallic or FRP composite, and the core can be metallic or polymer in either a solid, corrugated, honeycomb, or foam structure. Failure is typically observed to occur by way of either 1) peeling delamination at the core/face interface, 2) face wrinkling or buckling, or 3) face sheet indentation. The predominance of each failure mode depends upon several factors including the stiffness of the core, the face stiffnesses, the strength of the adhesive bond between the core and faces, and the severity of the load concentration.

Taking this approach in the analysis of structural shapes presents a new approach to treating FRP shapes and their assemblies (e.g. FRP decks) in assessing capacity. Sandwich theories offer a few distinct advantages over the high-order beam theories, which make them attractive for analysis of thin-walled composite beams. These include high-order kinematics, core compressibility, the ability to apply more realistic boundary conditions, and potentially, a simpler solution. As opposed to beam theories where the kinematic assumptions and boundary conditions are defined for the structure as a whole, these sandwich theories allow different assumptions for the core and faces. The features of a few basic and high-order sandwich theories are presented here. For a complete review of the state-of-the-art in sandwich modeling, the reader is referred to review papers by Noor et al. (1996) and Frostig (1992).

Characterizing and Predicting the Durability of FRP Structural Elements & Systems. The durability of FRP sections and assemblies remains a major question in their use as primary load bearing members and structures. Moreover, how capacity and structural stiffness will change over time present a central conundrum for designers when specifying allowable loads and reductions factors. It is safe to say that the community cannot wait, nor does it possess the resources, to qualify all full-scale structural elements and systems under the many varied service environment conditions possible nationally. To allow for sufficiently generalized descriptions of life, credible simulations that accurately describe the combination of synergism of load and environment must be used to ensure practical and efficient design guidelines for durability. These simulations must be robust and developed from reliable descriptions of material degradation descriptions and their interactions, which may include characterizations from accelerated testing to extend the validity of the predictions. Such simulations must be validated over a wide range of conditions at both the component and structural levels.

We also recognize that the traditional linear cumulative damage theory (Miner's Rule) is not sufficient to describe FRP life. Given these circumstances our group has taken the following approach to assess the useful life of an FRP component or structure.

- Establish how service life will be judged (e.g. loss of a certain percentage of stiffness, complete failure)
- Identify the governing failure mode(s) (coined the critical element) and possible shifts in failure mode(s)
- Assess the dominate stress state(s) that influence the failure mode(s)
- Postulate that remaining strength of the controlling failure mode may be used as the damage metric
- Track residual stiffness of the FRP material that influences the stress state which controls the failure mode(s) of interest

The culmination of these principles is embodied in a life prediction code called MRLife (Reifsnider et al. 1996). This code assumes that remaining strength may be determined (or predicted) as a function of load level and some form of generalized time. For a given load level, say S_a^i, a particular fraction of life corresponds to a certain reduction in remaining strength. We claim that a particular fraction of life at a second load level is equivalent to the first if and only if it gives the same reduction in remaining strength, as illustrated in Figure 1. In the case of Figure 1, time τ_1 at an applied stress level S_a^1 is equivalent to τ_2^0 time at stress level S_a^2 because it gives the same remaining strength, S_r^1 as compared to the initial ultimate strength S_{ult}. In addition, the remaining life at the second load level is given by the amount of generalized time required to reduce the remaining strength to the applied load level. In this way, the effects of several increments of loading may be incorporated into the analysis by adding their respective reductions in remaining strength. Moreover, tracking remaining strength can be accomplished by accounting for all of the degradation mechanisms that reduce the stiffness and strength of the critical element. This assumes that the kinetics and their synergism are understood and describable in terms of stiffness and strength reduction.

This method has been employed, with some validation, to assessing the fatigue life of a hybrid FRP shape (Senne 2000). Delamination within the hybridized region of the flange was shown to be the controlling failure mode in both quasi-static and fatigue loading. The life prediction scheme uses a simple beam theory to determine effective beam stiffnesses and then, assuming pure bending, calculates the curvature of the flanges. In-plane stresses are determined using classical laminate theory (CLT), and the model applies a web "smearing" simplification to model the double web beam as a rectangular beam, then utilizes a complementary energy solution for laminated plates to determine free-edge (out-of-plane) stresses. This stress analysis was then integrated into a residual-strength-based strength and life prediction methodology. A unique and practical feature of this analysis was that coupon level data was employed to predict the stiffness and strength reduction of the entire structure (Phifer 1999). It was shown that using flange delamination as the damage metric was reasonable for predicting life and that the coupon level data was sufficient to assess loss in stiffness of the beam and changes in the controlling stress state.

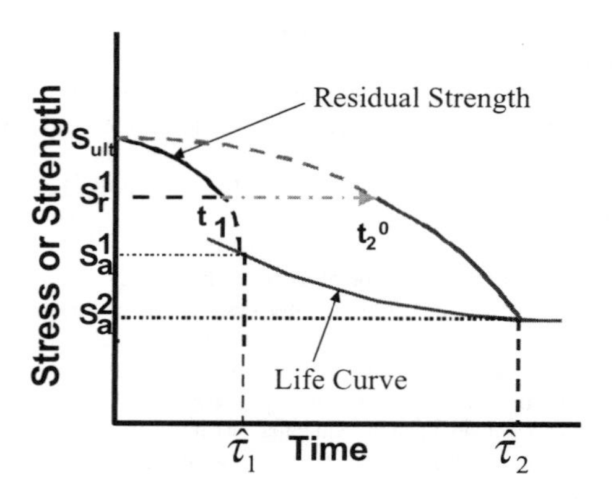

Figure 1. The use of remaining strength as a damage metric.

Advancing this analysis to describe actual bridge service environments requires the addition of strength and stiffness reduction features from the synergistic effects of load, moisture, UV, thermal (cycling), etc. The nature of the residual strength approach presented above allows for the inclusion of synergistic effects through their contribution to changing stiffness and strength. This however requires that the kinetics of strength and stiffness reduction can be described and combined. At present we do not possess the mechanics necessary to accomplish this for even general classes of materials. This is in part due to the plethora of material combinations and processing methods used to derive these composite material systems, as these variations can strongly influence their durability. The community at this point must identify reasonable and general kinetics, and methods to accurately accelerate and combine effects, to support realistic life prediction of FRP shapes and systems for long-term service environments.

Design Guidelines for FRP Shapes. Throughout this discussion we have touched on the idea of standardization and the need and use of design guidelines necessary for the routine use of the new materials systems in civil infrastructure. It is interesting to point out that the FRP step-ladder industry drove a key standardization of pultruded FRP shapes (Werner 1979). The development of B-basis allowable properties were established for pultruded channels through the guidance of the US Military Handbook 17 (MIL-HDBK-17 2001). Subsequent efforts by our group have lead to the development of a design guide for a 203-mm-deep double web beam (DWB) fabricated by Strongwell Corp (2000). A reliability-based approach is taken in conjunction with Weibull statistics to describe property variation and develop A and B basis allowables. The resulting guide contains recommendations on stiffness and capacity under bending including the effects of shear. Specifications for bearing capacity, bolted connections and lateral torsional stability are also described. A subsequent design guide is under development for a 914-mm version of the DWB.

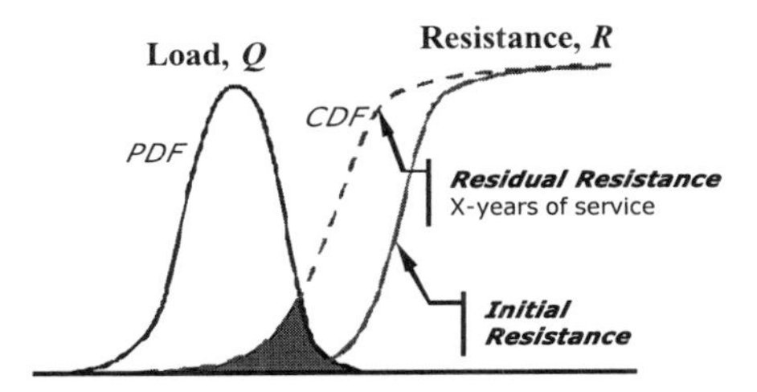

Figure 2. Effect of decrease in resistance due to service conditions on the probability of failure (red/shaded area) calculated using the LRFD design philosophy.

What these specifications lack is the inclusion of how long-term service will affect the validity of the derived allowables. The load and resistance factor design (LRFD) approach, we believe, represents an ideal place in which to develop a basis for the inclusion of durability. (The LRFD approach is presently gaining greater acceptance in the highway bridge design community.) This more than likely must be accomplished through simulation (potentially through the MRLife approach presented above). The engineer of record must assess how probability of failure will change over time by calculating the residual strength of the structural element. The strength reduction factor, ϕ, should account for the possibility of including strength reduction due to the specific in-service exposure imparted to the structure. Here the probability of failure is defined as the integral of the product of the loads probability density function (PDF) and the cumulative density function (CDF). The ability of our community to develop a set of reduction factors for various components, and service conditions (climate regions within the US and average daily traffic) could present the designer with some guidance on how a structure may be conceived reliably for reasonable service lives.

Summary and Remarks

The characterization of FRP shapes and systems must take a holistic approach to evaluating the materials, fabrications, design and durability if the bridge design community is to realize routine use of these new material systems in bridges and buildings. Recommendations are provided in this discussion for improving the materials used by designing resins for the purposes of infrastructure applications and for focusing further attention on the use of fiber sizings to improve durability. As many of the FRP shapes and structures employed in present applications resemble sandwich construction we propose looking to the mechanics literature for insight into strength. We also suggest further investigation into the use of residual-strength-based life prediction approaches for simulating durability and the inclusion of these procedures in the context of the LRFD philosophy.

References

Bakis, C. E., Bank, L. C., Cosenza, E., Brown, V., Davalos, J., Lesko, J. J., Machida, A., Rizkalla, S., Triantafillou, T., (2001) "FRP Composites in Construction – State-of-the-art Review," ASCE *Journal of Composites for Construction*, 150th Anniversary Edition, to be published

Bank, L.C. (1989) "Flexural and Shear Moduli of Full-Section Fiber Reinforced Plastic (FRP) Pultruded Beams," *Journal of Testing and Evaluation*, Vol. 17, No.1, 40-45.

Barbero, E. J., Lopez-Anido, R., and Davalos, J. F. (1993). "On the mechanics of thin-walled laminated composite beams." *J. Composite Materials*, 27(8), 806-829.

Broyles, N. S., Verghese, K. N. E., Davis, R. M., Lesko, J. J., & Riffle, J. S., (2001) "Pultruded AS-4 Carbon Fiber/Vinyl Ester Composites Processed with G', Phenoxy, and K-90 PVP Sizings Part I: Processing & Static Mechanical Performance," ASCE J. for Materials, to be published

Davalos, J. F., Qiao, P., and Barbero, E. J. (1996). "Multiobjective material architecture optimization of pultruded FRP I beams." *Intl. J. Composite Structures*, 35(3), 271-281.

Davalos, J.F., Salim, H.A., Qiao, P., Lopez-Anido, & Barbero, E.J., (1996), "Analysis and design of pultruded FRP shapes under bending," Composites: Part B, **27B**, 295-305.

Davalos, J. F., Qiao, P. Z., Xu, X. F., Robinson, J., and Barth, K. E. (2001). "Modeling and characterization of fiber-reinforced plastic honeycomb sandwich panels for highway bridge applications." Journal of Composite Structures, 52, 441-452.

Frostig, Y., Baruch, M., Vilnay, O., and Sheinman, I. (1992). "High-Order Theory for Sandwich-Beam Behavior with Transversely Flexible Core," *J. Eng. Mech.*, 118(5): 1026-1043.

Hayes, M.D., Schniepp, T., and Lesko, J.J., (2001), "Shear Effects in the Strength of FRP Structural Beams", to be published in American Society for Composites 16th Annual Technical Conference, September 9-12, Virginia Tech.

Henderson, M.P., "Evaluation of Salem Avenue Bridge Deck Replacement," ODOT Final Report MOT-49-1.634, PID N o. 17939, December, 2000.

Li, H., Rosario, A. C., Davis, S V., Glass, T., Holland, T. V., Davis, R. M., Lesko, J. J., & Riffle, J. S., "Network Formation of Vinyl Ester-Styrene Composite Matrix Resins," *Journal of Advanced Materials*, Vol. 28, No. 4, July 1997, pp. 55-62.

Madhukar, M. S. and Drzal, L. T. (1991) "Fiber-Matrix Adhesion and Its Effect on Composite Mechanical Properties: II. Longitudinal (0°) and Transverse (90°) Tensile and Flexure Behavior of Graphite/Epoxy Composites", J. of Comp. Matr., Vol. 25, pp. 958-991.

McGhee, K. K., Barton, F. W. and McKeel, W. T. (1991). "Optimum design of composite bridge deck panels." *Advanced Composite Materials in Civil Engineering Structures, Proceedings of the Specialty Conference,*, ASCE, Reston, VA, 360-370.

Meyer, R.W., 1985. *Handbook of Pultrusion Technology,* Chapman and Hall, NY.

MIL-HDBK17, 2001. Composite Materials Handbook. Available at <http://www.mil17.org>.

Noor, A.K., Burton, W.S., and Bert, C.W. (1996). "Computational Models for Sandwich Panels and Shells," 49(3): 155-198.

Phifer, S.P., (1999) "Quasi-Static and Fatigue Evaluation of Pultruded Vinyl Ester/E-Glass Composites," M.S. Thesis, Dept of Engineering Science & Mechanics, Virginia Tech.

Qiao P., Davalos, J. F., and Brown, B. (2000). "A systematic approach for analysis and design of single-span FRP deck/stringer bridges." *Composites Part B – Engineering,* 31(6-7), 593-610.

Reifsnider, K. L. (1994) "Modeling of the interphase in polymer-matrix composite material systems", Composites, Vol.25, Number 7, pp. 461-469.

Reifsnider, K. L., Iyengar, N., Case, S. W. and Xu, Y. L. (1996), "Damage Tolerance And Durability Of Fibrous Material Systems: A Micro-Kinetic Approach," Durability Analysis of Structural Composite Systems, A. H. Cardon (ed). A. A. Balkema (Rotterdam), 123-144.

Senne, J.L. (2000) "Fatigue Life of Hybrid FRP Composite Beams," M.S. Thesis, Dept of Engineering Science & Mechanics, Virginia Tech.

Strongwell. (2000), "EXTREN DWB™ Design Guide"

Verghese, K. N. E., Broyles, N. S., Lesko, J. J., Davis, R. M., & Riffle, J. S., (2001)"Pultruded, Hexcel AS-4 Carbon Fiber/Vinyl Ester Composites Processed with G', Phenoxy, and K-90 Sizing Agents Part II: Enviro-mechanical Durability," ASCE J. for Materials, to be published.

Werner, R. I., 1979. "Properties of pultruded sections of interest to designers," Proc. 34th Annual Technical Conference, Reinforced Plastics/Composites Institute, SPI, New York, Section 9-C, 1-7.

Bond to Concrete of FRP Rebars and Tendons

Amnon Katz[1]

Introduction

Fiber reinforced polymers (FRP) have recently been introduced into the construction market in various applications: rebars for concrete reinforcement, tendons for prestressed concrete, bolts for grouted anchors and more.

In all the applications, the composite technology is used to produce a material that is as strong as steel, durable- with high resistance to aggressive environment, high strength to weight ratio and easy for handling and installation. Fibers of high properties, such as carbon, glass and Kevlar, are used for this purpose and together with a good polymeric matrix they can meet the above requirements. However, a weak link exists in the system that transfers the loads from the carrying fibers to the surrounding concrete, namely 'bond'. In all the applications, the polymeric system at the surface of the rods is the one that transfers the loads between the two; therefore special demands are required from this system.

An extensive study has been carried out in the last decade to establish the bond properties of FRP rebars and tendons as well as the load-slip relationship during pullout, effect of rod and tendon configuration on the bond, bond at the anchors and more. Part of this extensive study is covered by a review prepared by Cosenza et al., 1997, hence later data will be presented here. The understanding of these parameters can allow a better design of new FRP materials and accurate prediction of the behavior of reinforced and prestressed concrete members.

The following paper presents a state of the art review on bond behavior of FRP rebars and tendons, the parameters affecting the bond with respect to rebars and tendon configuration, external parameters affecting the bond and models for bond behavior. Finally, points where lack of knowledge still exists and need further investigation in the future will be highlighted.

---

[1] National Building Research Institute, Department of Civil Engineering, Technion-Israel Institute of Technology, Haifa 32000, Israel

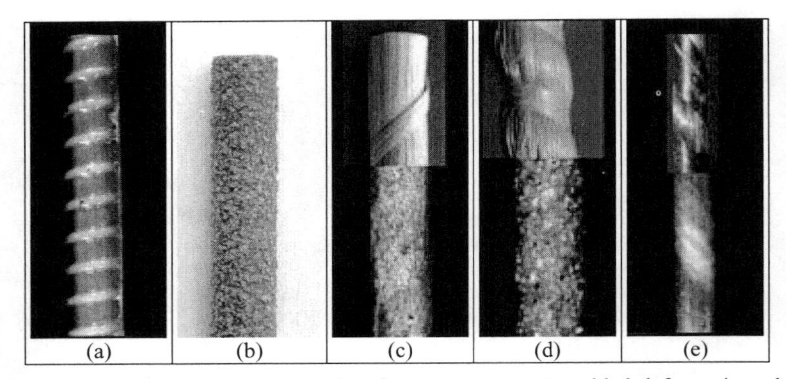

Figure 1: various configurations of surface treatments: a) molded deformations, b) sand coating, c) helix and sand coating, d) tight helix and sand coating, e) helix and excess resin. Exposed fibers are shown at the top of c, d and f.

Bond mechanism of FRP rebars and tendons

FRP rebars: The bond mechanisms of FRP rebars to concrete is significantly different from that of steel due to the differences in materials properties (both the fibers and the resin) and geometry. Generally, the resin used for the production of the rebars is water repellent and therefore the formation of chemical bonds between the concrete and the rebar is very limited unless special treatment is used. From experimental results published in the literature two mechanisms are seen: i) mechanical interlocking at different size scales; ii) friction due to surface roughness. In some cases the distinction between friction due to surface roughness and mechanical interlocking is difficult and depends on the roughening degree. Therefore in general, when referring to 'bond' the resistance to pullout is generally considered without a precise distinction between the two.

The bond of smooth FRP rebars to concrete is generally very low, therefore additional means to enhance the bond are needed.. Mechanical anchorage to the concrete is achieved by several means such as molded deformations, helical wrapping by fibers that lead to hump and dents in the rod's surface by using different methods of winding, sand coating, using excess of resin that leads to irregular humps, braiding of the fibers and a combination of some of them, as shown in Figure 1.

Most of the surface treatments improve the bond strength significantly and it increases to values close to those of deformed steel. Some values presented lately in the literature are shown in Table 1. The data in the table present various testing methods, embedment lengths, thickness of concrete cover, concrete strengths etc.; therefore they cannot be compared on the same basis. However, it can be seen that for most of the rod's configuration, the bond strength is in the same order as that of deformed steel.

Table 1: Bond data of various FRP rebars*

Surface treatment	Fiber type	Bond strength (MPa)	Reference
Smooth	Carbon	0.84	Kanakubo et al., 1993
Smooth (polyester)	Glass	1.0	Bank et al., 1998
Smooth (vinylester)	Glass	0.75	Bank et al., 1998
Smooth	Glass	1.0	Katz, 1999
Sand+2x helix	Glass	5.1-11.3	Tighiouart et al., 1998
Sand + helix	Glass	7.4-12.3	Tighiouart et al., 1998
Ribbed	Aramid	11.4-16.6	Wang et al., 1999
Spiral (helix)	Aramid	10.8-13.2	Wang et al., 1999
Strand	Aramid	2.1-4.2	Wang et al., 1999
Ribbed	Carbon	15.0-20.1	Wang et al., 1999
Braid	Carbon	16.3-20.7	Wang et al., 1999
Strand	Carbon	7.4-8.4	Wang et al., 1999
Spiral (helix)	Carbon	14.2-19.2	Wang et al., 1999
Ribbed steel	Steel	12.1	Katz, 1999

* note: tests were performed in different methods.

Katz (1999) distinguished between three modes of failure at the post-peak region (Figure 2). Mode I is characterized by a moderate shift from the pre-peak zone to the post-peak zone and a gradual drop of the load along approximately half of the embedment length. A rapid drop of the load after the peak was reached characterizes mode II and locations of increased pullout load as the rod slips out characterize Mode III. Katz explained these modes by the failure mechanism of the bond. Mode I is typical to rods having a relatively fine roughness, achieved by means like sand coating. The failure is mixed in the concrete and the rod, leading to a relatively gradual decrease of the pullout load after the peak. Mode II is typical to steel rods where a sudden shear of the concrete surrounding the rod occurs, but also to rods having relatively large deformations that shear off suddenly. The locations of increased pullout load during the slip of the rod as in Mode III are a result of entrapment of concrete crumbs or polymer residues from the sheared deformations between the rod and the concrete, exhibiting locations of increased resistance to pullout.

Wang et al. (1999) and Achillides et al. (1997) have studied several effects related to parameters such as rebar diameter, embedment length etc. Their findings, summarized below, confirm earlier studies (Cosenza et al. 1997).

Effect of rebar diameter: the average bond strength decreases as the diameter increases. This phenomenon was explained by the development of higher shear stresses near the surface of the rod, which is more noticed in larger diameters.

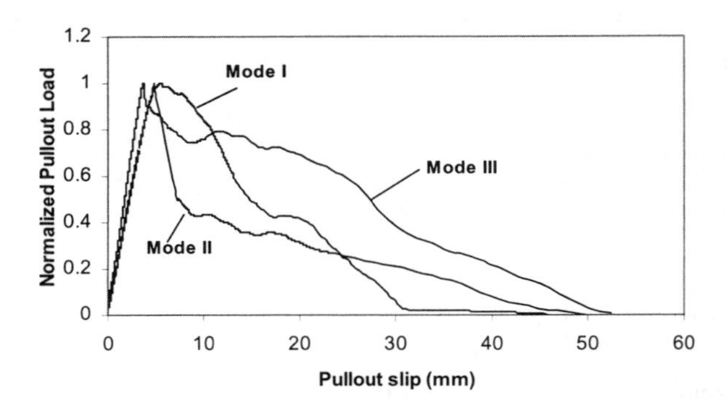

Figure 2: Three modes of bond failure.

Effect of embedment length: the average bond strength decreases as the embedment length increases, strengthening the assumption that the distribution of bond stresses along the embedment length is not linear.

Effect of modulus of elasticity: the comparison between rods having different modulus of elasticity is difficult due to differences in the surface configuration. However, it seems that there is a trend for higher bond strength with higher modulus. This finding needs further investigation.

Effect of concrete strength: it appears that bond strength is not affected by the concrete strength for concrete strength exceeding approximately 20 MPa. When considering the failure mechanism described before, it is clear that the weak link is located in the rod's surface after a minimum concrete strength is achieved. Therefore, concrete strength shows no effect on the bond strength in structural applications.

Hydrostatic pressure due to thermal incompatibility between the rod and the concrete (Gentry and Husain 1999, Aiello 1999) or shrinkage of the concrete can increase the surface friction, leading to an increase in the apparent bond. However, it may also crack the concrete cover, leading to loss of bond.

FRP tendons: The bond behavior of FRP prestressing tendons is quite similar to that of rebars described before. However, development length and bond at the anchors are of special interest in these applications. The effect of bond on the development length and transfer length of prestressing tendons was investigated by Ehsani et al. (1997), Domenico et al. (1998) and Lu et al. (2000) on a large scale test setup. The transfer and development lengths were determined by measuring the length changes along a prestressed beam. They found that the transfer length for several types of FRP tendons is shorter than that of steel tendons, leading to conservative recommendations in ACI 318 when applying them to FRP tendons. Lu et al. have found that the transfer length was identical for all the tested tendons despite the initial stress level or the tendon materials. Therefore they recommend on changes in the ACI equations. However, different findings were presented by Domenico et al. suggesting that additional study is required. Unlike the results for FRP rebars,

Domenico et al. found no relationships between the average bond strength and strand's diameter.

Anchoring FRP tendons requires special considerations due to the inhomogeneous properties of the tendons in different directions and the complicated state of stress at the anchor that combines lateral pressure and shear stresses on the surface, together with a high axial stress. Zhang et al. (2000) have reviewed the bond mechanism of FRP tendons to steel anchors. They found several methods of anchoring: clamp anchorage, bond anchorage and wedge-bond anchorage. Benmokrane et al. (2000) concluded that pullout behavior of grouted FRP tendons is similar to that of FRP rebars described above, with additional parameters that are typical to grouted anchors, such as the effect of anchor material, use of expansive agent etc. These results were used by Zhang et al. (2000) to develop a model to predict the pullout behavior of grouted anchors.

Bond under external effects

FRP is proposed as a replacement for steel in aggressive environment. Therefore its long-term performance should be dealt with very thoroughly. It should be considered that deterioration of the interface may lead to structural failure, while the individual constituents (concrete or FRP) can still carry loads, as in the case of high temperatures (see below).

Contradictory data have been reported in the literature regarding the effect of the alkaline environment of the concrete on bond. Bakis et al. (1998a) found that bond strength increased when the vinylester resin contained increasing amounts of polyester. Conard et al. (1998) tested the bond strength of several FRP rebars and tendons after conditioning in saturated $Ca(OH)_2$ solution at 80°C for 28 days. They found changes in the range of ±10% but a uniform trend could not be detected.

In the case of high temperatures, in the range of 100-150°C, the concrete and the bars can preserve a significant strength level but Katz et al. (1999, 2000) found a significant reduction (up to 90%) in the bond strength as the temperature exceeds the glass transition temperature (Tg) of the polymer at the surface. The loss of bond

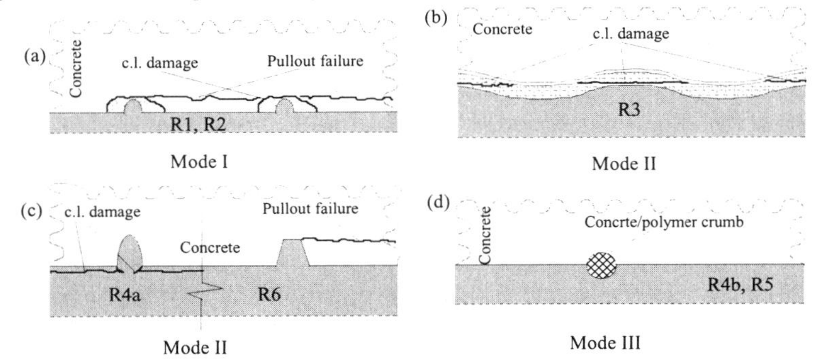

Figure 3: Schematic description of damage due to cyclic loading (Katz, 2000).

leads to structural failure while the individual constituents are still functioning.

Cyclic loading was found to damage the bar surface, leading to bond reduction that depends on the surface materials and configuration. Katz (2000) resolved several modes of damage due to cyclic loading (denoted "c.l. damage" in Figure 3), depending on the extent of damage and its location. In the cases where the damage to the surface layer due to cyclic loading is extensive (R4a in the figure) a reduction of more than 50% was seen in the bond strength.

Modeling the load-slip behavior

Several models have been developed in order to describe the bond behavior of FRP rebars to concrete and its effect on the bond-slip behavior during pullout. Malvar (1995) proposed a set of empirical equations to describe the bond strength, the displacement up to peak pullout load and the complete bond slip behavior. A large number of empirical constants is needed to be determined experimentally for each type of bar. Thus the pullout behavior cannot be predicted from the equations unless an extensive experimental work is done.

Cosenza et al. (1997) modified an earlier model, developed previously for reinforcing steel bars (Eligehausen et al. 1993), to describe the ascending and descending branches of the load-slip curve. The model was extended later (Cosenza et al. 1999, Focacci, et al. 2000) for further predictions of the load-slip behavior, but some parameters still need to be determined experimentally. Therefore, these models can be beneficial to predict the structural behavior of concrete elements reinforced with a particular type of bar, but the behavior of a new bar needs to be determined experimentally every time and cannot be predicted.

Achillides et al. (1997) used the finite element method to predict the load-slip behavior. The ascending portion of the load-slip curve was predicted quite well for both the loaded and unloaded ends. However, the peak load and the ascending curved were not predicted well and the authors reported that additional modifications are needed. Another finite element solution was presented by Bakis et al. (1998b) for smooth and lugged rods. The model predicted quite well the actual pullout results based on the behavior of a smooth rod and of a single lug.

Conclusions and research needs

1. It is clear that the bond strength of FRP rebars and tendons nowadays is of the same order as that of steel. Differences in the properties of FRP compared with steel prevent a direct use of the equations applicable for the design with steel. Moreover, the variety of FRP materials and configurations requires special consideration for each type of FRP product. Any conclusion regarding the bond properties should be restricted to the particular materials and configuration tested.

2. Many test methods to determine the bond strength were detected while conducting this review, as well as the effect of environmental conditioning. A standard test method that will take into account all variables related to FRP materials and configuration is needed. Alternatively, knowledge on the effect of testing procedure on the measured values, and the "correct" bond value for design is needed.

In addition, accelerated methods to assess environmental effects should be established.

3. New FRP products are being developed every year, improving the existing products. An analytical model that will allow the prediction of bond behavior as function of materials properties and surface treatment and configuration is needed. Such a model will enable a proper development of new products.

4. Environmental effects need more attention. FRP products are proposed as a replacement for steel as a non-corroding material; however, the durability of FRP in the aggressive environment of the concrete for the life-time of the structure is still in question. It should be noted that the properties of the polymer at the surface are the governing factors when assessing the durability of the bond. Therefore, special conditions that affect polymers should be studied, such as the behavior at temperatures in the range of 70-90°C that are likely to develop in bridge decks and pavements in hot climate, or special mechanical loadings like cyclic loading or impact.

References

Achillides Z., Pilakoutas K. and Waldron P., (1997), "Modeling of FRP rebar bond behaviour", Proceedings, Non-metallic Reinforcement for Concrete Structures, Supporo, Japan, Japan Concrete Institute, Vol. 2, pp. 423-430.

Aiello M.A., (1999), "Concrete cover failure in FRP reinforced beams under thermal loading", *Journal of composites for construction,* Vol. 3, No. 1, pp. 46-52.

Bakis C. E., Freimanis A. J., Gremel D. and Nanny A., (1998a), "Effect of resin material on bond and tensile properties of unconditioned and conditioned FRP reinforcement rods", Proc. *Durability of fiber reinforced polymer (FRP) composites for construction*, Benmokrane B and Rahman H. Eds. Dep. Of Civil Engineering, University of Sherbrooke, Canada, pp. 525-535.

Bakis C. E., Uppuluri V. S., Nanni A. and Boothby T. E., (1998b), "Analysis of bonding mechanisms of smooth and lugged FRP rods embedded in concrete", *Composites Science and Technology*, Vol. 58, pp. 1307-1319.

Bank L. C., Puterman M. and Katz A., (1998), "The effect of materials degradation on bond properties of fiber reinforced plastic reinforcing bars in concrete", *ACI Materials Journal*, Vol. 95, No. 3, pp. 232-243.

Benmokrane B., Zhang B. and Chennouf A., (2000), " Tensile properties and pullout behavior of AFRP and CFRP rods for grouted anchor applications", *Construction and Building Materials*. Vol. 14 No. 3, pp. 157-170.

Conard J. O., Bakis C. E., Boothby T. E. and Nanny A., (1998), "Durability of bond of various FRP rods in concrete", Proc. *Durability of fiber reinforced polymer (FRP) composites for construction*, Benmokrane B and Rahman H. Eds. Dep. Of Civil Engineering, University of Sherbrooke, Canada, pp. 299-310.

Cosenza E., Manfredi G. and Realfonzo R., (1997), "Behavior and modeling of bond of FRP rebars to concrete", *Journal of composites for construction,* Vol. 1, No. 2, pp. 40-51.

Consenza E., Manfredi G., Pecce M. and Realfonzo R., (1999), "Bond between glass fiber reinforced plastic reinforcing bars and concrete- experimental analysis", Proceedings, 4th Int. Symposium on *Fiber Reinforced Polymer Reinforcement for Reinforced Concrete Structures*, Dolan et al. Eds., ACI SP-188, pp. 347-358.

Domenico N. G., Mahmoud Z. I. and Rizkalla S. H., (1998), "Bond properties of carbon fiber composite prestressing strands", ACI Structural, Vol. 95, No. 3, pp. 281-290.

Ehsani M. R., Saadatmanesh H. and Nelson C. T., (1997), "Transfer and flexural bond performance of aramid and carbon FRP tendons", *PCI Journal*, Vol. 42, No 1, pp. 76-86.

Eligehausen R., Popove E.P. and Bertero V. V., (1983), "Local bond stress slip relationships of deformed bars under generalized excitation", Rep. No. 83/23, Earthquake Engineering Research Center EERC), University of California, Berkeley.

Focacci F., Nanni A. and Bakis C. E., (2000), "Local bond-slip relationship for FRP reinforcement in concrete", *Journal of Composites for Constructio*n, Vol. 4, No. 1, pp. 24-31.

Gentry T. R. and Husain M., (1999), "Thermal Compatibility of Concrete and Composite Reinforcements", *Journal of Composites for Construction*, Vol. 3, No. 2, pp. 82-86.

Kanakubo T., Yonemaru K., Fukuyama H., Fujisawa M. and Sonobe Y., (1993), "Bond performance of concrete members reinforced with FRP bars", in Proceedings of *Fiber Reinforced Plastic Reinforcement for Concrete Structures*, Nanni A. and Dolan C. W. Eds., American Concrete Institute (ACI) Special Publication SP-138, pp. 767-788.

Katz, A., (1999), “Bond Mechanism of FRP Rebars to Concrete”, *Materials and Structures*, Vol. 32, pp. 761-768.

Katz, A., Berman N. and Bank, L.C., (1999), “Effect of High Temperature on the Bond Strength of FRP Rebars”, Journal of *Composites in Construction*, Vol.3, No. 2, pp. 73-81.

Katz, A. and Berman N., (2000), “Modeling the Effect of High Temperature on the Bond of FRP Rebars”, Journal of *Cement and Concrete Composites*, Vol. 22 No. 6, pp. 433-443.

Katz, A., (2000), “Bond to Concrete of FRP Rebars After Cyclic Loading”, *Journal of Composites for Construction*, Vol. 4, No. 3, pp. 137-144.

Malvar, L. J., (1995), "Tensile and bond properties of GFRP reinforcing bars", *ACI Materials Journal*, Vol. 92, No. 3, pp. 276-285.

Tighiouart B., Benmokrane B. and Gao D., (1988), " Investigation of bond in concrete member with fibre reinforced polymer FRP bars", *Construction and Building Materials*, Vol. 12, pp. 453-462.

Wang Z., Goto Y. and Joh O., (1999), "Bond strength of various types of fiber reinforced plastic rods", Proceedings, 4th Int. Symposium on *Fiber Reinforced Polymer Reinforcement for Reinforced Concrete Structures*, Dolan et al. Eds., ACI SP-188, pp. 1117-1130.

Zhang B., Benmokrane B. and Chennouf A., (2000), "Prediction of tensile capacity of bond anchorages for FRP tendons", *Journal of Composites for Construction*, Vol. 4, No. 2, pp. 39-47.

Compatibility related problems for FRP and FRP reinforced concrete

Ralejs Tepfers[1]

Abstract

Among different compatibility problems for FRP the tensile strength reduction due to uneven stress distribution in elastic fibers is discussed. Uneven stress in fibers, not fully straight fibers, shear lag also accentuated by creep, influence the tensile strength. Fracture at tension and creep of hybrid fiber rods and rotation capacity of concrete members are discussed. For laminate type application the confinement efficiency decreases with increasing number of plies due to increased non-linearity of fibers. These problems are logical but not always notified in code clauses.

Introduction

Compatibility problems arise mostly when some kind of discontinuity appears in a structure e.g. differences in strain, stress or slip develop within a construction. Such problems may become serious when the materials are fully elastic without any possibility to redistribute forces. FRP fibers are such elastic materials. When the most stressed fibers start to break, the others have to be able to take over force, otherwise there is a zipper character failure. Such failures give no pre-warning and therefore should be avoided. There are many compatibility problems such as differences between materials in thermal elongation, moisture uptake, deformations caused by chemical reactions etc. However in this paper the interest will be directed towards the effects in FRPs when tensioned as rods also in concrete and when used as laminates for strengthening.

Tensile strength and ultimate elongation

The tensile strength can be determined for one fiber and expressed in stress if the area is precisely known. If a fiber filament is tested instead, its tensile strength will be lower because it is not possible to grip all fibers equally in the filament when fixing in testing machine. As the fibers are fully elastic up to fracture, there is no redistribution of load among them. When the first fiber reaches its breaking elongation it fails and the filament capacity decreases. Therefore the strength will decrease with increasing number of fibers in the reinforcing unit.

[1]Professor emeritus, Dept. of Building Materials, Chalmers University of Technology, SE-412 96 Göteborg, Sweden; phone +46 31 772 1991; Tepfers@bm.chalmers.se

Impregnating resin interconnects fibers in an FRP rod. This resin transfers shear force from bonded or gripped surface fibers to central fibers in the rod. The resin has certain shear deformations and it creeps more than fibers. The consequence is that the central fibers will therefore take less force as the surface fibers, which will start to break under tensile load. This phenomenon is known as shear lag, Figure 1. It is also possible that the same fiber in a rod is not gripped equally in both ends. Then, failure start may be the exceeding of resin shear strength between the fibers. Consequently size and shape of rods will have influence on tensile strength.

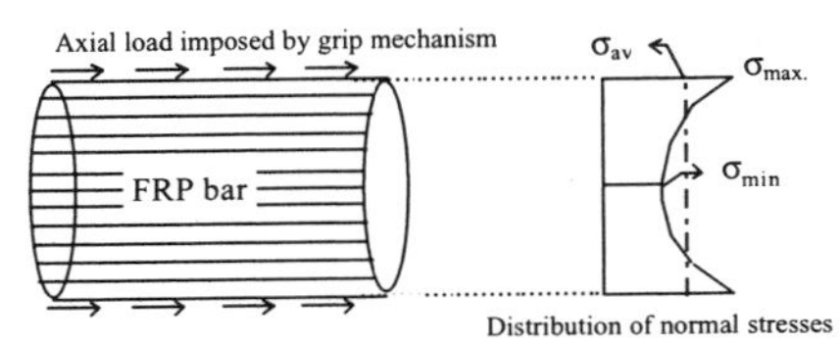

Figure 1. Distribution of stresses on a FRP bar cross-section subjected to axial load, Achillides (1998).

The resin has tensile breaking strain in excess of that of fibers when tested homogenous. However, in FRP rods a cracking up of resin happens before the fibers break. This because the resin in the rods has voids and many micro-cracks. At the crack tips stress concentrations exist and cause cracks to grow further, so the contribution of resin in taking tensile load in most cases has to be neglected.

The tensile failure starts with the most stressed fiber breaking and load overtaken by neighboring fibers. These soon are overloaded, break and redistribute load to other fibers. Finally, the remaining fibers are not able to carry the load and fail. In this way the failures are progressive and give lower failure load than if all fibers would carry the same load all the time. This means that the fiber nominal stress cannot be obtained in an FRP rod. The load capacity is further depending on the shape of the rod and how the force is transferred to and distributed in the rod. The pultrusion process is used to stretch fibers as equally as possible, so they take about the same load in an FRP rod. As the blocking devices in tests influence the tensile strength of the FRP rods, the gripping device should simulate the bond of concrete as much as possible. A Japanese proposal to arrange the gripping in an internally rough steel pipe, in which the rod is blocked with expanding mortar, seems to be for time being the best.

FRP flexural tensile reinforcement in concrete members with low height will have certain stress variation due to bending influence. This fact will result in earlier flexural reinforcement failures due to breaking of most stressed fibers. The calculated flexural member resistance, using mean tensile stress in the FRP rod, may overestimate the failure load.

Hybrid fiber FRP rods

Ductile properties of CFRP-AFRP or CFRP-GFRP hybrid composites are achieved by mixing the different fibers in an appropriate proportion. Under increasing tensile load the low breaking strain carbon fibers will break and the load will be overtaken by aramid or glass fibers. The hybrid fiber rods should be made so that aramid or glass fibers alone are able to carry at rupture of rod 1.1 times the load, which was taken by the carbon fibers together with aramid or glass fibers at rupture of carbon fibers. The effect of resin is disregarded. The shocks from the failing carbon fiber filaments, see Figure 2, will also rupture probably 5% of the neighboring aramid or glass fibers. Consequently the following formula, Apinis et al. (1999), can be set up to determine the relation between the area of carbon and area of aramid or glass fibers:

$$1.1 \cdot 1.05\,(\varepsilon_c \cdot E_c \cdot A_c + \varepsilon_c \cdot E_a \cdot A_a) = \varepsilon_a \cdot E_a \cdot A_a \qquad (1)$$

where A_a – area of aramid or glass fibers, A_c – area of carbon fibers, E_a – modulus of aramid or glass fibers, E_c – modulus of carbon fibers, ε_a – strain at failure of aramid or glass fibers, ε_c – strain at failure of carbon fibers.

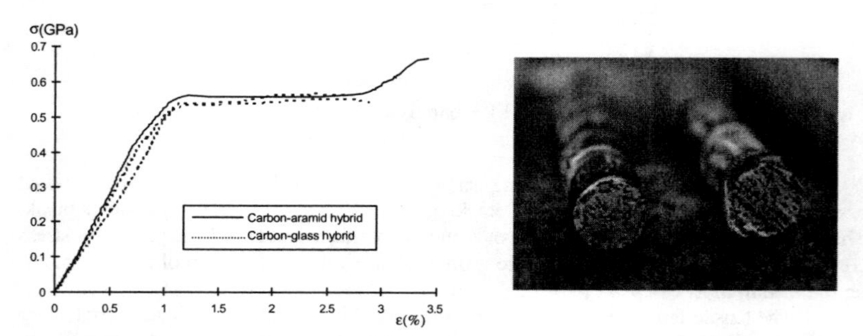

Figure 2. Carbon-aramid and carbon-glass fiber hybrid composite σ-ε-diagrams. Monotonic load. Load control. To the right hybrid carbon/glass fiber rod sections with carbon fiber filament sections visible. Apinis et al (1999).

In Figure 2 three distinct parts are clearly seen on the curve. In the initial stage, the rod deforms as elastic material. Having reached the critical strain, the carbon fibers rupture. In the second stage the strain grows quickly and reaches the value appropriate for the aramid or glass fibers at this load. In the third stage, the specimen elongates, up to the rupture, with the modulus determined by the amount of aramid or glass fibers left. At carbon fiber filament rupture elastic energy is released for a fiber length, which becomes debonded around the failed section. Therefore the performance of the hybrid rod is regarded as pseudo-ductile. If the rod is long, several carbon fiber failure zones will appear along rod. Tension stiffening effect will develop with carbon fibers being active between the failed zones. Due to economic reasons mono carbon fibers cannot be used, but filaments of a certain minimum size, Figure 2. Therefore the influence of filament failure on surrounding aramid or glass fibers will be more disastrous in small size bars.

Creep of fiber composites and hybrids

Carbon fibers practically do not creep. Aramid has considerable creep and glass has in between. When mixing fibers and imposing a certain strain, it must be kept in mind that creeping aramid or glass fibers transfer load to carbon fibers. These might be overloaded and fail. Then the full load has to be taken by aramid or glass fibers, which now creep more. This might even result in creep failure. In Figure 3 CFRP-AFRP hybrid creep development is shown at a level of 65% of short-term breaking load. Creeping aramid fibers cause carbon fibers to take load and to fail at a strain of 1.2%. Then the creep accelerates due to aramid fibers being taking higher load. At about the same starting breaking load level, 60%, the glass fibers are used up to about 25% of their load capacity in the hybrid CFRP-GFRP rod. At this load level, the creep of glass fibers is low and is stopped by carbon fibers. Glass fibers most likely do not suffer from stress corrosion when used to only 25% of tensile capacity.

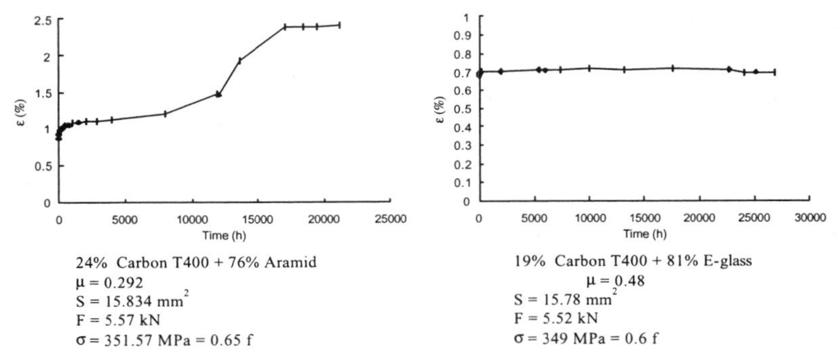

Figure 3. Creep of hybrid CFRP-AFRP rod to the left and CFRP-GFRP rod to the right. Apinis et al (2000), Tamuzs et al (2001).

Shear lag effect may be accentuated for FRP bars by creep. When these bars fail the fiber fracture will be progressive, starting with lateral fibers and spreading towards bar centre. The ultimate load will be lowered.

Rotation capacity

The following observations can be stated concerning rotation capacity for reinforced concrete beam. Bond of reinforcement is necessary to obtain many flexural cracks in concrete, which results in curved structure. Between cracks tension stiffening exists. If there is no bond (only end anchors), a singular crack appears in concrete member, a hinge is formed and no real rotation happens. When deformed steel bars yield in cracks, tension stiffening still exists between cracks. To obtain rotation it is necessary that the bond and tension stiffening break down. This happens when reinforcement in cracks goes into strain hardening region for steel and yield penetrates in between cracks. Then bond and consequently tension stiffening breaks down, rotation takes place and is followed by a tough failure.

FRP reinforcement in general has good bond, which gives many flexural cracks. Strain maximum is in cracks and between cracks is tension stiffening. No yield takes place and tension stiffening does not break down. Beam has consequently bad rotation and relatively brittle failure.

Two beam tests have been performed by Apinis et al (1999). The tested beams are identical with exception of tensile reinforcement. The first beam has 2 elastic CFCC

(Japanese Carbon Fiber Composite Cable) rods and the second 2 laboratory made hybrid CFRP-GFRP rods with stress-strain relation according to Figure 2. Both tensile reinforcements have the same load capacity and tensile modulus up to carbon fiber failure in hybrid rod. If there is a difference in performance between the beams it is caused by hybrid effect.

CFCC and hybrid FRP (pseudo ductile) reinforcement has good bond, which gives dense flexural cracking. Strain maximum for both type of rods exists in flexural cracks and between these cracks tension stiffening reduces the strain. CFCC reinforced beam has almost no real rotation and brittle failure. In hybrid rod, when carbon fibers fail in section of flexural cracks, hybrid strain increase happens only in cracks, Figure 4. (left) This widens cracks but gives almost no deflection increase, Figure 4 (right). In hybrid rod between flexural cracks high modulus fibers are active due to strain hardening effect in the FRP rod itself. No "yield" penetrates rod in between cracks. Tension stiffening does not break down. Beam has almost no real rotation and shows "brittle" failure. However some pre-warning of approaching failure is given by widening cracks. As the "yielding" of hybrid FRP is located only to the carbon fiber cracking zones in concrete cracks, this reinforcement should be taken as "pseudo" ductile. Rotation capacity can be obtained by intermittent bond, which reduces tension stiffening effect, Lees & Burgoyne (1996), but the bond might be reduced too much where it is necessary.

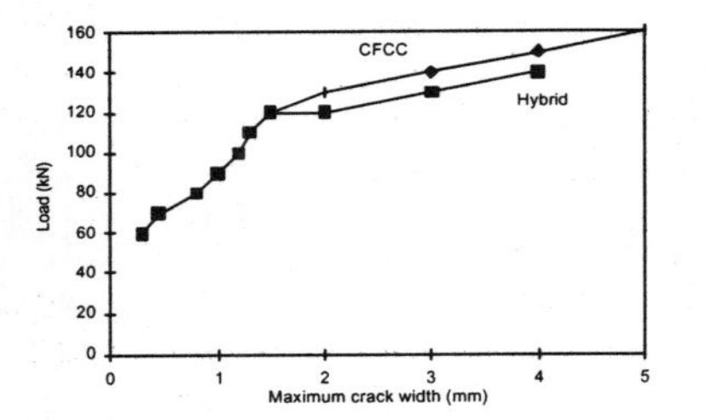

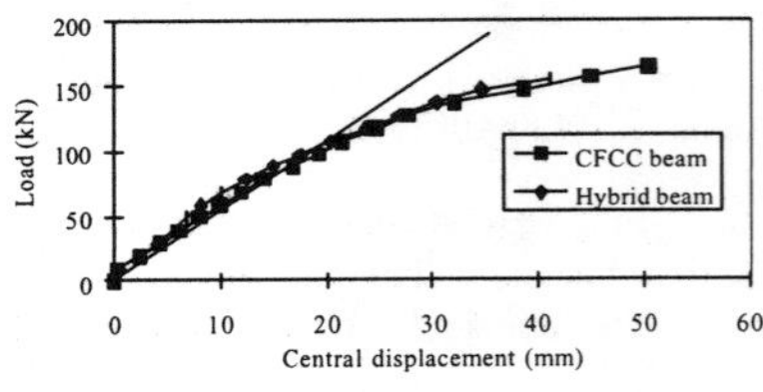

Figure 4. Measured maximum crack widths at level of tensile FRP reinforcement (left) and mid deflection curves (right) for two beams with equal tensile reinforcement capacity and modulus. Apinis et al (1999).

Compatibility between concrete and surrounding wraps

Tensile strength of carbon fibers is higher than that of carbon fiber sheets and this in its turn is higher than that of sheets on wrapped concrete members. This is due to the fact that it is difficult to stretch all fibers equally. The fibers in a sheet are somewhat wavy, Figure 5, and at tensioning do not transfer the same load. Further, fibers in the sheet take load transversally from confining pressure in addition to the tensile load. Moreover cracked concrete and the gravel size and sharpness might influence the breaking of carbon fibers due to uneven pressure from concrete against sheet.

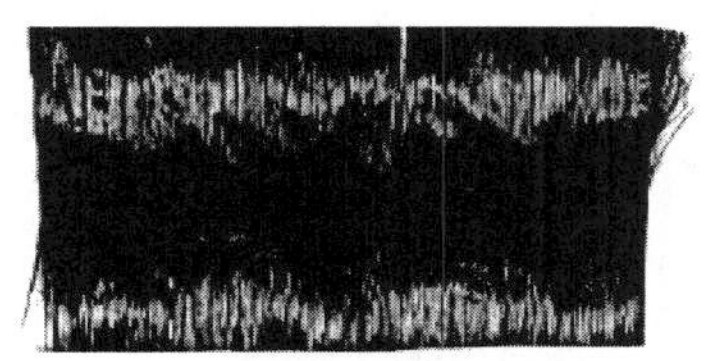

Figure 5. Fiber orientation in carbon fiber sheet. The fibers are somewhat wavy and not fully stretched. The fibers become more and more uneven in added wraps.

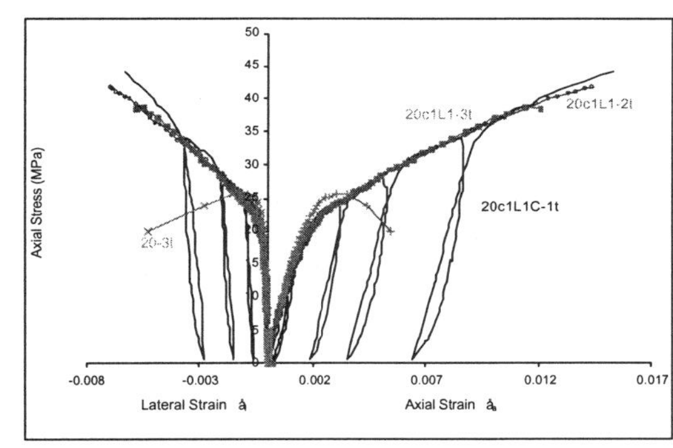

Figure 6. Compressive test of one layer carbon fiber sheet confined concrete cylinder with f_c=20 MPa. In diagram is shown axial strain to the right and lateral to the left. Unconfined cylinder 20-3t results are shown for comparison. Rousakis (2001).

In Figure 6 concrete cylinder wrapped with 1 layer of carbon sheet is shown after compressive failure. The concrete compressive strength is 20 MPa. I Figure 6 the stress-axial and lateral strain diagrams for 1 unwrapped and 3 wrapped concrete cylinders with compressive load repetitions for 1 cylinder are shown. In the diagram it can be seen that, at the load level when the plain concrete cylinder fails, wraps are activated and the load can be increased considerably. There is a reduction in modulus, but the stress-strain relation is linear, although giving residual deformations, which the unloading cycles show. The failure strain of the wraps is about 0.007, which is less than that of plain fibers 0.012. If the wraps are in several layers the modulus of elasticity of the compressive stress-strain relation for the cylinder increases, when wraps are activated. However the failure happens earlier at lower lateral strain, which becomes visible in Figure 7 as decreasing efficiency of wraps when their number increases. If the wrapping is done by somewhat prestressed filaments with stretch control it is possible that the failure load is less affected by progressive fiber rupture.

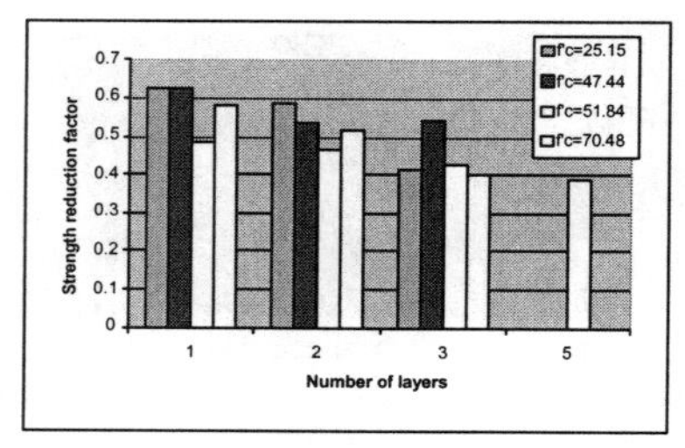

Figure 7. Strength reduction for every carbon sheet layer around concrete cylinders with different concrete strengths. De Lorenzis & Rousakis (2001).

The models produced to describe performance of confined concrete are related to the tested specimens and yields for these. Usually empiric constants are determined and used, Figure 8. The formulas produced for the experiments with cylinders can hardly be used directly in design, because the columns and compressed members in practice have bigger circular sections than 150mm diameter. There might be a size effect. Further if the section has different shape the efficiency of wrapping changes. For square section parts along the sides there is confinement along the side but not in perpendicular direction. On the other hand there is intense confinement in corners. The corners have to be rounded to prevent fiber breakage and to extend the confinement in corner regions. Therefore section parameters have to be introduced in a design formula. Finally the buckling problem for wrapped columns has to be considered, especially as the modulus of elasticity decreases when wraps are activated.

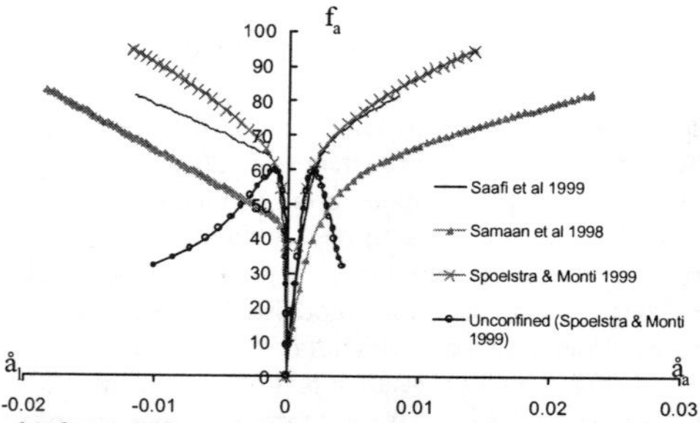

Figure 8. Results from different models simulating confining effect of carbon fiber wrapping of concrete cylinders. Rousakis (2001).

When FRP sheets are glued to the surface of the concrete the sheets are rolled and smoothened. However, all the fibers are not stretched equally, which makes some fibers to be more loaded. This results in a progressive failure of the fibers in the sheet, which happens at a lower mean strain in comparison with that which is achieved in a tensile test with all fibers taking the same load. The progressive failure is probably more pronounced for high modulus fibers, because the difference in fiber stretching is the same but high modulus and low ultimate strain for the fibers result in higher differences in load taking. In a design formula a fiber efficiency factor, which relates to the number of wraps, has to be introduced. It would be favorable if continuity in thinking could be developed in formulas from one and two dimensional surface fiber sheet strengthening to one and two dimensional confining wraps giving three dimensional effect.

Conclusions

The observed disturbances in tensile strength make up a challenge for code writers to evaluate results correctly and to produce the right clauses, test interpretations and reduction factors. Tensile strength is affected by uneven load taking by individual fibers and by shear lag also accentuated by creep. In hybrid rods creep might cause overloading and fracture of the not creeping carbon fibers. In spite of appropriate 3% elongation of hybrid rods concrete member rotation capacity is lacking because bond between cracks does not break down. Fiber stretching in sheet is uneven and this causes sheet tensile efficiency reduction with increasing number of wraps. Progressive fiber failures must be taken into account in code writing.

Acknowledgement

The investigation was sponsored by European Commission - TMR - Network, ConFibreCrete "Research leading to the Development of Design Guidelines for the Use of FRP in Concrete Structures" and Ralejs Tepfers Consulting.

References

Achillides Z. (1998). "Bond behavior of FRP bars in concrete." *(PhD Thesis),* Centre for Cement and Concrete, Dept. of Civil and Structural Eng., The University of Sheffield, UK.

Apinis R., Modniks J., Tamuzs V., Tepfers R. (1999). "Pull-out, flexural rotation capacity and creep tests using hybrid composite rods and CFCC rods for reinforcement in concrete." Chalmers University of Technology, Division of Building Technology, Work No 32. Publication 99:4. Göteborg, 1999. p. 49 + A1 to A5. *Contribution to Eleventh International Conference on "Mechanics of Composite Materials"* June 11-15, 2000 Riga, Latvia as abstract". 14-15.

Apinis R., Modniks J., Tamuzs V., Tepfers R. (2000). "Creep tests using hybrid composite rods for reinforcement in concrete." Chalmers University of Technology, Division of Building Technology, Work No 33. Publication 00:2. Göteborg, May 2000. p. 8. *3rd International Conference on Advanced Composite Materials in Bridges and Structures,* August 15-18, 2000 Ottawa, Canada.

Lees J.M., Burgoyne C.J. (1996). "Influence of bond on Rotation Capacity of Concrete Pre Tensioned with AFRP." *Advanced Composite Materials in Bridges and Structures. 2nd International Conference*. Montréal, Québec, Canada, August 11-14. 901-908.

Saafi M., Toutanji H.A., Li Z. (1999). 'Behavior of Concrete Columns Confined with Fiber Reinforced Polymer Tubes." *ACI Materials Journal,* V. 96, No. 4, July – August 1999. 500-509.

Samaan M., Mirmiram A., Shahawy M. (1998). "Model of Concrete Confined by Fiber Composites." *ASCE Journal of Structural Engineering, V. 124, No 9,* September 1998. 1025-1031.

Spoelstra M. R., Monti G. (1999). "FRP-Confined Concrete Model." *ASCE Journal of Composites for Construction*, V. 3, No. 3, August 1999. 143-150.

Tamuzs V., Maksimovs R., Modniks J. (2001). "Long-yterm creep of hybrid FRP bars". *FRPRCS-5 Conference in Cambridge* 16-18 July 2001.

Rousakis T., De Lorenzi L. (2001). "Ongoing Experimental Investigation of Concrete Cylinders Confined by Carbon FRP Sheets, under Monotonic and Cyclic Axial Compressive Load.". *Chalmers University of Technology, Division of Building Technology, Work No 44*. Göteborg, March 2001.

Debonding Failures in FRP-Strengthened RC Beams: Failure Modes, Existing Research and Future Challenges

J.G. Teng [1], J.F. Chen [2] and S.T. Smith [3]

Abstract

Both the flexural and shear capacities of reinforced concrete (RC) beams can be increased by bonding fibre-reinforced polymer (FRP) composites. A variety of debonding failure modes are possible in these strengthened beams, and for reliable strengthening measures, these debonding failures must be designed against. This paper provides a review of existing research on debonding failures in FRP-strengthened RC beams. Various failure modes are identified, and for each failure mode, existing research is summarised, with particular attention to studies on strength models. Issues which require future research are also outlined.

Introduction

Extensive research has been carried out in recent years on the flexural and shear strengthening of reinforced concrete (RC) beams using fibre-reinforced polymer (FRP) composites. This research has identified a variety of debonding failure modes as the limiting factor of the ultimate strength (Teng et al. 2001a). This paper provides a review of existing research on debonding failures in FRP-strengthened RC beams. Various failure modes are identified, and for each failure mode, existing research is summarised, with particular attention to studies on strength models. Issues which require future research are also outlined. Only beams strengthened with plates/sheets without prestressing or mechanical end anchorage are covered, as these are the more common and due to space limitation. Also due to space limitation, the review is relatively brief and the list of references is limited, but more details and a comprehensive list of references can be found in Teng et al. (2001a).

---

[1] Professor, Dept of Civil and Struct. Engrg, The Hong Kong Polytechnic Univ., Hong Kong, China. Tel: +852 2766 6012, Email: cejgteng@polyu.edu.hk

[2] Lecturer, School of the Built Environment, Nottingham Univ., Nottingham NG7 2RD, UK. Tel: +44 115 951 4889, Email: jianfei.chen@nottingham.ac.uk.

[3] Postdoctoral Fellow, Dept of Civil and Struct. Engrg, The Hong Kong Polytechnic Univ., Hong Kong, China. Tel: +852 2766 6060, Email: cesmith@polyu.edu.hk

Only debonding failures which occur in the concrete are treated here. Debonding at the adhesive-to-concrete interface and away from the concrete should be avoided by the use of strong adhesives and appropriate surface preparation. For such debonding failures in the concrete, studies on RC beams strengthened with a bonded steel plate are often as relevant as those on RC beams bonded with an FRP plate, and reference to the former are made in this paper when appropriate.

FRP-to Concrete Bond Strength

General. In the strengthening of RC beams, FRP plates or sheets are externally bonded to RC beams using adhesives. For the external FRP to be effective in enhancing the load-carrying capacity of the beam, effective stress transfers between the FRP and the concrete are required. The bond strength between FRP and concrete has thus been a major topic of research and is first discussed in this section.

Test Methods, Failure Mode and Effective Bond Length. Substantial experimental and theoretical work exists on the bond strength of FRP or steel plate-to-concrete joints (Chen and Teng 2001a, Teng et al. 2001a). Most of the experiments have been carried out using simple shear tests (Figure 1), on which the discussion here is based, but several other test set-ups have been used. A numerical investigation showed that significant differences exist between different test set-ups (Chen et al. 2001a). Provided strong adhesives are used, FRP and steel plate-to-concrete bonded joints fail within the concrete adjacent to the adhesive-to-concrete interface, starting from the critically stressed position in most cases (Chen and Teng 2001a).

An important feature of plate-to-concrete bonded joints is that the bond strength, in terms of the force in the bonded plate, cannot always increase with an increase in the bond length L (Figure 1), and that the ultimate tensile strength of an FRP plate may never be reached however long the bond length is. This leads to the concept of effective bond length, beyond which any increase in the bond length cannot increase the bond strength, as confirmed by many experimental studies (e.g. Täljsten 1994, Maeda et al. 1997) and fracture mechanics analyses (e.g. Holzenkämpfer 1994, Yuan et al. 2001). However, a longer bond length can improve the ductility of the failure process. This phenomenon is substantially different from the bond behaviour of internal reinforcement and must be accounted for in the development of bond strength models and design rules.

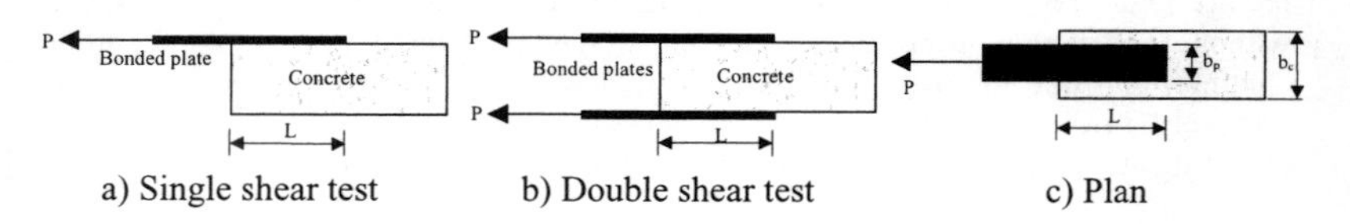

a) Single shear test b) Double shear test c) Plan

Figure 1. Single and double shear tests

Bond Strength Models. Several bond strength models (Maeda et al. 1997, Hiroyuki and Wu 1997, Tanaka 1996, Täljsten 1994, Neubauer and Rostásy 1997, Holzenkämpfer 1994, Yuan et al. 2001) and design proposals (van Gemert 1980, Khalifa et al. 1998, Chaallal et al. 1998) have been advanced in the last few years. Chen and Teng (2001a) recently examined these models by comparing them with experimental data collected from the literature. They further developed a strength model based on a fracture mechanics solution (Chen and Teng 2001a). This new model not only captures all the characteristic behaviour of an FRP-to-concrete bonded joint, but is also in better agreement with existing test data.

Debonding Failures in Flexurally-Strengthened Beams

Failure Modes. A number of debonding failure modes have been observed in numerous tests on RC beams flexurally-strengthened with an FRP soffit plate (Teng et al. 2001a), which are often referred to as FRP-plated RC beams. These may be broadly classified into two types: (a) plate end debonding failures, including concrete cover separation (Figure 2a) and plate end interfacial debonding (Figure 2b); and (b) intermediate crack induced interfacial debonding failures including those due to a major flexural crack (Figure 2c) or a flexural-shear crack (Figure 2d).

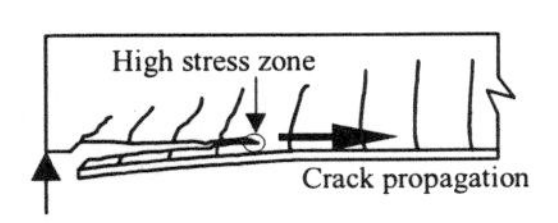

(a) Concrete cover separation

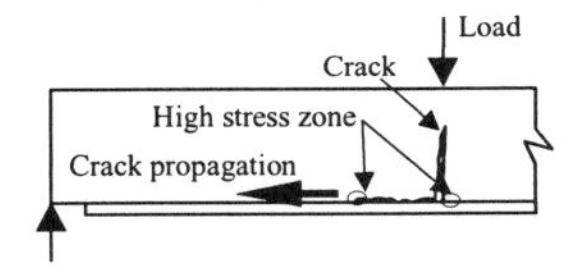

(c) Intermediate flexural crack induced interfacial debonding

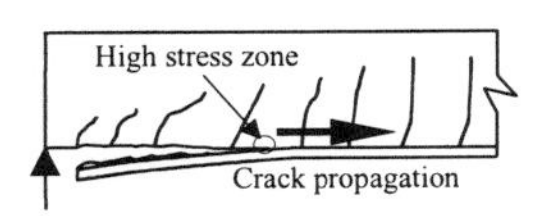

(b) Plate end interfacial debonding

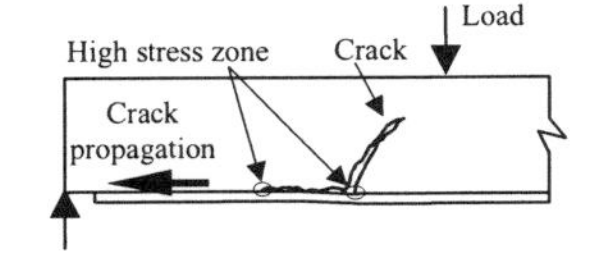

(d) Intermediate flexural-shear crack induced interfacial debonding

Figure 2. Debonding failure modes of FRP-plated RC beams

Interfacial Stresses. Plate end debonding failures are due to high interfacial stresses near the plate end, and a number of approximate closed-form solutions for these stresses suitable for direct exploitation in design have been formulated (Vilnay 1988, Roberts 1989, Roberts and Haji-Kazemi 1989, Liu and Zhu 1994, Täljsten 1997, Malek et al. 1998, Smith and Teng 2001a). All these solutions are based on the assumption that the shear and normal stresses are uniform across the adhesive layer thickness, and have been found to give closely similar results for RC beams bonded with a thin plate (Smith and Teng 2001a). Higher order solutions for interfacial stresses (Rabinovich and Frostig 2000, Shen et al. 2001) have also been obtained as well as finite element results (Teng et al. 2000). These studies showed that the distributions of interfacial stresses near the plate end are much more complicated than are predicted by simple approximate solutions such as that of Smith and Teng (2001a), although the latter does provide a useful approximation.

Strength Models for Plate End Debonding Failures. Several strength models for plate end debonding have been developed for FRP-plated RC beams in the last decade (Varastehpour and Hamelin 1997, Saadatmanesh and Malek 1998, Wang and Ling 1998, Ahmed and van Gemert 1999, Tumialan et al. 1999, Raoof and Hassanen 2000). In addition, a number of strength models have also been developed for plate end debonding in steel plated beams (Oehlers 1992, Ziraba et al. 1994, Jansze 1997, Raoof and Zhang 1997).

These debonding strength models can be divided into three categories in terms of approaches, namely (a) shear capacity based models (Oehlers 1992, Jansze 1997, Ahmed and van Gemert 1999) in which the debonding failure strength is related to the shear capacity of the beam; (b) concrete tooth models (Raoof and Zhang 1997, Wang and Ling 1998, Raoof and Hassanen 2000) in which the behaviour of a concrete "tooth" formed between two adjacent cracks deforming like a cantilever under the action of lateral shears applied at the beam-plate interface is considered; and (c) interfacial stress based models (models I and II of Ziraba et al. 1994, Varastehpour and Hamelin 1997, Saadatmanesh and Malek 1998, Tumialan et al. 1999) in which predictions of interfacial stresses are made use of, generally in conjunction with a concrete failure criterion. Model II of Ziraba et al. (1994) in fact combines approaches (a) and (c). In addition, it should be mentioned that Oehlers' (1992) model is not purely based on the shear capacity, as it also takes into account the interaction between shear and bending.

An assessment of the performance of these models against existing test data of FRP-plated beams have been recently undertaken (Smith and Teng 2001b, c, Teng et al. 2001a) which shows that Oehlers' (1992) model is the best in terms of providing safe and close predictions for design use for the plate end debonding mode, despite the fact that it was developed based on test data of steel-plated beams. Smith and Teng (2001c) also proposed a new simple model by modifying Oehlers' (1992) model. This new model is superior to other existing models in terms of providing safe and close predictions for direct application in design (Smith and Teng 2001c, Teng et al. 2001a).

Strength Model for Intermediate Crack Induced Debonding Failures. The intermediate crack induced debonding mode involves interfacial debonding in the concrete adjacent to the adhesive-to-concrete interface which initiates at a flexural or flexural-shear crack and propagates towards one of the plate ends. Limited experimental data show that the debonding failure strength of this mode can be closely predicted by Chen and Teng's (2001a) bond strength model with a simple modification (Teng et al. 2001a, 2001b).

Debonding Failures in Shear-Strengthened Beams

Debonding Failure Mode. The common methods for shear strengthening using FRP composites include bonding FRPs on the sides of a beam only, bonding U jackets to cover both sides and the soffit, and wrapping FRPs around the cross-section. A detailed description of these strengthening schemes is given in Teng et al. (2001a). When an FRP-strengthened RC beam fails in shear, the two common failure modes are shear failure due to FRP rupture and that due to FRP debonding. It may be noted that, in the shear failure mode due to FRP rupture, FRPs are likely to have debonded before they rupture but this has little effect on the shear capacity of the beam.

Figure 3. FRP debonding failure of a U jacketed RC beam

For test beams controlled by shear failure, available test data show that all beams with FRPs bonded on sides only, and many bonded with U jackets, failed by debonding of the FRP from the concrete. In this mode, once the FRP starts to peel off, the beam fails very quickly in a brittle manner. Bond strength between FRP and concrete thus plays the key role here. Figure 3 shows the shear failure due to FRP debonding of a U jacketed beam.

Strength Models. A number of design proposals have been presented for the shear capacity of shear-strengthened RC beams (Chaallal et al. 1998, Triantafillou 1998, Khalifa et al. 1998, Triantafillou and Antonopoulos 2000). In all these proposals, the shear capacity of a shear-strengthened RC beam is expressed as the sum of the contributions from the concrete, the steel shear reinforcement and the bonded FRP.

Chaallal et al. (1998) treated the FRP as conventional shear reinforcement. Debonding is dealt with by limiting the design average shear stress between the FRP and the concrete to half the value expected at debonding. However, the debonding strength model used by them does not match well with experimental data and the effective bond length is not considered (Chen and Teng 2001a).

Triantafillou (1998) proposed to limit the strain in the FRP to an effective strain which was obtained from regression of experimental data. The model does not distinguish between different strengthening schemes and failure modes. Triantafillou and Antonopoulos (2000) extended Triantafillou's (1998) model so that different effective strain expressions are given for CFRP wrapping and other strengthening schemes, but no distinction is made between side bonding and U jacketing.

Khalifa et al. (1998) proposed a modification to Triantafillou's (1998) effective strain model in which the ratio of effective stress (or strain) in the FRP to its ultimate strength (or strain) is used instead of the effective stress (or strain) itself. In particular, they proposed a bond mechanism design approach based on the bond strength model of Maeda et al. (1997). However, the bond strength model of Maeda et al. (1997) adopted in deriving the design proposal cannot correctly predict the effective bond length (Chen and Teng 2001a).

Recognising the deficiencies of the above models, Chen and Teng (2001b) derived a rational debonding strength model for shear-strengthened RC beams based on their FRP-to-concrete bond strength model (Chen and Teng 2001a). The new model is in better agreement with test data than all other models.

Future Challenges

General. The above review shows that significant advances have been made in the last few years in achieving a better understanding of debonding failures in FRP-strengthened RC beams. Rational strength models are now available for most of these debonding failure modes. However, a great deal of further work is required on a number of aspects. Some of the more obvious are discussed below.

FRP-to-Concrete Bond Strength. Although many test results are now available, there is still the need for more tests, particularly those carefully conducted, instrumented and documented. For example, more experimental data on effective bond length are required.

Flexurally-Strengthened RC Beams. For plate end debonding failures, the recently proposed Smith and Teng (2001c) model considers only the effect of plate end shear force and ignores any interaction between shear force and bending moment. Test results can be several times higher than the predictions of their model, which indicates that there is room for significant improvement to the accuracy of the model. The lack of interaction between shear force and bending moment is another aspect which needs attention. While this lack of interaction reflects the trend of the existing test database, it has been attributed to the limitation of this database in a subsequent study (Smith and Teng 2001d). Smith and Teng (2001d) showed that interaction between shear force and bending moment at the plate end is important through a series of experiments and proposed a bi-linear approximation to describe this interaction based on these experimental results. A great deal of further work is thus required on plate end debonding failures so that more accurate models can be developed.

For intermediate crack-induced debonding failures, existing test results are rather limited, so additional tests covering a wide range of dimensions are required. In particular, Teng et al. (2001b) assumed that the same strength model is appropriate for both failures due to a major flexural crack or a major flexural-shear crack. Further studies are required to confirm the validity of this assumption over a wide range of parameters.

Shear-Strengthened RC Beams. Most experiments have been conducted on beams with a small range of shear-span/depth ratios and beam sizes. Further research is needed to establish a better understanding of the failure process, verify existing strength models in practical ranges of parameters and develop improved strength models if necessary.

Numerical Simulations and Large Scale Experiments. Most existing studies on debonding failures have been experimentally based, or on the exploitation of experimental results to develop relatively simple strength models. Numerical simulations of debonding failures, to predict both failure processes and failure loads, have lagged behind. Efforts in this direction should be encouraged. Similarly, tests have been commonly conducted on small to medium sized beams, while those on larger beams, due to the considerable cost, have been relatively few.

Conclusions

For both flexurally-strengthened and shear-strengthened RC beams, the main debonding failure modes have been identified and are reasonably well understood. Rational strength models are now available for these failure modes, although a great deal of further work is required both to gain better insight into failure mechanisms and to develop more accurate strength models.

Acknowledgements

The authors wish to thank The Hong Kong Polytechnic University for the financial support provided to the project "Retrofitting of RC Structures using FRP Composites" through the Area of Strategic Development in Advanced Buildings Technology in a Dense Urban Environment and for a postdoctoral fellowship for the third author.

References

Ahmed, O. and van Gemert, D. (1999). "Effect of longitudinal carbon fiber reinforced plastic laminates on shear capacity of reinforced concrete beams.", *Fiber Reinforced Polymer Reinforcement for Reinforced Concrete Structures, Proceedings of the Fourth International Symposium,* edited by C.W. Dolan, S.H. Rizkalla and A. Nanni, Maryland, U.S.A., 933-943.

Chaallal, O., Nollet, M.J., Perraton, D. (1998). "Strengthening of reinforced concrete beams with externally bonded fibre-reinforced-plastic plates: design guidelines for shear and flexure.", *Canadian Journal of Civil Engineering*, 25(4), 692-704.

Chen, J.F. and Teng, J.G. (2001a). "Anchorage strength models for FRP and steel plates attached to concrete.", *Journal of Structural Engineering, ASCE*, 127(7), 784-791.

Chen, J.F. and Teng, J.G. (2001b). "Shear strengthening of RC beams by external bonding of FRP composites: a new model for FRP debonding failure.", *Proceedings of the Ninth International Conference on Structural Faults and Repair*, London, U.K.

Chen, J.F., Yang, Z.J. and Holt, G.D. (2001). "FRP or steel plate-to-concrete bonded joints: effect of test methods on experimental bond strength.", *Steel and Composite Structures - An International Journal*, 1(2), 231-244.

Hiroyuki, Y. and Wu, Z. (1997). "Analysis of debonding fracture properties of CFS strengthened member subject to tension.", *Non-Metallic (FRP) Reinforcement for Concrete Structures, Proceedings of the Third International Symposium*, Sapporo, Japan, 287-294.

Holzenkämpfer, O. (1994). *Ingenieurmodelle des Verbundes geklebter Bewehrung für Betonbauteile*, Dissertation, TU Braunschweig.

Jansze, W. (1997). *Strengthening of RC Members in Bending by Externally Bonded Steel Plates*, PhD Thesis, Delft University of Technology, Delft, The Netherlands.

Khalifa, A., Gold, W.J., Nanni, A. and Aziz, A. (1998). "Contribution of externally bonded FRP to shear capacity of RC flexural members.", *Journal of Composites for Construction, ASCE*, 2(4), 195-203.

Liu, Z. and Zhu, B. (1994). "Analytical solutions for R/C beams strengthened by externally bonded steel plates.", *Journal of Tongji University* (in Chinese), 22(1), 21-26.

Maeda, T., Asano, Y., Sato, Y., Ueda, T. and Kakuta, Y. (1997). "A study on bond mechanism of carbon fiber sheet.", *Non-Metallic (FRP) Reinforcement for Concrete Structures, Proceedings of the Third International Symposium*, Sapporo, Japan, 279-285.

Malek, A.M., Saadatmanesh, H. and Ehsani, M.R. (1998). "Prediction of failure load of R/C beams strengthened with FRP plate due to stress concentration at the plate end.", *ACI Structural Journal*, 95(1), 142-152.

Neubauer, U. and Rostásy, F.S. (1997). "Design aspects of concrete structures strengthened with externally bonded CFRP plates.", *Proceedings of the Seventh International Conference on Structural Faults and Repair*, edited by M.C. Ford, Edinburgh, U.K., 109-118.

Oehlers, D.J. (1992). "Reinforced concrete beams with plates glued to their soffits.", *Journal of Structural Engineering, ASCE*, 118(8), 2023-2038.

Rabinovich, O. and Frostig, Y. (2000). "Closed-form high-order analysis of RC beams strengthened with FRP strips.", *Journal of Composites for Construction, ASCE*, 4(2), 65-74.

Raoof, M. and Zhang, S. (1997). "An insight into the structural behaviour of reinforced concrete beams with externally bonded plates.", *Proceedings of the Institution of Civil Engineers, Structures and Buildings*, 122, 477-492.

Raoof, M. and Hassanen, M.A.H. (2000). "Peeling failure of reinforced concrete beams with fibre-reinforced plastic or steel plates glued to their soffits.", *Proceedings of the Institution of Civil Engineers: Structures and Buildings*, 140, 291-305.

Roberts, T.M. (1989). "Approximate analysis of shear and normal stress concentrations in the adhesive layer of plated RC beams.", *The Structural Engineer*, 67(12), 229-233.

Roberts, T.M. and Haji-Kazemi, H. (1989). "Theoretical study of the behaviour of reinforced concrete beams strengthened by externally bonded steel plates.", *Proceedings of the Institution of Civil Engineers,* 87(2), 39-55.

Saadatmanesh, H. and Malek, A.M. (1998). "Design guidelines for flexural strengthening of RC beams with FRP plates.", *Journal of Composites for Construction, ASCE,* 2(4), 158-164.

Shen, H.S., Teng, J.G. and Yang, J. (2001). "Interfacial stresses in beams and slabs bonded with a thin plate.", *Journal of Engineering Mechanics, ASCE*, 127(4), 399-406.

Smith, S.T. and Teng, J.G. (2001a). "Interfacial stresses in plated beams.", *Engineering Structures,* 23(7), 857-871.

Smith, S.T. and Teng, J.G. (2001b). "FRP-strengthened RC beams-I: Review of debonding strength models.", *To be published.*

Smith, S.T. and Teng, J.G. (2001c). "FRP-strengthened RC beams-II: Assessment of debonding strength models.", *To be published.*

Smith, S.T. and Teng, J.G. (2001d). "Plate end debonding failures in FRP-plated RC beams.", *Proceedings of the Ninth International Conference on Structural Faults and Repair, London, U.K.*

Täljsten, B. (1994). *Plate bonding. Strengthening of existing concrete structures with epoxy bonded plates of steel or fibre reinforced plastics*, Doctoral Thesis, Luleå University of Technology, Sweden.

Täljsten, B. (1997). "Strengthening of beams by plate bonding.", *Journal of Materials in Civil Engineering, ASCE*, 9(4), 206-212.

Tanaka, T. (1996). *Shear resisting mechanism of reinforced concrete beams with CFS as shear reinforcement*, Graduation Thesis, Hokkaido University, Hokkaido, Japan.

Teng, J.G., Zhang, J.W. and Smith, S.T. (2000). "Finite Element Interfacial Stresses in RC Beams Bonded with a Soffit Plate.", *Proceedings of International Symposium on High Performance Concrete*, Hong Kong and Shenzhen, China, 499-505.

Teng, J.G., Chen, J.F., Smith, S.T. and Lam, L. (2001a). *FRP-strengthened RC structures*, John Wiley and Sons, Chichester, U.K., *in press.*

Teng, J.G., Smith, S.T., Yao, J. and Chen, J.F. (2001b). "Intermediate crack induced interfacial debonding in FRP-plated RC beams and slabs.", *to be published.*

Triantafillou, T.C. (1998). "Shear strengthening of reinforced concrete beams using epoxy-bonded FRP composites.", *ACI Structural Journal*, 95(2), 107-115.

Triantafillou, T.C. and Antonopoulos, C.P. (2000). "Design of concrete flexural members strengthened in shear with FRP.", *Journal of Composites for Construction, ASCE*, 4(4), 198-205.

Tumialan, G., Belarbi, A. and Nanni, A. (1999). "Reinforced concrete beams strengthened with CFRP composites: failure due to concrete cover delamination.", *Report No. CIES-99/01*, Department of Civil Engineering, Center for Infrastructure Engineering Studies, University of Missouri-Rolla, U.S.A.

van Gemert, D. (1980). "Force transfer in epoxy-bonded steel-concrete joints.", *International Journal of Adhesion and Adhesives*, 1, 67-72.

Varastehpour, H. and Hamelin, P. (1997). "Strengthening of concrete beams using fibre-reinforced plastics.", *Materials and Structures*, 30, 160-166.

Vilnay, O. (1988). "The analysis of reinforced concrete beams strengthened by epoxy bonded steel plates.", *The International Journal of Cement Composites and Lightweight Concrete*, 10(2), 73-78.

Wang, C.Y. and Ling, F.S. (1998). "Prediction model for the debonding failure of cracked RC beams with externally bonded FRP sheets.", *Proceedings of the Second International Conference on Composites in Infrastructure (ICCI 98)*, Arizona, U.S.A., 548-562.

Yuan, H., Wu, Z.S. and Yoshizawa, H. (2001). "Theoretical solutions on interfacial stress transfer of externally bonded steel/composite laminates.", *Journal of Structural Mechanics and Earthquake Engineering, JSCE*, in press.

Ziraba, Y.N., Baluch, M.H., Basunbul, I.A., Sharif, A.M., Azad, A.K. and Al-Sulaimani, G.J. (1994). "Guidelines for the design of reinforced concrete beams with external plates.", *ACI Structural Journal*, 91(6), 639-646.

FRP-Glulam Structures: From Material and Processing Issues to a Performance-Based Evaluation Methodology

Roberto Lopez-Anido[1], Lech Muszynski[2], Douglas Gardner[3]and Barry Goodell[3]

Abstract

While significant advances in the development of cost-effective and durable fiber-reinforced polymer (FRP) composite systems for reinforcing glulam structures have been achieved, there is a need to address relevant material, processing, and durability issues. The objective of this paper is to present a performance-based material evaluation methodology for FRP-glulam structures. Challenges and opportunities in the development of hybrid FRP-glulam composites as structural materials are analyzed in the context of specific construction applications. The ongoing research effort will develop a set of simple and integrated material stiffness, strength, toughness and durability test methods with associated performance limits, and will provide data and recommendations necessary to draft performance-based material specifications.

Introduction

Fiber-reinforced polymer (FRP) composites, which are been increasingly used in civil infrastructure (Lopez-Anido 2000), offer new opportunities to repair or strengthen existing wood structures in buildings and bridges (Plevris 1992, Gardner 1994). However, the use of FRP composites for wood reinforcement is limited because of the relatively high cost of materials and fabrication processes. There are various systems for FRP reinforcement of wood structural members that have been developed, as follows: a) FRP wrapping of wood members with E-glass and carbon fabric reinforcement using the wet layup method (Sonti 1995, GangaRao 1997), (Lopez-Anido 2000); b) FRP adhesive bonding of E-glass pultruded plates to glulam beams (Davalos 1992, Dagher 1996, Tingley 1997, Dagher 1998); and c) FRP tendons for prestressing of laminated wood decks (Dagher 1997). Examples of FRP-wood structural applications in civil infrastructure are: a) FRP-glulam beams for bridge beams (Davalos 1992, Davalos 1994), Dagher 1999); b) FRP-glulam panels for bridge decks (Lopez-Anido 2000); c) Reinforced railroad ties (Sonti 1996, Davalos 1999); and d) FRP-glulam beams integral with a concrete slab (Brody 2000). In parallel with construction applications, methods of analysis have been developed for FRP-wood structural members. For example, a layer-wise theory for failure analysis of FRP reinforced glulam beams was proposed (Davalos 1994), and a

[1]ASCE Member, Assistant Professor, Dept. of Civil & Environmental Engineering, Advanced Engineered Wood Composites (AEWC) Center, University of Maine, Orono, ME 04469-5711.

[2] Assistant Scientist, AEWC Center, University of Maine, Orono, ME 04469-5793.

[3] Professor, Wood Science & Technology, AEWC Center, University of Maine, Orono, ME 04469-5793.

progressive failure analysis was developed (Kim 1996). More recently, a non-linear stochastic layered moment-curvature analysis of glulam beams was developed (Lindyberg 2000).

While physical and mechanical properties of wood and several fiber-reinforced polymer (FRP) materials are already well recognized, properties of hybrid FRP-wood composites and, in particular, FRP-wood interfaces need to be better understood and fully characterized. Furthermore, the durability of FRP reinforcement in the presence of wood preservatives, moisture and temperature fluctuations; and subjected to harsh environmental conditions combined with cyclic or permanent load regimes is of concern for the application of FRP-wood composites in construction.

Basic physical and mechanical FRP composite properties, such as: glass transition temperature; tensile strength and modulus, compressive strength, interlaminar shear strength; and FRP-wood interfacial properties, such as: shear strength and interlaminar fracture toughness, are commonly regarded as long-term performance indicators of FRP-wood hybrid materials. Performance indicators are computed for intact and exposed materials to evaluate residual properties and assess durability. Materials are exposed under accelerated conditions according to operational requirements, e.g., temperature, moisture, wood preservatives and treatment pressures, ultraviolet radiation exposure. Once relevant performance indicators are identified for FRP-wood materials, associated performance limits can be determined.

A performance-based material evaluation methodology has been developed at the University of Maine to allow prediction of FRP-glulam structural properties with acceptable tolerances. The methodology includes evaluation of short-term mechanical response and long-term durability properties, simple accelerated exposure methods and associated performance limits that apply to FRP systems for glulam reinforcement. Where possible, test methods are taken from accepted standards and modified as necessary. Material and processing issues that influence the development of hybrid FRP-glulam composites as structural materials for construction are analyzed in the context of current applications. Examples of field implementation of hybrid FRP-glulam composites are presented and future needs and trends are discussed.

Performance-Based Material Evaluation Methodology for FRP-Glulam Beams

Methodology. The objectives of the performance-based material evaluation methodology are: a) to critically define one set of simple and integrated material stiffness, strength, toughness and durability test methods with associated performance limits, and b) to provide data and recommendations necessary to develop performance based material specifications. The first stage in the performance-based material evaluation methodology is based on adapting ASTM D2559-00 test protocol "Standard Specification for Adhesives for Structural Laminated Wood Products for use under Exterior (Wet Use) Exposure Conditions", to FRP reinforcement bonded to glulam. The proposed protocol encompasses three standard test methods: a) Resistance to Shear by Compression Loading; b) Resistance to Delamination During Accelerated Exposure; and c) Resistance to Deformation Under Static Loading

Exposure. Other stages of the performance-based material evaluation methodology (preservative treatments, environmental resistance, cyclic loads, creep rupture and fracture) are also being developed.

FRP Systems. Four commercially available FRP reinforcement systems exhibiting exterior durability potential for glulam reinforcement have been selected to validate the material qualification protocol (See Figure 1). Material A is E-glass/vinyl ester applied by the resin infusion or vacuum assisted resin transfer molding (VARTM) process with hydroxymethylated resorcinol (HMR) coupling agent as a wood primer. Material B is E-glass/urethane sheet fabricated by the pultrusion method bonded with one-part moist-cure urethane adhesive.

Material C is E-glass/epoxy sheet fabricated by a continuous lamination process bonded with epoxy adhesive using HMR coupling agent as a wood primer. Material D is a Carbon/Phenolic sheet with phenolic impregnated paper (CF/P) bonded using a standard wood adhesive. The four FRP materials represent a broad spectrum of fiber reinforcement, matrix and adhesive systems, as shown in Table 1.

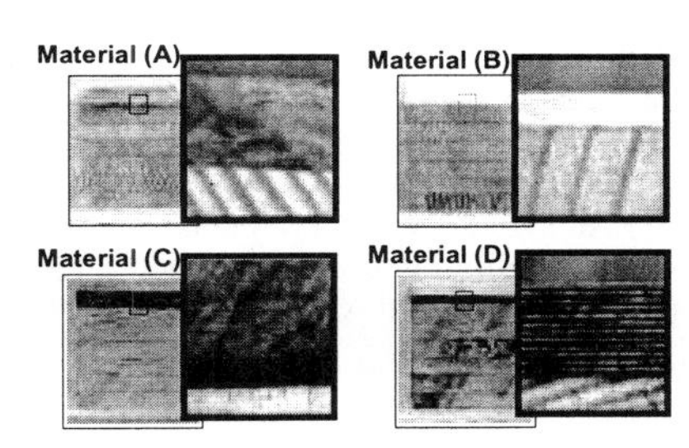

Figure 1 – FRP Systems Evaluated

Table 1. FRP Systems: Materials and Processes

Material	Fiber reinforcement	Polymer matrix	FRP fabrication	Bonding process
A	E-glass	vinyl ester	Resin infusion (VARTM)	Fabric reinforcement bound and bonded by resin infusion
B	E-glass	urethane	pultrusion	Pre-consolidated sheet bonded with urethane adhesive
C	E-glass	epoxy	continuous lamination	Pre-consolidated sheet bonded with epoxy adhesive
D	Carbon	phenolic	continuous lamination	Pre-consolidated sheet/paper bonded with phenolic adhesive

Glulam Materials. Two wood species, which are representative of US glulam production, are considered in this study: southern pine and Douglas-fir. Furthermore, two different oil-borne preservative treatments are evaluated: a) Creosote, which is a coal tar distillate with a complex mixture of hydrocarbons that is widely used in conventional timber bridge construction; and b) Copper naphthenate, which is a chemical that has been in use for many years and has a good track record with regard to long-term protection of wood from biodeterioration (Tascioglu 2000). Water-borne preservatives were not considered because of their deleterious effect on the dimensional stability on large exterior-use timbers.

Controversial issues. Relevant issues have been identified during the research work. The current effort is directed toward answering the following issues: What is the effect of wood preservative chemicals on FRP materials? What is the correct bond model for FRP-wood interfaces (stress-based or fracture mechanics-based)? What are the most durable and cost-effective adhesives (wood adhesives or FRP adhesives)? What is the most efficient type of fiber reinforcement (carbon or E-glass)? What is the most efficient bonding process (bonding pre-consolidated plates, wet lay-up of fabrics or resin infusion of fabrics)? What is the most efficient procedure for primer and adhesive curing (room temperature, elevated temperature or radio frequency)?

Fatigue of FRP-Wood Interface. The proposed criteria for developing material interface fatigue tests and interface bond models consider the following aspects: Multiaxial stresses (shear and peeling), Stress concentrations (ends, notches, gripping areas), Ease of specimen preparation, Ease of testing, Reproducibility (Coefficient of Variation), Varying failure modes (independent of system integrity) and Sensitivity (capacity to discriminate). Two test methods are being applied to evaluate the fatigue performance of FRP-wood interfaces, as shown in Figure 2:

a) Fracture mechanics test based on the double-cantilever setup (Mode I); and
b) Strength test based on the single-lap shear by tension loading setup.

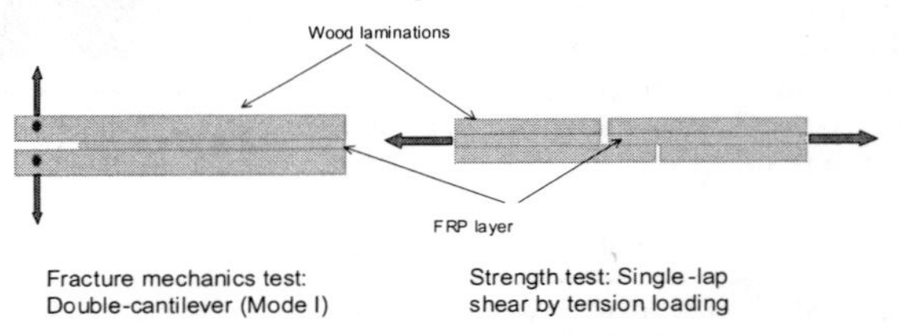

Figure 2 – FRP-Wood Interface: Fatigue Tests Setup

Fabrication Procedures

There are three general procedures to reinforce wood with FRP composites, as shown in Table 2: a) Bonding of consolidated laminates, b) Wet lay up of fabrics, and c) Resin infusion of fabrics. Adhesive bonding of consolidated FRP composites to wood has been studied by several researchers (Gardner 1994, Vick 1995). In the wet lay up process, the polymer resin has a double function of adhesive for wood/FRP interface and matrix for the fabric reinforcement (Sonti 1996, Lopez-Anido 2000). The wet lay up method has been used with phenol-resorcinol-formaldehyde (PRF) adhesive to bond E-glass woven fabric to wood beams (Foster 1998). However, the PRF adhesive is not formulated as a resin system for wetting glass fibers and binding them together into a composite. Recently, research conducted at the Forest Products Laboratories (Vick 1995, 1997) showed that epoxies develop durable bonds to wood when a coupling agent, such as HMR, is utilized. Although epoxies are high performance resins for composites, their application in developing panels is restricted by the high material cost, fabrication complexity and post curing requirements.

In the resin infusion or VARTM process (SCRIMP™ license), reinforcing fabric is placed on the glulam member and the entire member is encapsulated in a vacuum bag. Resin is then infused into the assembly with compaction taking place under vacuum pressure. Unlike the wet layup process this is a closed process and the infusing resin can fill wood cracks and voids as well. Vinyl ester resin systems are used extensively in the production of FRP composite materials because of their excellent chemical resistance and tensile strength combined with low viscosity and fast curing (Lopez-Anido 2001). These resin systems have been used to infuse E-glass fabric reinforcement for glulam beams in our work.

Table 2- Hybrid FRP-Glulam Composite Fabrication Systems

System	Aspects	Issues / Concerns
Wet Lay-up of fabric	Use of impregnator Flexible Ease of use in restricted areas Compaction with vacuum	Quality control Entrapped air Compaction & fiber wrinkling Environmental issues
Adhesive Bonding of prefabricated Plates	Use of prefabricated sections Adhesive bonding Rapid procedure Ease of fabrication	Shear lag effect Durability of the adhesive Two-step fabrication
Resin Infusion (VARTM)	Placement of dry fabric Infusion under vacuum High compaction Fill wood cracks	Fiber wet-out Difficulty of holding vacuum Vacuum bag, flow media & conduits

Glued Laminated Wood in Transportation

Wood as a traditional construction material has been widely used in transportation infrastructure. Glued laminated (glulam) wood material has been used in bridges for approximately 30 years *(Wipf 1990)*. Glulam panels have also been used in bridge decks supported either on steel or glued laminated stringers. Pioneer research work on the application of glulam panels to bridge decks was conducted at the Forest Products Laboratory (McCutcheon 1973, 1974). Stress laminated timber decks have also been applied for highway bridges (Barger 1993, Crews 1998). A method for designing glulam bridge decks, which is available in the Timber Bridge Design Manual (Ritter 1992), led to the construction of glulam bridge decks. For example, the West Seboeis Stream Bridge in Maine was built with FRP-glulam beams and a glulam deck (Dagher 1999). FRP-glulam beams with sufficient tension reinforcement not only exhibit significant strength increases, but also they can develop wood ductile compression failure, rather than the typical brittle tension failure of wood (Dagher 1996), (Dagher 1999). It has been shown that FRP-glulam beams performed satisfactorily in creep tests under controlled temperature and moisture conditions (Davids 1999).

Opportunity: FRP-Glulam Beams Integral with Concrete Slabs

Traditional glulam beam designs, while economical and durable, often have difficulty producing adequate stiffness for longer spans. Addition of FRP tension reinforcement has been shown to improve stiffness to a limited extent, but not nearly as much as it improves strength. Significant research efforts have been dedicated to developing effective composite action between concrete slabs and timber or glulam beams (Ahmadi 1993, Mantilla Carrasco 1999). However, relatively little work has been done on FRP-glulam beams with integral concrete slabs.

To address the stiffness issue, a structural system with an FRP-glulam beam integral with a concrete slab was developed (Brody 2000). On the compression face of the glulam beam, lag screws were driven before pouring the concrete slab to make the glulam beam and concrete slab integral. The potential benefits of this method of construction include greatly increased bending stiffness due to composite action coupled with greater strength and ductility provided by the FRP reinforcement.

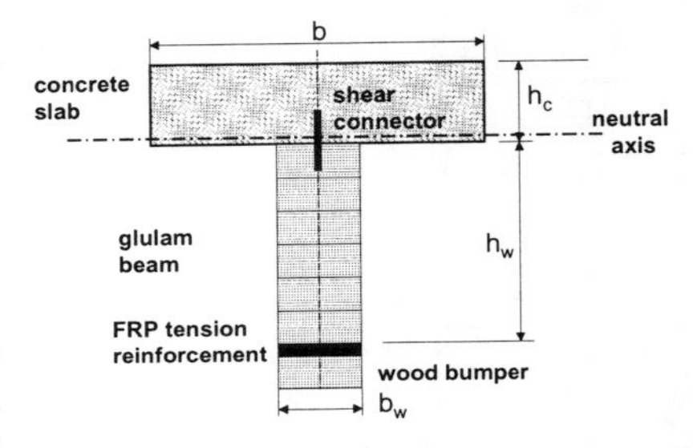

Figure 3. FRP-Glulam Beam Integral with a Concrete Slab

Opportunity: FRP-Glulam Panels for Bridge and Pier Decks

The need for replacement of aged decks in highway bridges and waterfront piers has motivated the development of new durable construction systems. Advances in research and engineering design during the last 25 years, led to the development of glulam timber panels for bridge deck construction (Ritter 1992, Zwerneman 1995). Glulam decks are constructed of glulam panels that are typically 130 to 220 mm (5-1/8 to 8-3/4 inches) thick and 0.91 to 1.50 m (3 to 5 feet) wide (Ritter 1992). In a bridge deck, glulam panels are placed transverse to the supporting beams and traffic loads act parallel to the wide face of the laminations. Compared to reinforced concrete slabs, glulam panels have limited maximum deck spans, which constrains their application in bridge deck replacement projects.

One promising bridge deck system was investigated at the University of Maine by reinforcing glulam panels with fiber-reinforced polymer (FRP) composites using the concept of sandwich construction (Lopez-Anido 2000). The main advantages of FRP composites to strengthen glulam panels for decking are: 1) High tensile strength, 2) Environmental protection of glulam panels, 3) Ease and flexibility of installation, and 4) Reduced weight. The FRP reinforcing faces provide tensile strength in both the positive and negative moment regions of the deck and protect the wood core from environmental effects. In addition, the top FRP composite face sheet can be bonded to a polymer concrete overlay that serves a wearing surface. The wood glulam core resists shear forces, preventing buckling of the FRP face sheets under compression and contributes to the bending stiffness and strength of the deck.

The development of a durable bond between the FRP face sheets and the glulam panel treated with preservative chemicals is a challenge to the advancement of this new technology. The first demonstration of this deck replacement system has been scheduled for the Skidmore Bridge, Maine (USA) in September 2001 (See Figure 4).

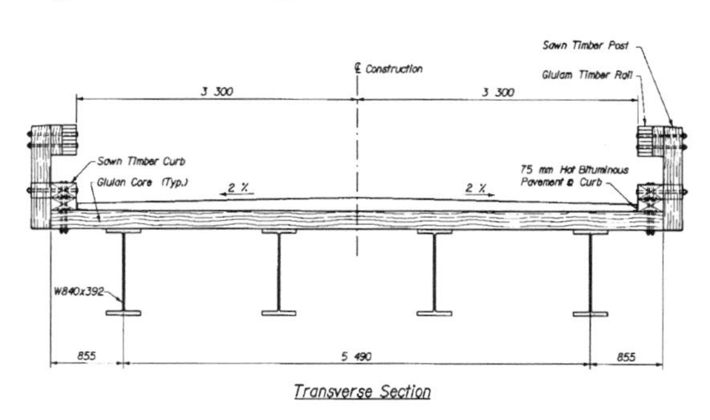

Figure 4. FRP-Glulam Deck over Steel Beam Bridge

Summary and Conclusions

The evaluation of hybrid FRP-glulam structures is a three-fold task that encompasses the elaboration of design guidelines and specifications, the understanding of long-term performance, and the development of construction technologies. While the work presented focuses on transportation applications, there are also opportunities of FRP-glulam construction in buildings (e.g., large span roof structures and architectural sensitive projects). The extent of FRP-glulam composites applications will depend in large part on the resolution of outstanding critical issues that include: a) Durability and fire resistance; b) Reparability of FRP-glulam structural elements; c) Development of validated specifications, standards, and guidelines of use to civil engineers; d) Development of practical design methods and cost-effective manufacturing processes that optimize the use of the material; and e) Provision of an appropriate level of quality assurance and control both during manufacturing and installation by contractors. The first step proposed by the authors to overcome the existing barriers and contribute to the widespread acceptance and use of FRP-glulam construction is to develop a set of performance-based specifications and to qualify representative commercially available FRP reinforcing systems.

Acknowledgements

The research presented in this paper is sponsored through the FHWA Contract DTFH61-99-C-00064- FRP Reinforced Glulams for Bridge Applications which is sponsored by the Federal Highway Administration.

References

Ahmadi, B. H. a. S., M.P. (1993). "Behavior of Composite Timber-Concrete Floors." Journal of Structural Engineering **119**(10): 3111-3130.

Barger, L. S. J., Lopez-Anido, R., and GangaRao, H.V.S. (1993). "Experimental Evaluation Of Stressed Timber Bridge Systems." Transportation Research Record(1426): 57-61.

Brody, J., Richard, A., Sebesta, K., Wallace, K., Hong, Y., Lopez Anido, R., Davids, W. and Landis, E. (2000). FRP-Wood-Concrete Composite Bridge Girders. Structures Congress 2000 - Advanced Technology in Structural Engineering, Philadelphia, PA, ASCE Press.

Crews, K. (1998). International Guidelines for Design of Stress Laminated Timber Bridge Decks. 5th World Conference on Timber Engineering, Montreux, Switzerland.

Dagher, H. J., Abdel-Magid, B., Ritter, M. and S. Iyer. (1997). GRP Prestressing of Wood Decks. Structures Congress XV, Portland, OR, ASCE Press.

Dagher, H. J., Kimball, T., Abdel-Magid, B., and Shaler, S.M. (1996). Effect of FRP Reinforcement on Low-Grade Eastern Hemlock Glulams. National Conference on Wood Transportation Structures, Madison, WI, NWTIC, Morgantown, WV.

Dagher, H. J., Poulin, J., Abdel-Magid, B., Shaler, S.M., Tjoelker, W., and Yeh, B. (1998). FRP Reinforcement of Douglas Fir and Western Hemlock Glulam Beams. International Composites Expo '98, Nashville, TN.

Dagher, H. J. a. L., R. (1999). FRP Reinforced Wood in Bridge Applications. 1st RILEM Symposium on Timber Engineering, Stockholm, Sweden, RILEM.

Dagher, H. J. a. L., R. (1999). West Seboeis Stream FRP-Glulam Highway Bridge. 1999 International Composites Expo, Cincinnati, OH.

Davalos, J. F., Barbero, E., and Munipalle, U. (1992). Glued-laminated timber beams reinforced with E-glass/polyester pultruded composites. Structures Congress X, San Antonio, TX, ASCE.

Davalos, J. F., H.A. Salim, and U. Munipalle (1992). Glulam-GFRP composite beams for stress-laminated T-system timber bridges. 1st International Conference on Advanced Composite Materials in Bridges and Structures, Sherbrooke, Quebec, Canada, CSCE-CGC.

Davalos, J. F., Kim, Y., and Barbero, E.J. (1994). "Analysis of laminated beams with a layer-wise constant shear theory." Composite Structures **28**(3): 241-253.

Davalos, J. F., Zipfel, M.G. and Qiao, P. (1999). "Feasibility study of prototype GFRP-reinforced wood railroad crosstie." Journal of Composites for Construction **3**(2): 92-100.

Davids, W. G., Dagher, H.J., and Breton, J. (1999). Experimental and Numerical Study on Creep of FRP Reinforced Glulam Beams. 1st RILEM Symposium on Timber Engineering, Stockholm, Sweden, RILEM.

Foster, F. (1998). Flexural Repair and Strengthening of Timber Beams Using Fiber Reinforced Polymers. Dept. of Civil and Environmental Engineering. Orono, ME, University of Maine.

GangaRao, H. V. S. (1997). "Sawn and Laminated Wood Beams Wrapped with Fiber Reinforced Plastic Composites." Wood Design Focus: 13-18.

Gardner, D. J., Davalos, J. F. and Munipalle, U. M. (1994). "Adhesive Bonding of Pultruded Fiber-Reinforced Plastic to Wood." Forest Products Journal **44**(5): 62-66.

Kim, Y., Davalos, J.F. and Barbero, E.J. (1996). "Progressive Failure Analysis of Laminated Composite Beams." Journal of Composite Materials **30**(5): 536-539.

Lindyberg, R. (2000). A Nonlinear Stochastic Analysis of Reinforcement Glulam Beams in Bending. Dept. of Civil and Environmental Engineering. Orono, ME, University of Maine.

Lopez-Anido, R., and Karbhari, V.M. (2000). Chapter 2: Fiber Reinforced Composites in Civil Infrastructure. Emerging Materials for Civil Engineering Infrastructure - State of the Art. R. Lopez-Anido, and Naik, T.R. Reston, VA, ASCE Press: 41-78.

Lopez-Anido, R., and Wood, K. (2001). Environmental Exposure Characterization of Fiber Reinforced Polymer Materials used in Bridge Deck Systems, Advanced Engineered Wood Composites Center, University of Maine.

Lopez-Anido, R., Gardner, D.J. and Hensley, J.L. (2000). "Adhesive Bonding of Eastern Hemlock Glulam Panels with E-Glass/Vinyl Ester Reinforcement." Forest Products Journal **50**(11/12): 43-47.

Mantilla Carrasco, E. V. a. O., S.V. (1999). Behavior of Composite Timber-Concrete Beams. 1st RILEM Symposium on Timber Engineering, Stockholm, Sweden, RILEM.

McCutcheon, W. J., and Tuomi, R. L. (1973). Procedure For Design Of Glued-Laminated Orthotropic Bridge Decks. Madison, WI, Department of Agriculture, Forest Service, Forest Products Laboratory.

McCutcheon, W. J., and Tuomi, R. L. (1974). Simplified design procedure for glued-laminated bridge decks. Madison, WI, Department of Agriculture, Forest Service, Forest Products Laboratory.

Plevris, N. a. T., T.C. (1992). "FRP-Reinforced Wood as a Structural Material." Journal of Materials in Civil Engineering **4**(3): 300-317.

Ritter, M. (1992). Timber Bridges: Design, Construction, Inspection and Maintenance. Washington, DC, USDA-FS.

Sonti, S. S., GangaRao, H.V.S. and Talakanti, D.R. (1996). Accelerated aging of wood-composite members. 41st International SAMPE Symposium, Society for the Advancement of Material and Process Engineering, Covina, CA.

Sonti, S. S. a. G., H.V.S. (1995). Strength and Stiffness Evaluations of Wood Laminates with Composite Wraps. 50th Annual Conference, Composites Institute, Cincinnati, OH.

Sonti, S. S. a. G., H.V.S. (1996). Banding Timber Crossties using Composite Fabrics for Improving their Performance. Materials for the New Technology, Washington, DC, ASCE Press.

Tascioglu, C., Goodell, B., Lopez-Anido, R., and Abdel-Magid, B. (2000). Effects of Preservative Treatment on FRP Reinforcement for Wood. 54th Annual Meeting of the Forest Products Society, Lake Tahoe, Nevada, Forest Products Society.

Tingley, D. A., Gai, C. and Giltner, E. (1997). "Testing methods to determine properties of fiber reinforced plastic panels used for reinforcing glulams." Journal of Composites for Construction **1**(4): 160-167.

Vick, C. B. (1995). Hydroxymethylated Resorcinol Coupling Agent for Enhanced Adhesion of Epoxy and Other Thermosetting Adhesives to Wood. Wood Adhesives - Proceedings No. 7296, Forest Products Society, Madision, WI.

Vick, C. B. (1997). "More durable epoxy bonds to wood with hydroxymethylated resorcinol coupling agent." Adhesives Age **40**(8): 24-29.

Wipf, T. J., Klaiber, F.W. and Funke, R.W. (1990). "Longitudinal Glued Laminated Timber Bridge Modeling." Journal of Structural Engineering **116**(4): 1121-1134.

Zwerneman, F. J., and Huhnke, R.L. (1995). "Performance of Glued-laminated Timber Panels for Bridge-Deck Replacement." Journal of Performance of Constructed Facilities: 231-241.

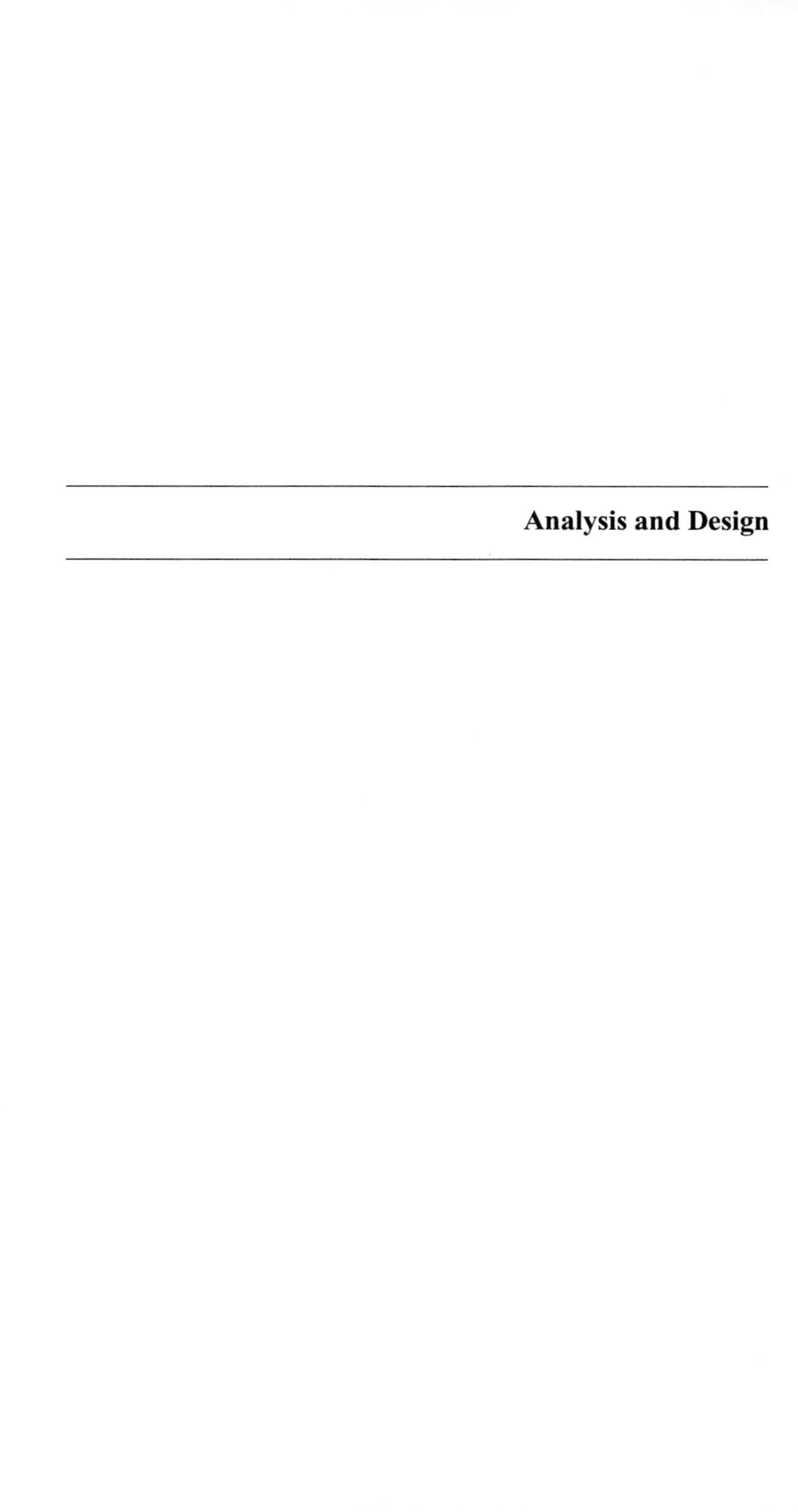

Analysis and Design

Flexural Performance of RC Elements with FRP Reinforcement

Edoardo Cosenza[1], Roberto Realfonzo[2]

Abstract

In the field of fiber reinforced polymer (FRP) reinforced concrete (RC) structures most of the available research works relate to the flexural behaviour. This paper covers the main issues affecting the flexural behaviour of FRP reinforced members: the most significant experimental results obtained from several researchers are summarized and provisions reported by the international codes are discussed. Finally, some research needs are suggested.

Introduction

Composite materials have a wide range of physical and mechanical properties: in fact they can be designed, choosing the type and quantity of fibers and matrix, and commercial products can be realized with different characteristics. In addition FRP rebars with different outer surface treatment vary the performances in terms of bar-concrete bond, that influence many behavioural aspects such as crack opening, deflections, anchorage length etc.

The main characteristics of FRP bars are the high ratio of strength to elasticity modulus and the linear behaviour up to failure; the former property determines that serviciability controls the design of most of such elements, while the latter results in a brittle behavior of the member, where concrete becomes the ductile component. Therefore, the flexural design philosophy of the upcoming guidelines (i.e., ACI 440.1R-01 and *fib* bulletin) aims at achieving concrete failure at ultimate; this can be obtained by placing an amount of FRP reinforcement higher than that producing balanced strain conditions.

These concepts are introduced in the recent design methods and codes for FRP RC structures have been developed in Japan (JSCE, 1997), Canada (CHBDC, 1996), USA (ACI 440.1R.01, 2001) and Europe (Eurocrete 1998, IStructE, 1999). These design guidelines are mainly modifications of existing steel RC codes of practice; they take into account the unconventional mechanical properties of FRP reinforcement and are often based on empirical conservative equations matching experimental results.

Flexural Behaviour and Design Code Provisions: ultimate limit state

Flexural capacity. Experimental results confirmed that the basic assumptions used for the section analysis of steel RC sections are still valid for FRP reinforced elements. Figure 1 allows to observe that plane section remains plane (i.e., Bernoulli assumption) also in the case of FRP reinforced sections.

[1] Full Professor, Dept. of Structural Analisys and Design, University of Naples Federico II, Via Claudio 21, 80125 Naples, Italy; Ph: +39-81-7683489; e-mail: cosenza@unina.it;

[2] Assistant Professor, Dept. of Structural Analisys and Design, University of Naples Federico II, Via Claudio 21, 80125 Naples, Italy; Ph: +39-81-7683485; e-mail: robrealf@unina.it

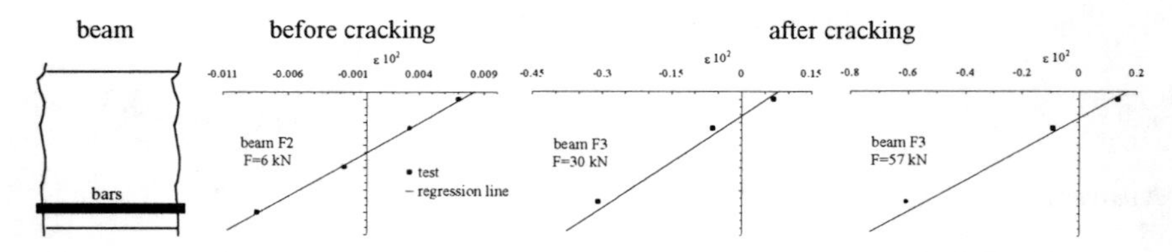

Figure 1. Stresses along the section depth (Pecce et al. 2000)

Therefore, the flexural strength can be calculated using the usual methods based on the strain compatibility, the internal force equilibrium and the controlling mode of failure. The failure of FRP reinforced flexural members can be by either concrete crushing or FRP rupture. Both failures mode are acceptable, but the concrete crushing is marginally more desirable, because at the collapse the elements show some plastic behavior.
Different codes suggest that the design properties of FRP rebars have to take into account the effect of the long-term exposure, while some experimental tests show a reduction of the tensile strenght of FRP rebars in flexure. However in general the serviceability criteria control the design and influence the load capacity.
In Figure 2, some experimental cases (Faza and Gangarao, 1991; Buyle Bodin et al., 1995; Alsayed et al., 1996; Al Salloum et al., 1996; Benmokrane et al., 1996; Benmokrane and Masmoudi, 1996; Almusallam et al., 1997; Theriault and Benmokrane, 1998; Cosenza et al., 1997) are shown, considering four criteria for evaluating the serviceability load (Pecce et al. 2001): the theoretical ultimate load F_u reduced with a coefficient equal to 1.5; the load corresponding to the deflection for a deflection/span ratio of 1 to 250; the load corresponding to the crack width equal to 0.5 mm; the load corresponding to the maximum stress in concrete equal to 45% of the cylinder compression strength or to the maximum stress in FRP equal to 25% of its ultimate tensile strength.
The specimens are defined by the the geometrical slenderness *L/H* and the ratio between the geometrical percentage of reinforcement ρ and the reinforcement percentage corresponding to the balanced failure ρ_b evaluated considering the partial safety factors for the materials.

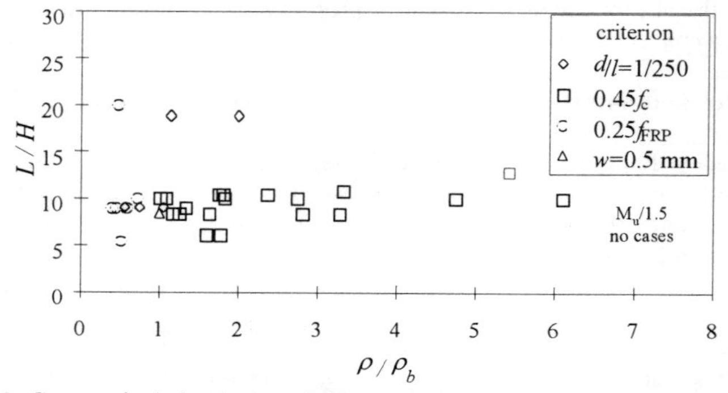

Figure 2. Geometrical slenderness *L/H* vs ratio between geometrical percentage of reinforcement ρ and the percentage of reinforcement corresponding to the balanced failure ρ_b (Pecce et al., 2001)

In 80% of the beams the serviceability load corresponds to the criterion of the materials limiting stresses (65% for the concrete stress since the reinforcement percentage is high); in the remaining beams, the deflection criterion mainly governs the design (for beams with high slenderness and low reinforcement percentage). The ultimate strength of the beams is never the criterion. These observation confirm that verification at the serviceability conditions control the load capacity in all the beams, and in most cases the stress criterion is the one which governs. The evaluation of the serviceability load has been simplified, but the conclusions are reliable

Ductility. According to a "classic" definition the structural behavior should be considered brittle, because an almost elastic response up to failure is followed by sudden collapse. However, from an engineer stand-point this is acceptable because the high deformability and wide cracks allow to see warnings of upcoming collapse; in this sense such structures provide a certain level of ductility. This does not apply to cases such as seismic applications where the redistribution is very important.
Large deformations and wide cracking normally accompany flexural failure of FRP RC elements. In addition, FRP RC elements carry higher loads as the deformations increase and warning of impending failure will be given soon after the design load is applied. In fact, failure of FRP reinforced elements is only likely to occur at deflections much greater than those expected from steel reinforced sections at steel yielding.
The relatively high deformability of FRP RC beams can serve as a warning of failure for structural purposes. It can also be utilized in situations where deformability is not a disadvantage, such as in many ground structures and elements that are likely to be subjected to impact or explosive loading.
When structural ductility is of highest importance, alternatives techniques can be used to enable FRP reinforced elements to develop a pseudo-ductile behaviour, such as increase the ultimate strain of concrete in compression with lateral confinement or using FRP hybrid rods or combinations of FRP rods with different material characteristics.

Flexural Behaviour and Design Code Provisions: serviceability limit state

The design according to the Limit State Method requires the check of the structures under serviceability loading conditions to guarantee the functionality performances. The mechanical characteristics of FRP make the serviceability conditions governing most of the design of the concrete structure reinforced with FRP bars.

Deflections. FRP RC members are expected to undergo larger deformations than steel RC members. Large deflections may cause damage to partitions or other attached members, and jeopardize the structural functionality. The overall deflection limitations are related to the importance of a given structural member (primary or secondary span), type of action (static or dynamic, permanent or 'live' loads) and the type of structure being considered (building, frame, bridge).
The approach to the problem of the deflections limitation is the same that for steel reinforced concrete structures; however the structure deformability is usually very higher when FRP bars are used, thus it becomes a critical aspect of the design.
Several researches about the short-term behaviour are still in progress, but even for this problem the effect of time had to be analysed, and a few knowledge is available.

The deflection limitations are always expressed as ratio between deflection (f) and length (L) of the beam and the restrictions are related to the cause of the limitation.
For example, in case of aesthetic and functionality conditions Eurocode 2 provisions fix a value of 1/250 for the ratio f/L. The same value is provided by the Japanese code for precast FRP reinforced concrete slabs.
Simplified methods for obtaining the design values of short- and long-term deflections are adopted in different codes of practice.
The short-term deflection of a cracked beam can be obtained by applying the standard linear-elastic approach for beams in flexure (Branson 1966, 1977) representing its effective (I_1) and reduced (I_2) second moment of area.
ACI 440.1R-01 has proposed an analytical model for the evaluation of the effective moment of inertia for FRP reinforced cracked sections; it is based on the following formula:

$$I_e = I_g \cdot \beta_d \cdot \left(\frac{M_{cr}}{M_a}\right)^3 + I_{cr} \cdot \left[1 - \left(\frac{M_{cr}}{M_a}\right)^3\right]$$

where M_{cr} is the cracking moment and M_a is the maximum bending moment under imposed service loads. The coefficient β_d is given by:

$$\beta_d = \alpha_b \cdot \left(\frac{E_f}{E_s} + 1\right)$$

where, α_b is a bond-dependent coefficient, which has been estimated to be equal to 0.5 for GFRP bars. No data are available for other FRP bar types; ACI 440.1R-01 suggests to adopt the value of 0.5 for all FRP bar types. E_f and E_s are the modulus of elasticity of FRP and steel, respectively.
The draft ISIS (Intelligent Sensing for Innovative Structures, 1999) Network Canada design manual suggests the following expression of the effective moment of inertia:

$$I_e = \frac{I_g I_{cr}}{I_{cr} + \left[1 - 0.5 \cdot \left(\frac{M_{cr}}{M_a}\right)^2 (I_g - I_{cr})\right]}$$

Due to the insignificant role of tension-stiffening in FRP reinforced concrete members and to their simple moment-curvature relation, the deflection of FRP reinforced concrete members can be calculated using a rational and general method, irrespective of the loading and boundary conditions of the beam.
The deflection due to flexure at any point along the beam at any load can be calculated as long as the true moment-curvature relationship and the bending moment variation are known.
Razaqpur et al (2000) used the moment-area method to develop closed-form deflection equations for several common types of loading and support conditions. For example, for a beam under four point bending, the maximum deflection is given by:

$$\delta_{max} = \frac{PL^3}{24EI_{cr}} \cdot \left[3 \cdot \left(\frac{a}{L}\right) - 4 \cdot \left(\frac{a}{L}\right)^3 - 8 \cdot \eta \cdot \left(\frac{L_g}{L}\right)^3\right]$$

where $\eta = (1 - I_{cr} / I_g)$, P is the concentrated load, a is the distance of the load from the support, and L_g is the length of the uncracked portion of the beam from the support.

EC2 and CEB model code have adopted the following approach to the calculation of long-term deflection δ:

$$\delta = \delta_1 \cdot \gamma + \delta_2 \cdot (1-\gamma)$$

where, the ratio between cracking and maximum bending moment under service loading is taken into account using the following equation:

$$\gamma = \beta \cdot \left(\frac{M_{cr}}{M_{max}}\right)^m$$

In the above expressions δ_1 and δ_2 are calculated assuming that the sectional second moment of area is constant along the beam for cracked and non-cracked section respectively. Values for coefficients β and m, which take into account the tension stiffening effect, recommended by two European codes of practice are given in Table 1. For FRP bars the coefficients β and m should be evaluated experimentally.

Table 1. Values for coefficients β and m (Eurocode 2)

	β	m
EC2	1	2
CEB	0.8	1

In Figure 3 a comparison between experimental values and deflections obtained by using the code models is presented.

A statistical analysis of experimental results developed in (Pecce et al. 2001) suggests that, using C-Bars™ as reinforcement, the average deflection can be evaluated using the EC2 formulation with β=1 and *m*=1.4. A factor *m*=2 will be on safe side for all the FRP bars considered and the characteristic deflection (5% fractile) can be evaluated by introducing a partial safety factor γ_v =1.75.

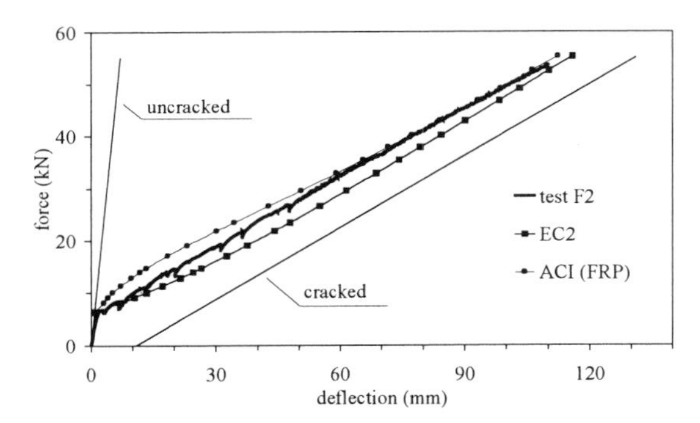

Figure 3: Experimental and theoretical evaluation of deflection. (Pecce et al. 2000)

Cracks width. This section deals with issues related to flexural cracks caused by flexural stress, without considering cracks caused by tensile stress, temperature, shrinkage, shear and torsion. The long-term effects will only be considered briefly for what concerns long-term deformations.

Major variables which affect the development and characteristics of cracks are the concrete tensile strength, concrete cover, bond between concrete and reinforcement, concrete tensile area, reinforcement ratio, position of bars and bar size. The analysis of

crack development is very important for understanding the global deformation behaviour.
The check of crack width is essential for the durability of steel reinforced concrete structures, due to the correlation with the corrosion of the steel bars. Conversely the use of FRP bars avoids corrosion, so that the penetration of the aggressive agents through the cracks is not influent; however, in FRP reinforced structures, the cracks width has to be evaluated and limited to satisfy the performances of aesthetic and psychological comfort; furthermore, cracks could question the functionality of the structure (i.e. when the water penetration has to be avoided). For this reason, it has been accepted in many codes of practice that maximum allowed crack widths for FRP reinforced concrete members may have greater values than those defined for steel ones. Maximum values for design crack widths in FRP reinforced concrete members, extracted from several design codes of practice are given in Table 2.
To evaluate the cracks width w, the ACI provisions suggests the following formulation obtained as a modification of the Gergely–Luz relationship:

$$w_{max} = \frac{2.2}{E_f} \cdot \beta \cdot k_b \cdot f_f \sqrt[3]{d_c \cdot A} \times 10^{-3} \text{ [mm]}$$

where $\beta = \frac{H - x_c}{h - x_c}$ is approximate equal to 1.2, E_f is the FRP Young modulus, f_f is the stress in the FRP reinforcement in tension, d_c is the concrete cover and A is the effective area of concrete for each bar.

Table 2. Crack width limitations for FRP and steel reinforced concrete members

Code	Exposure	w_{max} [mm]
JSCE (1997)		0.51
CSA	Interior	0.71
CSA	Exterior	0.51
Eurocrete (1998)	Tightness	0.2
Eurocrete	Aesthetics	0.3-0.5
Eurocrete	Structural integrity	0.5-1.0
ACI 440.1R-01, (2001)	Exterior	0.5
ACI 440.1R-01, (2001)	Interior	0.7

The coefficient k_b is a bond coefficient that take into account the bond properties of the rebars. For FRP bars having similar bond behaviour to steel k_b=1.0; for FRP bars having inferior bond behaviour k_b>1.0 and for FRP bars having superior bond behaviour k_b<1.0. These values indicate that bond characteristics of FRP bars can be either superior and inferior than that of steel. Further research is needed to verify the effect of bond strength on crack width. Test data should be obtained for commercially available rod. If k_b is unknown, the value 1.2 is suggested for deformed rods.
The Eurocode crack width equation is also strain based and can be adopted directly for the crack width determination of FRP RC elements. The approach adopted is also sophisticated enough to allow both for different bond characteristics (via parameter β1) and for long-term stress (via parameter β2).
Eurocode 2 (ENV, 1992) suggests calculating cracks width for steel rebars as:

$$w_{max} = \varepsilon_{FRP,m} \cdot s_{r,m} \text{ [in mm]}$$

The average strain of the bars is given by:

$$\varepsilon_{FRP,m} = \frac{\sigma_{FRP,e}}{E_{FRP}}\left[1-\beta_1\beta_2\left(\frac{\sigma_{FRP,r}}{\sigma_{FRP,e}}\right)^2\right]$$

where β_1 is a bond coefficient, β_2 is a coefficient depending on the term load (equal to 1 for short-term loads), $\sigma_{FRP,r}$ is the stress in the bars when the first crack occurs.
The average distance between cracks $s_{r,m}$, expressed in mm, is given by:

$$s_{r,m} = 50 + 0.25 \cdot k_1 k_2 \cdot \frac{D}{\rho_r}$$

where k_1 is a bond coefficient, k_2 is a coefficient equal to 1 for pure tension or 0.5 for bending stress, D is the bar diameter and ρ_r is the ratio between the total area of the tensile bars and the effective area of concrete.
The coefficient β_1 and k_1 have to be experimentally determined for FRP bars.
For the evaluation of crack widths, similar provisions for FRP bars are given by the Japanese Society of Civil Engineering and by the ACI Provisions. JSCE uses the following equation to obtain the maximum crack width:

$$w_{max} = k \cdot [4c + 0.7(c_t - D)] \cdot \left[\frac{\sigma_{FRP,e}}{E_{FRP}} + \varepsilon'\right]$$

where k is a bond coefficient, c_t is the centroidal distance between the bars and ε' is the mean strain due to the effects of creep, duration of loading and shrinkage.

Stresses in the concrete. The stress level of concrete in serviceability conditions influences the creep behaviour and the durability. If the stress in the concrete exceeds $0.4 \div 0.5\ f_c$, then much larger creep deformations could take place. Even though the centrality of the argument is clear and the need of the concrete stress limitation is still discussed in many groups of researchers, the solution is difficult due to the lack of the experimental results; the necessary tests are particularly sophisticate due to the necessity of simulating the time and environmental action. However, when the concrete stresses are lower than 40%÷50% of the strength, the creep linear model is considered reliable.
Stress limitations in concrete for satisfying serviceability limit states are provided in all codes of practice, for steel reinforced concrete members.
Since the flexural capacity of FRP RC sections was shown to depend on concrete crushing, a strong limitation for stress of concrete in FRP r/c members would lead to very conservative estimations of the capacity. Until now, no research or recommendations are available in this field.

Stresses in the FRP rebars. For the FRP materials various considerations have to be done about the stress limitation:

- glass FRP (GFRP) and aramid FRP (AFRP) composites can develop significant creep deformations when subjected to high tensile stresses;
- the FRP material generally show high durability, however glass fibers can be damaged by an alkaline solution and only the resin can assure the protection from this chemical attack; therefore microcracking in the resin due to the high level of stress can reduce this protection.
- fatigue (cycling) loading can reduce the strength of FRP structural elements together

with their bonding properties.

These considerations and the uncertainties about FRP suggest having low stresses in the bars. This requirement is usually indirectly satisfied by the design due to the high deformability of the material: the Young modulus ratio between FRP and concrete is low, so that the bars can be strongly stressed only when the reinforcement percentage is very low.

The stress limits for creep rupture suggested in the ACI provisions (ACI 440.1R-01) are summarized in table 3. This limits are comparable to the stress limitation given by the Japan Society of Civil Engineers. Limits based on fatigue capacity are more conservative. All these values have to be considered as indicative and need of the validation by more experimental data.

Table 3. Stress limits in FRP reinforcement (ACI 440.1R-01)

Fiber type	Glass FRP	Aramid FRP	Carbon FRP
Stress limit	$0.20f_u$	$0.30f_u$	$0.55f_u$

Conclusive remarks and research needs

Examining the available experimental results in case of FRP RC beams in bending, the following conclusions can be drawn: the Bernoulli hypothesis – i.e. plane sections remain plane - has been verified; models proposed by new codes appear reliable in evaluating beam deflections; for what concerns the evaluation of crack widths experimentally measured code models are probably suitable, but a reliable evaluation of some foundamental parameters which such models depend on is needed. For this purpose, further well established experimental tests have to be performed.

It appears to be crucial for the development of this technology that a standard procedure for determining the strength of FRP rebars is defined. Along with a unified methodology of testing, it will be necessary to outline standard rules for their quality control. This could allow to characterize also the bond properties of each type of rebars, lowering the safety factor affecting the bond coefficients for the calculation of deflection and crack width. As stated, serviceability checks generally govern the design; therefore, an optimisation of such parameters could have important consequences on FRP rebars applications.

The definition of safety coefficients would be another important step toward a wide application market for FRP rebars as internal concrete reinforcement. The present lack of knowledge induces researchers and engineers to strongly reduce the nominal properties of FRP rebars in order to cover the uncertainties. The formulation of a consistent safety theory would allow to apply lower reduction factors and then fully use the actual properties of such rods. This would have important influence also on the economics of this materials; in fact, high reduction factors impose strong limitations on FRP and concrete stresses in service and then require more amount of material. The definition of reliable coefficients could be important for reducing also the cost of structures reinforced with FRP rebars.

References

Abdalla,H., El-Brady, M.M., and Rizkalla, S.H., (1996). "Deflection of Concrete Slabs Reinforced with Advanced Composite Materials", *Proceedings of 2nd International*

Conference: Advanced Composite Materials in Bridges and Structures, Canadian Society for Civil Engineering, Montreal, Quebec, Canada, 201-208.

ACI 440.1R-01, (2001). "Guide for the Design and Construction of Concrete Reinforced with FRP Bars", *ACI Committee 440* , American Concrete Institute, Detroit, Michigan.

Almusallam T., Al Salloum Y., Alsayed S., and Amjad A., (1997). "Behaviour of concrete beams doubly reinforced by FRP rebars", *Proc. of 3rd Int. RILEM Symposium: Non Metallic (FRP) Reinforcement for Concrete Structures*, Sapporo Japan, 471-477

Al-Salloum, Y.A., Alsayed, S.H., Almusallam, T.H., and Amjad, M.A., (1996). "Evaluation of Service Load Deflection for Beams Reinforced by GFRP Bars", *Proc. of 2nd Int. Conference: Advanced Composite Materials in Bridges and Structures*, Canadian Society for Civil Engineering, Montreal, Quebec, Canada, 165-172

Alsayed, S.H., AL-Salloum, Y.A., Almusallam, T.H., and Amjad, M.A., (1996). "Evaluation of Shear Stresses in Concrete Beams Reinforced by FRP Bars", *Proc. of 2nd Int. Conference: Advanced Composite Materials in Bridges and Structures*, Canadian Society for Civil Engineering, Montreal, Quebec, Canada, 173-179

Benmokrane, B., and Masmudi, R., (1996). "FRP C-Bars as Reinforcing Rod for Concrete", *Proc. of 2nd Int. Conf.: Advanced Composite Materials in Bridges and Structures"*, Canadian Soc. for Civil Engineering, Montreal, Quebec, Canada, 181-185.

Benmokrane, B., Challal, O., and Masmoudi, R., (1996). "Flexural response of Concrete Beams reinforced with FRP Reinforced Bars", *ACI Structural J.*, Vol.93, No.1, 46-55.

Branson, D.E., (1966). "Deflections of Reinforced Concrete Flexural Members", *ACI Committee Report, ACI-435.2R-66,* (re-approved 1989), Detroit.

Branson, D.E., (1977). "Deformation of Concrete Structures", *McGraw-Hill,* New York, pp. 537.

Buyle, Bodin, F., Benhouna, M., and Convain, M., (1995). "Flexural Behaviour of Jitec-FRP Reinforce Beams", *Proceedings of 2nd International RILEM Symposium: Non Metallic (FRP) Reinforcement for Concrete Structures*, E&F Spoon, 235-242

CEB, (1993). "CEB-FIP Model Code 1990*", CEB Bulletin d'Information 213-214,* Comite International du Beton, Thomas Telford Service Ltd., London, pp.437.

CHBDC, "*Canadian Highway Bridge Design Code, Section 16: Fibre Reinforced Structures*", Final Draft, July 1996, pp.25.

Cosenza, E., Greco, C., Manfredi, G., and Pecce, M., (1997). "Flexural Behaviour of concrete beams reinforced with fibre reinforced plastic (FRP) rebars*", Proc. of 3nd Int. RILEM Symp.: Non Metallic (FRP) Reinf. for Concrete Structures*, Sapporo, Japan.

CSA, (1996). "S806-97 - Design and construction of building components with fibre reinforced plastics (Draft 2)", *Draft issue of CSA S806*, CSA Technical Committee on

FRP Components and Reinforcing Materials for Buildings, Canadian Standards Association, Canada, pp. 89.

Duranovic, N., Pilakoutas, K., and Waldron, P., (1997). "Tests on concrete beams reinforced with glass fibre reinforced plastic bars", *Proc. of 3rd Int. RILEM Symp.: Non Metallic (FRP) Reinforcement for Concrete Structures*, Sapporo, Japan, 479-486.

ENV 1992-1-1, (1992). "Eurocode 2 – Design of Concrete Structures, Part 1-6:General Rules and Rules for Buildings", European Prestandard, *European Committee for Standardisation*, pp.253.

Eurocrete (1998). "Modifications to NS3473 when using fiber reinforced plastic (FRP) reinforcement." Cement and Concrete, Sintef, Trondheim, Norway.

Faza, S.S., GangaRao, H.V.S., (1991). "Bending Response of Beams Reinforced with FRP Rebars for Varying Concrete Strengths", *Proc. of Advanced Composite Materials in Civil Engineering Structures*, American Society of Civil Engineers, 262-270.

Faza, S.S., GangaRao, H.V.S., (1991). "Bending Response of Beams Reinforced with FRP Rebars for Varying Concrete Strengths", *Proc. of Advanced Composite Materials in Civil Engineering Structures*, American Society of Civil Engineers, 262-270.

ISIS, (1999), "Design Guidelines for FRP", draft, Private Communication, ISIS Canada.

IstructE, (1999). "Interim Guidance on the Design of Reinforced Concrete Structures Using Fibre Composite Reinforcement", *The Institution of Structural Engineers*, pp.116.

JSCE, (1997). "Recommendation for design and construction of concrete structures using continuous fibre reinforcing materials", *Japan Society of Civil Engineering*, Concrete Engineering Series N°23, Tokyo, Japan.

Nanni, A., (1993). "Flexural Behaviour and Design of RC Members Using FRP Reinforcement", *Journal of Structural Engineering*, ASCE, Vol.119, No.11, 3334-3359.

Pecce, M., Manfredi G., and Cosenza E., (2000). "Experimental Response and Code Models of GFRP RC Beams in Bending", *Journal of Composites for Construction*, ASCE, Vol.4, No.4, 182-190.

Pecce, M., Manfredi G., and Cosenza E., (2001). "A probabilistic assessment of deflections in FRP RC beams", *Proceedings of FRPRCS-5 Conference*, Cambridge, UK, 16-18 July, 2001.

Razaqpur, A.G., Svecova, D., Cheung M.S. (2000), "A Rational Method for CalculatiNg Deflection of FRP Reinforced Concrete Beams," *ACI Structural Journal*, V.7, No. 1, Jan.-Feb, pp.175-184.

Theriault, M., and Benmokrane, B., (1998). "Effects of FRP Reinforcement Ratio and Concrete Strength on the flexural Behaviour of Concrete Beams", *Journal of Composites for Construction*, ASCE, Vol. 2, No. 1, February 1998, 7-16.

Shear of FRP RC: a Review of the State-of-the-Art

Kypros Pilakoutas[1], Maurizio Guadagnini[2]

Abstract

The work reported in this paper reviews modifications to existing codes to incorporate Fibre Reinforced Polymer (FRP) reinforcement in reinforced concrete (RC) structures focusing on shear resistance. Various approaches are examined ranging from the first recommendations proposed by the Japan Society of Civil Engineers to the latest work published by the British Institution of Structural Engineers and the American Concrete Institute. The inadequacies of these approaches are exposed and a new approach developed at the University of Sheffield based on experimental work is presented.

Introduction

The shear capacity of RC elements still remains an issue of scientific debate and even the behaviour of conventionally RC structures is not fully understood in all circumstances.

Various theories have been developed since the simple truss model analogy was first proposed by Ritter in 1899 but, because of the sudden nature of shear failures and the difficulties in formulating reliable mathematical models, code writers have tended to concentrate on predicting shear performance on an empirical basis. Unfortunately this trend does not allow an easy extension of existing design rules to cover changes in practice and extensive new testing often needs to be carried out before new materials or systems are adopted. This was the case with high-strength steel and concrete, shear reinforcement systems, such as studrail and shearband, and again now with the introduction of advanced composite materials in the construction industry.

Work undertaken by the Japan Society of Civil Engineers, the Canadian Standards Association, the American Concrete Institute (ACI) and, in Europe (by the EUROCRETE project) has lead to recommendations to incorporate non-ferrous reinforcement in existing codes of practice. All these proposals are based on the same philosophy by keeping the same equations suggested by the existing codes of practice for steel and adjusting them according to the "strain approach" firstly introduced and adopted by the Japanese (Nagasaka T. et al. 1993).

During the EUROCRETE project, Clarke adopted this concept and proposed modifications to the British Standard, Eurocode and the Norwegian code (Clarke J. L. et al. 1996). The ACI Committee 440 (ACI Committee 440 2001), adopted the basic equations, but not the spirit, of the strain approach and proposes similar modifications for the ACI code.

In the following, the concepts and philosophies adopted by the various codes and researchers are considered and the modified equations reported and commented.

[1] Centre for Cement and Concrete, Dept. of Civil and Struct. Eng., University of Sheffield
[2] Centre for Cement and Concrete, Dept. of Civil and Struct. Eng., University of Sheffield

Main philosophies for incorporating FRP reinforcement

Due to the different mechanical properties of FRP reinforcement and especially their inherent lack of plasticity, FRP RC structures have a peculiar behaviour generally governed by brittle and undesirable modes of failure (Figure 1). Based on these considerations, it appears evident that both construction techniques and design philosophy need to be carefully reassessed (Pilakoutas K. 2000) when dealing with these new materials. However, such a new philosophy has not yet been adopted by the codes.

As far as shear performance is concerned, all the shear-resisting mechanisms provided by conventional steel RC elements such as aggregate interlock, tooth bending and dowel action are expected to decrease when using FRP reinforcement due to the reduced neutral axis depth and expected larger tensile strains. The mechanical properties of the longitudinal reinforcement have been demonstrated experimentally to influence the amount of concrete shear resistance as well as the overall deflections. Cracks in the tensile region are anticipated to be larger than for steel RC and, hence, the aggregate interlock effect is predicted to be reduced in FRP RC. Furthermore, not much dowel strength is expected from the more flexible and anisotropic FRP materials.

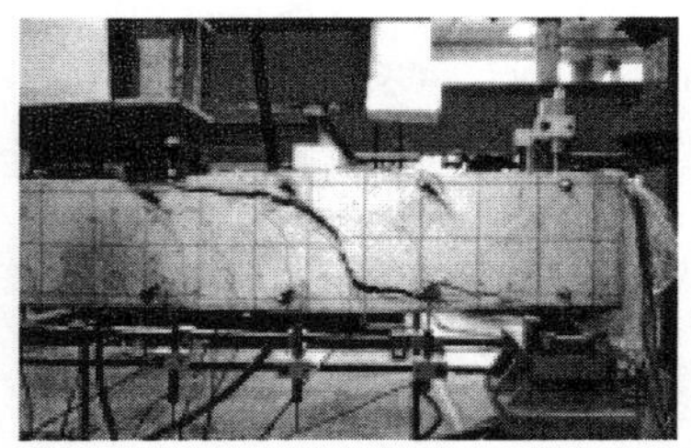

Figure 1. Shear failure of a FRP RC without shear reinforcement and concrete crushing failure

As far as shear reinforcement is concerned, due to their anisotropic properties, FRP links not only need large strains to develop their full strength but can not develop their full tensile potential and failure is expected at the corner anchorages.

The main aim of researchers working in this field has been to provide simple design equations. Thus, the trend so far has been the proposal of modification factors for inclusion in existing predictive equations. This also has the advantage that the code committees are more likely to accept these modifications rather than adopt fundamental changes.

The strain approach

The principle of this approach is that (assuming perfect bond) the concrete section experiences forces and strains which are independent of the type of reinforcement utilised. Hence, if the design using FRP reinforcement maintains the same strain as conventional steel ($\varepsilon_{FRP} = \varepsilon_s$) when the same design forces are developed ($F_{FRP} = F_s$), then that design will by definition should lead to the same safe result. Hence, according to this approach an equivalent area of steel (A_e) is introduced to evaluate the concrete shear resistance by multiplying the actual area of FRP reinforcement (A_{FRP}) by the modular ratio of FRP to that of steel, as shown below.

$$F_{FRP} = \varepsilon_{FRP} \cdot E_{FRP} \cdot A_{FRP} = \varepsilon_S \cdot E_S \cdot A_S = F_S \tag{1}$$

$$A_e = A_{FRP} \cdot \frac{E_{FRP}}{E_S} \tag{2}$$

When the applied shear stress exceeds the design concrete shear strength, shear reinforcement is required. The provision of such reinforcement is also subjected to strain restrictions when using FRP reinforcement. The strain approach (Clarke J.L. et al. 1996) maintains the strain limit of 0.0025, at which conventional steel bars are considered to yield. However, this strain limit implies that FRP links will only be stressed to a fraction of their potential.

The stress approach

In this approach developed for research purposes at Sheffield University (El Ghandour A. W. et al. 1998) the forces derived from the strength of the materials, F_s and F_{FRP}, are considered to be the same, but this time no restriction is imposed on the value of the strain (i.e. $\varepsilon_{FRP} \neq \varepsilon_s$), hence, the effective area becomes:

$$A_e = A_{FRP} \cdot \frac{\sigma_{FRP}}{\sigma_S} \tag{3}$$

This approach gives a shear resistance which is normally higher than the experimental one and is used only for research purposes as an upper bound approach and has never been incorporated into design recommendations.

Other contributions

An alternative approach was proposed by Alsayed et al. (1997) on the basis of experimental results to modify the ACI code equation. Various specimens with different kinds of longitudinal and shear reinforcement were tested and a different reduction in shear capacity was noticed for the specimens with both flexural and shear FRP reinforcement or with a mixture of FRP and steel reinforcement. They proposed to consider half of the expected concrete shear resistance (V_c) if flexural FRP reinforcement was used and half of the expected shear resistance offered by the vertical reinforcement (V_s) if FRP links were used. The equations proposed are given as in Eq. 4 – 6. Clearly this approach is too simplistic, since it does not recognise the type or amount of reinforcement.

$$V = \tfrac{1}{2} V_c + V_s \tag{4}$$

$$V = V_c + \frac{1}{2} V_s \tag{5}$$

$$V = \frac{1}{2} V_c + \frac{1}{2} V_s \tag{6}$$

Modifications to existing codes

In the following, the modifications to existing codes proposed by The British institution of Structural Engineers and the American Concrete Institute are discussed.

British Institution of Structural Engineers

The Institution of Structural Engineers adopted much of the initial work carried out as part of the EUROCRETE Project and in 1999 published an "Interim guidance on the design of reinforced concrete structures using fibre composite reinforcement". This guide is in the form of suggested changes to the British design Codes BS8110: "Structural use of concrete Parts 1 & 2" and BS5400: Part 4 "Code of practice for the design of concrete bridges", but the same philosophy may apply to the EC 2 equations.

The suggested modifications are in line with the strain approach and propose the use of the modification factor given in Eq. 2. Hence, the modified BS8110 equation for the concrete shear strength, v_c, is given in Eq. 7 (for notation see BS8110, 1985).

$$v_c = 0.79 \cdot \left(\frac{100}{b_w \cdot d} \cdot A_s \cdot \frac{E_r}{200} \right)^{\frac{1}{3}} \cdot \left(\frac{400}{d} \right)^{\frac{1}{4}} \cdot \left(\frac{f_{cu}}{25} \right)^{\frac{1}{3}} \tag{7}$$

As far as the shear strength resisted by the vertical shear reinforcement is concerned, this can be evaluated using the usual formulation derived by the strut and tie theory as reported for steel, but controlling the maximum strain developed in the vertical bars, as given by the strain approach.

Following these recommendations, the shear strength offered by the web reinforcement is given in Eq. 8 (E_{rv} is the elastic modulus of shear FRP reinforcement).

$$v_s = \frac{0.0025 E_{rv} \cdot A_{sv}}{b_w \cdot s} \tag{8}$$

The minimum amount of reinforcement has been assumed not to be a function of the type of reinforcement and, hence, no special provision is given.

American Concrete Institute

The ACI Committee 440 (ACI 440 2001), adopted the result, but not the spirit, of the strain approach and, after various revisions, proposes to modify the concrete shear resistance according to Eq. 9.

$$V_{c,f} = \frac{\rho_f E_{FRP}}{\rho_s E_S} V_c \tag{9}$$

where ρ_s can be taken as $0.375\rho_b$ (balanced flexural reinforcement ratio) and Eq. 9 can be rewritten in the form

$$V_{c,f} = \frac{\rho_f E_{FRP}}{75\beta_1 f_c'} V_c \tag{10}$$

where f'_c is the specified compressive strength of concrete

The value of V_c can be evaluated according to ACI 318-99 using Eq. 11 or the more sophisticated Eq. 12.

$$V_c = \frac{1}{6}\sqrt{f_c'}\, b_w d \tag{11}$$

$$V_c = \left(0.16\sqrt{f_c'} + 17\rho_w \frac{V_u d}{M_u}\right) b_w d \le 0.29\sqrt{f_c'}\, b_w d \tag{12}$$

The contribution of FRP stirrups is taken into account using the same method as for steel stirrups, but adopting a value of FRP tensile strength taken as the smallest of $0.002\ E_f$ and the strength f_{fb} of the bent portion of FRP stirrups according to the design recommendations first proposed by the Japan Society of Civil Engineers.

The recommended minimum amount of FRP shear reinforcement is given in Eq. 13.

$$A_{fv,\min} = 0.345 \frac{b_w s}{f_{fv}} \tag{13}$$

where f_{fv} is the shear reinforcement at ultimate for use in design (Eq. 14).

$$f_{fv} = 0.002 E_{FRP} \le f_{fb} \tag{14}$$

Evaluation of the code modifications and a new approach

Figure 2 shows the concrete shear resistance for a predefined RC section with a target concrete strength of 45 MPa versus longitudinal reinforcement ratio as predicted by the different codes. It is clear that not all codes take into consideration the effect of reinforcement and, hence, the strain approach can not be used directly. To compensate for that, different modifications are proposed by researchers at Sheffield for each code (El Ghandour A. W. et al. 1999, Guadagnini M. et al. 1999).

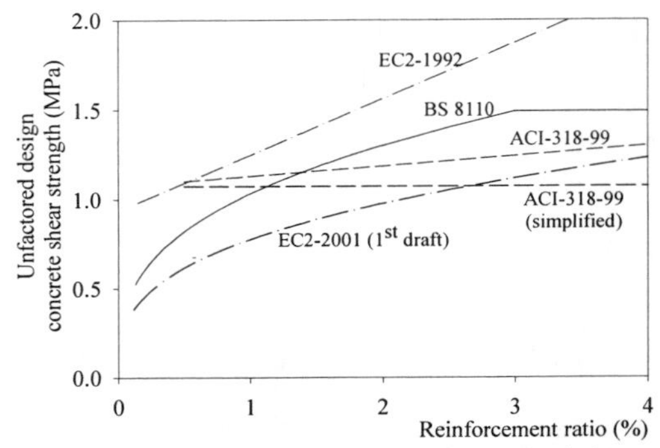

Figure 2. Concrete shear resistance against flexural reinforcement ratio

The modifications proposed, referred to as the Sheffield approach, are based on a stiffness approach along the lines of the strain approach, but allow larger strains to be developed both in the flexural and shear reinforcement. The modifications to the following codes are proposed when deriving the concrete shear resistance:

BS 8110 $$A_e = A_{FRP} \cdot \frac{E_{FRP}}{E_S} \cdot \phi_\varepsilon \qquad (13)$$

ACI-318-99 $$V_C = V_C \cdot \left(\frac{E_{FRP}}{E_S} \cdot \phi_\varepsilon \right)^{1/3} \qquad (14)$$

EC 2 $$A_e = A_{FRP} \cdot \frac{E_{FRP}}{E_S} \qquad V_C = V_C \cdot \left(\frac{E_{FRP}}{E_S} \cdot \phi_\varepsilon \right)^{1/3} \qquad (15)$$

where ϕ_ε represents the ratio between the maximum strain allowed in the FRP reinforcement and the yield strain of steel.

The design strength of FRP shear reinforcement is assumed to be $0.0025\ E_{FRP}\ \phi_\varepsilon$.

Comparison

The modified codes of practice were used to predict the shear capacity of sets of beams taken from the literature (Vollum R. 1993, Zhao W. et al. 1995, Alsayed, S. H. 1997, Duranovic N. et al. 1997, Guadagnini M. et al. 2001b) and results are shown in Figure 3. Different types of bars were used having different fibre composition. The approach applied to the BS and ACI equations is the one suggested by their respective committees and the same modifications where applied to the Eurocode equations as suggested in the Eurocrete report (Clarke J. et al. 1996).

The graphs, which were derived without the use of safety factors, show that the proposals are in general conservative and hence, provide a suitable starting point.

However their large scatter and high degree of conservatism can become unacceptable for large structures, where the amount of reinforcement and the price of the whole structure become relevant.

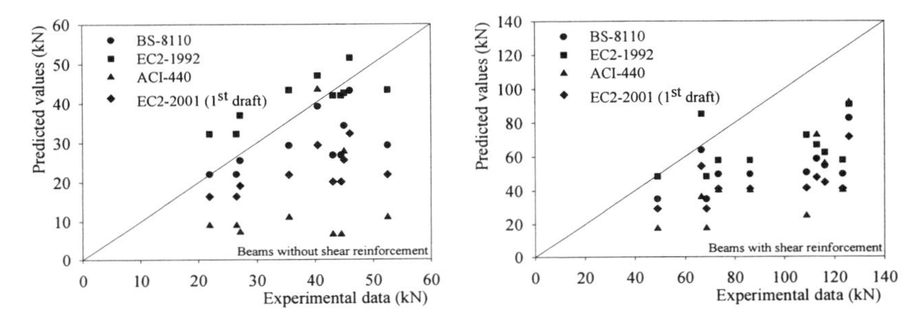

Figure 3. Comparison of analytical predictions using different modified code equations

Experimental tests carried out at the University of Sheffield provided evidence that the stresses in the FRP reinforcement at failure are usually much higher than suggested by the strain approach reaching average values of at least 80% higher than assumed. The ϕ_ε factor, considered equal to 1.8 for calculation purposes, was proposed based on this evidence and similar observations from punching shear experiments at Sheffield (El-Ghandour et al. 1998). More testing has been performed on beams with and without shear reinforcement (Guadagnini et al. 2001a and 2001b) and previous findings were confirmed reaching maximum strains in the shear links up to 300% higher than the one adopted by the strain approach.

Figure 4 and 5 show the results from a recent experiment. Strain on the flexural reinforcement exceeded 8000 µε whilst the strain on the shear link was just less than 8000 µε or 3 times the value allowed by the strain approach. Since the strain was recorded on externally applied reinforcement, this value can be considered to be conservative since such reinforcement is more likely to debond. In the same graph it can be seen that the strain on the corner (C59) of the link is lower and in fact failure of links is always observed at the corner. This supports the argument that the stress on the corner of the link should also be limited.

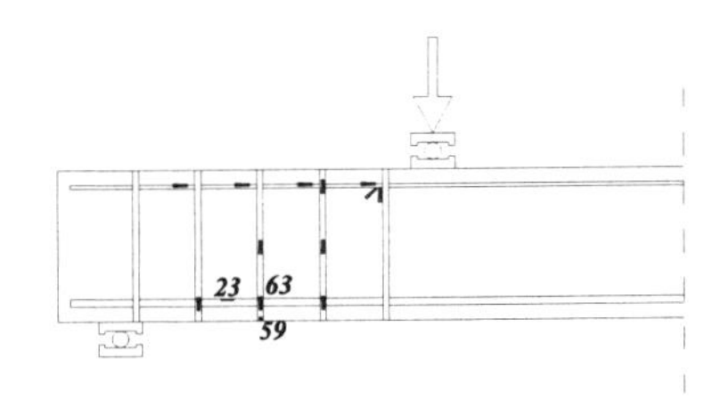

Figure 4. Location and numbering of strain-gauges (left) and shear failure for GB 44

Figure 6 shows the analytical predictions of the Sheffield Approach for the same sets of beams. It is evident that the results can be predicted with less scatter and a better accuracy.

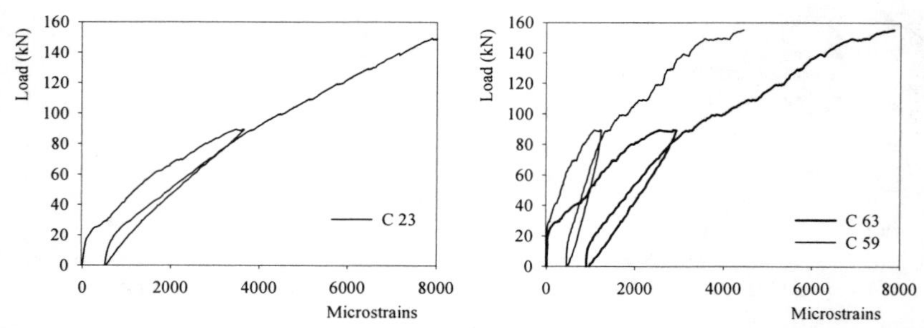

Figure 5. Maximum strains measured in the GFRP flexural reinforcement (C 23) and in the CFRP shear links (C59 and C 63)

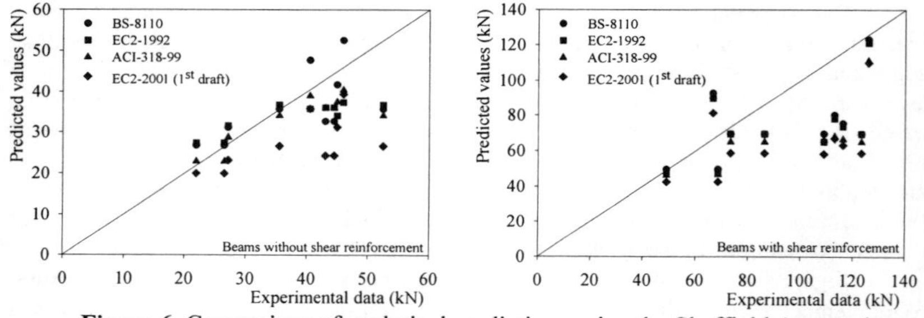

Figure 6. Comparison of analytical predictions using the Sheffield Approach

Conclusions

The main approach adopted by the codes of practice when dealing with FRP reinforcement is the strain approach, where the strains in the FRP are kept at the same level as for steel RC. This approach is shown to be conservative.

Not all codes have shear prediction equations that are suitable for modifications based on the strain approach and, hence, even more conservative results are given when the equations are modified inappropriately.

The proposed Sheffield approach can predict the shear behaviour in a more accurate and less conservative manner.

Further research is needed in this field so to improve the predictive models by making them less reliant on empirical data and reduce the conservatism.

Acknowledgement

The authors wish to acknowledge the European Commission for funding the EU TMR Network "ConFibreCrete".

References

ACI Committee 318 (1995), *Building Code Requirements for Reinforced Concrete*, ACI-318-99, Detroit, 1995.

ACI Committee 440 (2001), *Guide for the Design and Construction of Concrete Reinforced with FRP Bars*, ACI 440.1R-01, ACI, Farmington Hills, MI, USA

Alsayed, S. H., Al-Salloum Y.A., Almusallam T.H. (1997), "Shear Design for Beams Reinforced by GFRP Bars", Non-Metallic (FRP) Reinforcement for Concrete Structures, Proceedings of the Third International Symposium, Vol. 2, pp. 285-292

British Standards Institution (1985), *BS 8110. Structural use of Concrete. Part 1: 1985. Code of practice for design and construction*, BSI

Clarke J. L., O'Regan D. P. and Thirugnanenedran C. (1996), "EUROCRETE Project, Modification of Design Rules to Incorporate Non-Ferrous Reinforcement", EUROCRETE Project, Sir William Halcrow & Partners, London

Duranovic N., Pilakoutas K. and Waldron P. (1997), "Tests on Concrete Beams Reinforced with Glass Fibre Reinforced Plastic Bars", Non-Metallic (FRP) Reinforcement for Concrete Structures, Proceedings of the Third International Symposium, Vol. 2, pp. 479-486

El-Ghandour, A. W., Pilakoutas, K. and Waldron, P. (1998), "Use of FRP Reinforcement for Concrete Plate Elements", Proceedings of the International Conference on Advanced Composites ICAC 98, Hurghada, Egypt

El-Ghandour A.W., Pilakoutas K. and Waldron P. (1999),"New Approach for the Punching Shear Capacity Prediction of FRP RC Flat Slabs", approved for publication in the American Concrete Institute's Special Publication for the Fourth International Symposium on Fiber Reinforced Polymer for Reinforced Concrete Structures (FRPRCS-4)

ENV 1992-1-1 (1992), *Eurocode 2. Design of Concrete Structures, Part 1, General Rules and Rules for Buildings*, European Committee for Standardisation

Gdoutos E. E., Pilakoutas K., Rodopoulos C.A. (2000), *Failure Analysis of Industrial Composite Materials, McGraw-Hill*

Guadagnini M., Pilakoutas K., Waldron P. (1999), "Shear Design for FRP RC Elements", Conference proceedings of the Fourth International Symposium on Fiber Reinforced Polymer for Reinforced Concrete Structures (FRPRCS-4)

Guadagnini M., Pilakoutas K., Waldron P. (2001a), "Investigation on Shear Carrying Mechanisms in FRP RC Beams ", approved for publication in the Proceedings of the

Fifth International Symposium on Fiber Reinforced Polymer for Reinforced Concrete Structures (FRPRCS-5)

Guadagnini M., Pilakoutas K., Waldron P. (2001b), *Experimental Investigation on Shear Carrying Mechanisms in FRP RC Beams: Internal Report*, Centre for Cement and Concrete, The University of Sheffield, UK

Institution of Structural Engineers (1999), "'Interim guidance on the design of reinforced concrete structures using fibre composite reinforcement", Published by SETO Ltd, 116 pages.

Nagasaka T., Fukuyama H. and Tanigaki M. (1993), "Shear Performance of Concrete Beams Reinforced with FRP Stirrups", pp 789-811, Fiber Reinforced-Plastic Reinforcement for Concrete Structures (Ed. A. Nanni and C. W. Dolan), American Concrete Institute SP-138

Updated Design Philosophy For Structural Strengthening Of Concrete Structures With Fiber Reinforced Polymers (FRP)

Paul L. Kelley, Michael L. Brainerd, and Milan Vatovec[1]

Abstract

This paper discusses a general design philosophy for strengthening of concrete structures with FRP, which is in alignment with the proposed ACI 440F recommendations. The goal of the paper is to convey the main FRP design issues and several cautionary remarks to the engineering audience in the US. The paper focuses on several key topics: minimum required pre-strengthened strength criterion to prevent collapse if the CFRP system is compromised due to uncontrollable events (fire, impact, etc.); limits on strength enhancement to maintain "pseudo-ductile" behavior; and appropriate ϕ ("reliability") factors and limits on FRP design strength.

Introduction

Promised cost reductions in composite materials, long popular in the aerospace industry, are the catalyst of innovative uses for Fiber Reinforced Polymers (FRP) in civil engineering applications. One popular innovative concept is the use of FRP for external strengthening of existing reinforced concrete structures. Before the increased availability of FRP, external strengthening of concrete was limited to bonded steel plates, which in many instances were not feasible because of weight and splice difficulties. Figure 1 illustrates the simple constructability of external FRP flexural strengthening.

The introduction of FRP to the US Civil Engineering community has been driven by several European, Japanese, and American manufacturing firms. Each of the manufacturing groups and designers offered different design philosophies. This created difficulties with providing competitive systems and contractual bidding. The American Concrete Institute (ACI) Committee 440-F recently completed a document (ACI 440-F-01, 2001) which consolidates the differing design recommendations for the use of FRP as concrete post-strengthening reinforcement.

This paper presents the writers' perspectives on several key issues related to the ACI 440F design methodology, including the following:

- Threshold strength of structures: minimum required pre-strengthened strength criterion to prevent collapse if the CFRP system is compromised due to uncontrollable events (fire, vandalism, impact, etc.)

[1] Authors are Principal, Principal, and Senior Staff Engineer, respectively, at Simpson Gumpertz & Heger, Inc. Engineering Consulting Firm in Arlington, Massachusetts; phone 781-643-2000

- Minimum ratio to ensure "pseudo-ductile" behavior: limits on strength enhancement to maintain "pseudo-ductile" behavior.
- Appropriate ϕ ("reliability") factors and limits on FRP design strength.

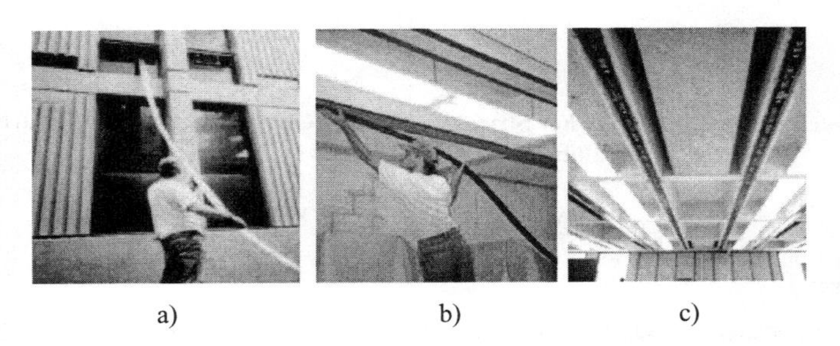

a) b) c)

Figure 1. a) Lightweight CFRP strips loaded into building by hand; steel plate strengthening would probably require a crane. b) Installation of CFRP strips utilizes a thixotropic adhesive paste to temporarily suspend and later bond the strips to the existing concrete structure. Steel plate bonding would require a larger installing crew and temporary suspension bolts. c) Completed flexural strengthening with CFRP strips does not change the structure's appearance. This work was later painted to match the ceiling.

Threshold Strength Of Structures

A major obstacle in the development of externally bonded strengthening design concepts is related to the potential loss of FRP effectiveness due to uncontrollable events. The direct risk is damage due to fire. Increased temperatures will cause the fixing adhesive to flow plastically causing a loss of load transfer to the FRP. Typically, the critical temperatures for the epoxy are in the range of 120°F to 200°F (approximately 50 °C to 95°C). Traditional fireproofing materials and systems cannot protect the adhesives in these low temperature ranges.

In recognition of the temperature risks, members to be strengthened with FRP should have an unstrengthened or pre-strengthened capacity that provides a positive factor of safety against collapse without FRP.

For rare load events like fire, typical load factors (1.4DL, 1.7LL) can be justifiably lowered to deliver a probability of failure in the expected short term period between fire and repair (<1 year) similar to the probability of failure commonly provided for a 50-year exposure to typical variation in dead and live loads. After long deliberation, ACI 440 F determined that the nominal strength of the unstrengthened system must resist the following:

$$\phi S_n \quad 1.2\, S_D + 0.85\, S_L + \ldots, \text{ where}$$

$\phi S_n =$ Nominal strength of an unstrengthened member
S_D, S_L = Demand due to dead and live load.

There has been discussion within ACI 440F about accounting for material degradation due to fire (in accordance with procedures defined in ACI 216R) to arrive at ϕS_n. The authors oppose this; an ACI 216R evaluation is not familiar to design engineers and these issues and risks are better and more traditionally accounted for by appropriate load factors and reliability factors.

In instances where CFRP loss is more likely than in fire, or where detection of CFRP loss might require longer than 1 year, higher load factors are prudent. Therefore, a mathematical model that relates the probability of accident, time to discovery, non-transient dead and live load variability, and transient live load variability to current ACI 50-year failure probability is needed to justify the use, or allow for modification of ACI 440 F load factors.

While the unstrengthened threshold concept can restrict the strengthening range of the FRP strengthening technique, the following must be considered:

- Safety factors are necessary to account for the probability of the coincidence of 1) unintended load, 2) understrengthed material, 3) unintended construction influences, and 4) unintended environmental influences. The entire reserve of strength implied by the "safety factor" cannot be consumed by one demand; an FRP-strengthened structure, compromised by FRP loss in a fire, can also be coincidentally overloaded and compromised by hidden original-construction anomalies.
- A load increase from 1.2 DL+ 0.85LL to 1.4DL+1.7LL is still significant. This limit still offers opportunities for meaningful strengthening. For example, a typical office slab with a dead load of 125 psf and an original design live load of 50 psf can be strengthened within this limit to accept a new live load of 130 psf.
- Until fire-protection methods are developed that will limit the CFRP exposure temperatures to within the150° F to 200°F range, a design-based protection remains the fire-protection rationale.
- Experience with FRP is limited; many performance issues are still incompletely tested and many environmental exposures have not yet passed the test of time. Caution is prudent.

As will be demonstrated below, other strengthening limits to ensure ductility, bond, anchorage, and strain compatibility provide similar restrictions.

Minimum Reinforcement Ratio To Ensure "Pseudo-Ductile" Behavior

Conventionally reinforced concrete members are ductile due to the presence of ACI-prescribed and controlled steel reinforcement ratio. This feature is especially pronounced and utilized in conventionally reinforced concrete flexural members; ACI prescribes limits to the amount of reinforcement to promote a ductile tension failure instead of a brittle compression failure. These limits are expressed in terms of the ratio of the area of reinforcement to the area of concrete ($\rho_{max} = A_s/bd \leq 0.75\ \rho_{bal}$). This requirement provides that a flexural element, if overloaded, will exhibit readily visible deformation as a warning prior to collapse.

One common measure of ductility is the "ductility index," which is defined as the ratio of reinforcing-steel strain at limit state (failure of the element) to the yield strain of the steel. For example, a simple steel rod loaded in tension has a ductility index of 25 (0.05 rupture strain/0.002 yield strain).

FRP is not a ductile material. Most FRP composites exhibit nearly linear stress-strain behavior when loaded to failure in tension. However, experimental tests show that steel-reinforced concrete flexural members strengthened with FRP can exhibit ductile behavior when loaded to failure. This behavior is known as "pseudo-ductile" behavior.

Idealized moment-rotation curves for FRP-enhanced, steel-reinforced concrete beams are multi-linear as shown in Figure 2. The portion of the curve between steel yielding and failure has an upward slope, compared to the characteristic horizontal slope for unstrengthened beams. The slope of the upper portion of the moment-rotation curve is a result of the linear behavior of FRP and depends upon the ratio of FRP reinforcement to the steel reinforcement.

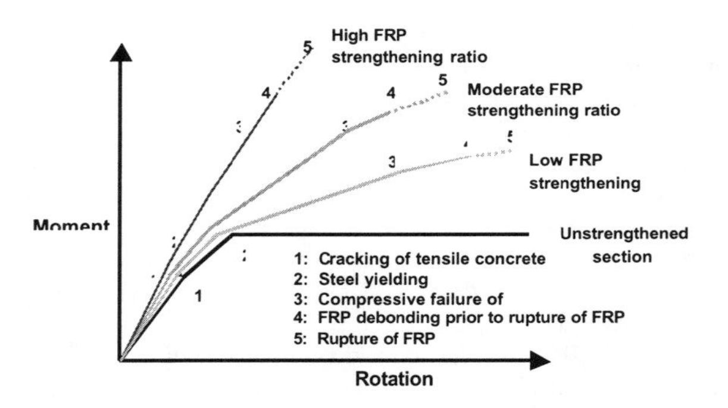

Figure 2. Moment Curvature Curves for Beams with and without FRP

The designer must evaluate whether the strengthened member possesses sufficient ductility. One method used to evaluate ductility of a flexural member is to plot an idealized moment-rotation curve for the post-strengthened member and compare it to the idealized curve for the unstrengthened member geometry reinforced to $0.75\ \rho_{bal}$. The level of ductility of the strengthened member is acceptable if the ratio of rotation at failure to rotation at steel yielding for the strengthened member ("ductility index") is equal to or greater than that of the unstrenghtened member reinforced to $0.75\rho_{bal}$. Some design-rule developers have suggested addressing ductility below this level with revised strength reduction factors.

To assess ductile performance, the designer must examine the behavior of the strengthened member at failure. This examination requires not only traditional evaluations like examination of limits associated with the rupture of the steel or FRP, or the crushing of the concrete, but bond-related FRP-system failure modes should also be evaluated. Bond-related failure modes such as delamination are typically sudden; delamination is related to the strain incompatibility of the reinforced concrete substrate and the FRP composite. The designer must understand the behavior of the member prior to and after the strengthening, and should ensure that the member contains enough steel to provide ductile behavior at or near the ultimate loads.

ACI 440-F (1999) states that adequate ductility of flexural members is achieved if the strain in the mild steel reinforcement at failure (regardless of the failure mode) is at least 2.5 times the strain in the steel at yield (0.002). For sections that do not meet this requirement, low ductility must be compensated with a higher reserve of strength. For flexure, this additional reserve strength is achieved by applying an overall strength reduction factor (ϕ) of 0.7 for brittle sections, as opposed to 0.9 for ductile sections. Linear-interpolation values between 0.7 and 0.9 are to be used for steel strains at ultimate limit state (failure) between 0.002 and 0.005.

Design Philosophy

Background

Concrete reinforced with steel is a composite material. The behavior and performance are determined by the composite action of the two materials. Each material contributes in a distinct way to the overall behavior of the composite system. The steel reinforcement provides tensile strength and ductility to the system. The concrete provides compressive strength, spatial form, and stability. Together, concrete and steel form a strong and ductile structural system.

In the United States, ACI 318 provides design requirements for reinforced concrete. In general, the ultimate strength design approach (USD) is used. In USD, the ϕ factor adjusted nominal strength of the reinforced-concrete composite is compared to the structural effects of factored loads imposed on the element.

For concrete elements reinforced with mild steel, ACI prescribes ϕ reduction factors to be applied to the composite strength of the element provided by the concrete and steel. This value varies with the type of element and with the failure mode under consideration; a ϕ value of 0.9 is used for flexure, 0.7 for compression, 0.85 for shear, etc. Only one strength reduction factor is applied to the entire composite element for a given design condition, even though the steel and concrete contribute differently to the strength and ductility of the member, and even though the two are affected differently by construction variations and the simplifying assumptions used in strength computations.

The Canadian Standards Association (CSA) publication A23.3-94, Design of Concrete Structures, recommends a slightly different approach. The basis of Canadian design is Ultimate Limit State Design (ULSD). In ULSD, strength-reduction factors are replaced by material resistance factors; a different value for ϕ is applied to each material (i.e., concrete = 0.60, steel rebar = 0.85, prestressing tendons = 0.90). The factored resistance of the whole composite is then calculated based on the formulations of equilibrium and strain compatibility. Material properties used in those formulations are adjusted by the appropriate values of ϕ. The factored resistance is compared to the appropriate combination of factored loads for ultimate limit states, as specified by the National Building Code of Canada.

Design Philosophy for Concrete Structures Strengthened with FRP

General. A reinforced concrete element strengthened with FRP is a composite consisting of three materials. The principles of equilibrium and strain compatibility still apply, and similar strength or resistance equations can be defined. However, the application of FRP as a strengthening tool requires engineers to address the interaction of three materials concrete, steel, and FRP; each material has different material properties and statistical reliability. New design formulations addressing each material's contribution to the overall strength of the element, as well as the approach for incorporating the ϕ factors into the overall strength formulation, must be established. These formulations can be based on either suitable individual ϕ factors applied separately to each of the materials constituting the composite (concrete, steel and FRP), a combined ϕ factor applied to the entire composite, or a combination of the two where both individual and combined ϕ factors are applied to the system.

Using a multifactor approach allows the designer to reflect the state of knowledge concerning the in-situ materials, as well as the FRP system. Depending on the information gathered about the in-situ concrete strength and condition, for instance, or the position of the steel reinforcement in a member being strengthened, an adjustment in the overall ϕ factor, either up or down, may be justified.

ACI 440 F proposes a combination of the composite ϕ factors (USD) and individual-material ϕ factors (ULSD). It proposes the use of traditional or reduced ACI ϕ factors, such as $\phi = 0.9$ for flexure and $\phi = 0.85$ for shear, applied to the overall element strength, in combination with a special FRP reduction factor ψ. The general philosophy is as follows:

$\phi\, S_n = \phi$ [steel/concrete + ψ (FRP strength)]; where:

ϕ = Typical ACI strength reduction factor.
ψ = Special reduction factor related to FRP.

Usable FRP Strength. The use of FRP for strengthening concrete structures is a new concept, and it may be used in applications where FRP will be exposed to harsh environments–

wide temperature fluctuations, variable wet/humid/ dry conditions, etc. Also, contact with highly alkaline concrete may affect the long term performance of FRP. Like with concrete in harsh environments, time will tell the obstacles for FRP performance in similar surroundings. While many FRP manufacturers show tremendous FRP strength and stiffness data, ACI 440F recommends limits to the ultimate FRP strength to account for the following potential issues:

- Environmental exposure
- Strain compatibility in bond zone
- Creep rupture
- Fatigue

Environmental Considerations. Environmental conditions uniquely affect mechanical properties of FRP materials. Different FRP materials (glass, aramid, carbon) react differently to certain environments, such as alkalinity, salt water, chemicals, UV light, temperatures, freezing and thawing cycles, etc. The performance of the FRP system in these environments will depend on the properties of the fiber itself and the matrix material (essentially homogeneous resin or polymer material in which the fiber system of an FRP composite is embedded). Carbon fiber (CFRP) is resistant to both alkali and acid environments, while glass fiber (GFRP) can degrade if exposed for long periods of time.

To account for expected in-service environmental conditions, ACI 440 F requires that the initial FRP material properties reported by the manufacturers (such as the guaranteed ultimate tensile strength) be adjusted by an environmental design factor.

For example, the ultimate tensile strength of an FRP system for use in design is:

$$f_{fu} = C_E f^*_{fu}$$

where f^*_{fu} is the guaranteed tensile strength guaranteed by the manufacturers (mean minus three standard deviations) and C_E is the environmental reduction factor. Similarly, the design ultimate tensile strain is:

$$\varepsilon_{fu} = C_E \varepsilon^*_{fu}$$

The design modulus of elasticity is the same as the manufacturer's reported modulus of elasticity.

The C_E factors proposed by ACI 440F are:

Exposure Condition		**Environmental Reduction Factor, C_E**
Interior Exposure	Carbon/epoxy	0.95
	Glass/epoxy	0.75
	Aramid/epoxy	0.85
Exterior Exposure (bridges, piers, unenclosed parking garages, etc.)	Carbon/epoxy	0.85
	Glass/epoxy	0.65
	Aramid/epoxy	0.75
Aggressive Environment (chemical plants, waste water treatment plants, etc.)	Carbon/epoxy	0.85
	Glass/epoxy	0.50
	Aramid/epoxy	0.70

Strain Compatibility. There may be limits on full FRP-strength utilization related to the strain compatibility of the substrate or parent member material. ACI 318 defines a concrete strain limit of 0.003, to define crushing failure of the concrete. In addition, it is commonly reported that aggregate interlock in concrete becomes unreliable at concrete strains in excess of 0.006. While CFRP systems in some applications can overcome these strain limits, the strength of the substrate will limit the utilization of the full FRP strength potential in some failure modes.

Since FRP research is still in the early stages and many practical variables are still to be studied and evaluated, simple strain limits are appropriate to define empirically developed limit states. A recent study by Seim et al. (2001) entitled "External FRP Post Strengthening of Scaled Concrete Slabs" provides test data indicating that CFRP strip systems (1.19 mm thick) debonded at strains on the order of 0.0065 versus an ultimate CFRP material strain of 0.0122; much thinner CFRP wraps achieved strains closer to the material limit.

ACI 440F proposes an "effective FRP strain" to preclude bonding failures. In flexural applications, for example, the factor k_m (see Figure 3), which is applied to limit the strength contribution by the FRP, is a function of FRP material thickness and stiffness.

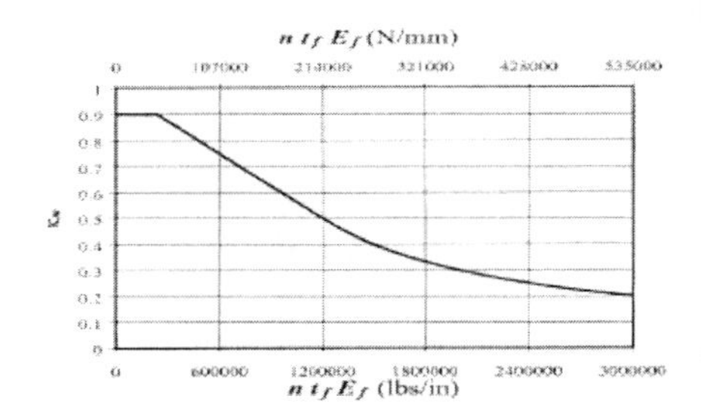

Figure 3. k_m factor as a function of FRP thickness and stiffness.

For example, for a 1.19 mm thick CFRP strip having E = 198 MPa; application of Figure 5 yields $k_m = 0.45$. The resulting effective strain is then $\varepsilon_{fe} = 0.45\ (0.0122) = 0.0055$; this strain is similar to but slightly less than the experimentally obtained failure strain of 0.0065.

Creep Rupture and Fatigue Stress Limits. Some FRP materials used for concrete post strengthening are more susceptible to loss of strength due to specific loading exposures. Glass Fiber Reinforced Polymers (GFRP), for instance, are more sensitive than other fiber materials to creep rupture and fatigue failure due to sustained and cyclic loads, respectively. Carbon FRP systems are reported to be highly resistive to both types of failure.

A designer must ensure that stresses in the FRP under sustained and/or cyclic loads are maintained at levels that do not present a risk of creep rupture and/or fatigue failures. The ACI

440 F document states that, because these concerns are typically encountered within the elastic range of the member, the stresses in the FRP can be computed through an elastic analysis. These stresses should be calculated based on all sustained loads (dead loads and the sustained portion of the live load).

To maintain safety, the service-load stress in the FRP at the service load levels should be limited to the values shown in the following table (ACI 440F).

	Fiber Type		
Stress Type	**Glass FRP**	**Aramid FRP**	**Carbon FRP**
Creep rupture and fatigue stress limit	$0.20\ f_{fu}$	$0.30\ f_{fu}$	$0.55\ f_{fu}$

where f_{fu} is the design ultimate strength of FRP.

Flexural Strengthening with FRP

In the ACI 440F approach, the flexural strength equation for a conventionally reinforced member strengthened with FRP becomes:

$$\phi M_n = \phi\ [(d\text{-}a/2)f_y A_S + (h\text{-}a/2)(\psi C_e f_{fe} A_f)];\ \text{where}$$

$\phi =$ Traditional ACI factor for flexure (varies between 0.7 and 0.9 based on the overall ductility at limit state, see Section 3),

$\psi =$ Special FRP factor based on type of application (flexure, shear, etc.)

$f_{fe} =$ Effective strength of FRP = $\varepsilon_{fe} E_f$, ksi (MPa)

$\varepsilon_{fe} =$ Effective strain, depends on strain formulations at ultimate state, but always $\varepsilon_{fe} \leq k_m\ \varepsilon_{fu}$

$A_f =$ Cross sectional area of FRP, in^2 (mm^2)

$a =$ Location of the resultant compressive force in concrete, in. (mm). It is a function of f'_c, b, and concrete stress-strain properties. When the governing mode of failure is not the crushing of concrete, it should be determined by means of force equilibrium and analysis of concrete nonlinear behavior (Whitney stress block can be inaccurate).

The above equation provides the flexural strength, based on an idealized condition where the FRP fails prior to concrete crushing and after yielding of the mild reinforcement. However, a designer needs to demonstrate that other limiting conditions do not control the flexural design.

ψ Factors for Strengthened Structures

A potential shortcoming of the proposed ACI 440F methodology is a lack of research supporting the proposed reliability factors for FRP. Without this research, one cannot interpret whether the currently proposed ψ factors, in combination with other environmental, strain compatibility, creep rupture and fatigue factors, are conservative or liberal.

Due to the emerging status of post-strengthening FRP systems, the existence of numerous FRP materials and variations, and the current lack of comprehensive standardized test data, the assignment of specific values to individual ψ factors is inexact, regardless of the method used. ACI 440F recommends a ψ factor of 0.85 for flexure, and a ψ factor of either 0.95 or 0.85 for shear, based on whether the section is completely or partially wrapped. A ψ factor of 0.95 is recommended for confinement applications. These numbers are not based on research data. They are solely another "safety measure" to account for the unknowns associated with FRP applications.

Prior to the development of ACI 440F guidelines, Karbhari and Sieble (1997) presented an approach for determining the FRP reduction factors for FRP properties alone (not including the application type, governing failure mode, etc.). According to Karbhari and Sieble, FRP materials have incomplete data records and must be treated as emerging materials. The ψ factors for each new system must reflect the variability in properties, quality control, and workmanship. The ψ factors must also address distinct responses of various FRP systems to different applications, to different governing failure modes, and to hostile environments. Karbhari and Sieble suggest the following philosophy for determining the ψ factors:

$$\psi_{FRP} = \psi_{MAT} * \psi_{PROC} * [(\psi_{CURE} + \psi_{LOC})/2] * \psi_{DEGR} \quad \text{, where}$$

- ψ_{MAT} is used to account for the deviation and/or level of uncertainty of material properties from the specified characteristic values (properties derived from tests as compared to those derived from theory);
- ψ_{PROC} is used to account for variation due to the processing method used (autoclave cure, wet winding, pultrusion, spray-up);
- ψ_{CURE} is used to account for variation in properties due to degree of cure achieved (autoclave cure, elevated temperature controlled cure, ambient cure);
- ψ_{LOC} is used to account for uncertainty in performance level due to the location of processing (controlled factory environment, field environment); and
- ψ_{DEGR} is used to account for changes in material properties over time and due to environmental effects (glass transition temperature, creep rupture).

Karbhari and Sieble suggest that values for ψ_{MAT}, ψ_{PROC}, ψ_{CURE}, ψ_{LOC}, ψ_{DEGR} vary in the range of 0.3 to 1. These values are much lower than the single ψ factors that ACI 440F recommends for use in strengthening design. Also, Karbhari and Sieble include factors such as ψ_{LOC} and ψ_{DEGR} in their approach, which ACI 440F approach handles separately. However, even though the final values for ψ factors determined as proposed by Karbhari and Sieble may be overly conservative, an engineer must approach this issue with due conservatism and obtain in-depth information about the strengthening system prior to comfortably using the ACI-440F recommended ψ values.

Conclusions

Strengthening of concrete structures with FRP is an emerging tool for which engineers will soon have design guidance from ACI. Although the ACI 440F document lays out detailed requirements for design with FRP systems, engineers should use their own judgment to make reasonable design assumptions. Unique characteristics of FRP, such as the lack of the high-temperature resistance and the linear-elastic performance up to failure, require special considerations uncommon to reinforced concrete design.

As the collective efforts of manufacturers, academia, and practicing design engineers continue to develop the knowledge base and expertise for designing with FRP, the practical guidelines will become more reliable; in the interim, caution is prudent.

Acknowledgements

The authors, who freely cited material from the draft of ACI 440F Document "Guide for the Design and Construction of Externally Bonded FRP Systems for Strengthening of Concrete Structures", wish to acknowledge all contributions by the members of ACI 440F Committee involved in preparation of that document.

References

ACI 440F-01, 2001. Guidelines for the Selection, Design, and Installation of Fiber Reinforced Polymer (FRP) Systems for Externally Strengthening Concrete Structures. Reported by ACI Committee 440, American Concrete Institute, Detroit, Michigan 48219.

Karbhari, V.M. and Sieble, F. 1997. Design Considerations for the Use of Fiber Reinforced Polymeric Composites in the Rehabilitation of Concrete Structures.

Kelley, P.L., Brainerd, M. L., Vatovec, M., 2000. Design Philosophy for Structural Strengthening with FRP. Concrete International, February 2000, Vol 22., No 2.

Seim, W. Herman, M. Karbhari, V., Seible, F. 2001, External FRP Post strenghtening of Scaled Concrete Slabs. ASCE Journal of Composites for Construction, V.5, No. 2, May 2001.

Van Gemert, D. 1992. Special Design Aspects of Adhesive Bonding Plates. ACI SP-165-2: Repair and Strengthening of Concrete Members with Adhesive Bonded Plates. pp.25-41.

Flexural Strengthening with Externally Bonded FRP Reinforcement

Thanasis C. Triantafillou,[1] Member, ASCE, and Stijn Matthys[2]

Abstract

Reinforced concrete members may be strengthened in flexure through the use of fiber reinforced polymer (FRP) composites epoxy-bonded to their tension zones, with the direction of fibres parallel to that of high tensile stresses (member axis). In this paper the authors present an overview of the design approach for flexural strengthening using FRP, according to recent developments of the working group within *fib* Task Group 9.3 on externally bonded reinforcement (EBR). Both ultimate limit state (including full composite action as well as bond failure modes) and serviceability limit state aspects are presented. Moreover, some special cases are discussed briefly: strengthening of post- or pre-tensioned concrete members, EBR in compression and strengthening with prestressed FRP. Finally, recommendations for future work related to flexural strengthening are given.

Introduction

Reinforced concrete members, such as beams, slabs and columns, may be strengthened in flexure through the use of FRP composites epoxy-bonded to their tension zones, with the direction of fibres parallel to that of high tensile stresses (member axis). The analysis for the ultimate limit state in flexure for such elements may follow well-established procedures for reinforced concrete structures, provided that: (a) the contribution of the external FRP reinforcement is taken into account properly; and (b) special consideration is given to the issue of bond between the concrete and the FRP.

[1] Assoc. Prof., Dept. of Civil Engrg., Univ. of Patras, Patras 26500, Greece. Email: ttriant@upatras.gr

[2] Post-Doctoral Res. Assoc., Dept. of Struct. Engrg., Ghent Univ., Ghent B-9052, Belgium. Email: stijn.matthys@rug.ac.be

In this paper the authors present an overview of the design approach for flexural strengthening using FRP, according to recent developments of the working group within *fib* Task Group 9.3 on externally bonded reinforcement (EBR). Both ultimate limit state and serviceability limit state aspects are presented. Moreover, some special cases are discussed briefly: strengthening of post- or pre-tensioned concrete members, EBR in compression and strengthening with prestressed FRP. Finally, recommendations for future work related to flexural strengthening are given.

Ultimate Limit State

For the ultimate limit state (ULS), classical calculation methods for RC elements still apply as long as **full composite action** between the FRP and the concrete may be assumed (Figure 1). The initial strain distribution (initial situation) during strengthening should be taken into account, based on elastic analysis of the (typically cracked) cross section. The following two failure modes may be considered as acceptable and most desirable: yielding of the tension steel reinforcement followed by concrete crushing, while the FRP is intact; and yielding of the tension steel reinforcement followed by FRP tensile fracture. The latter mode is rather unlikely to occur due to premature debonding, but it may be activated if very low FRP quantities are employed or if special care is taken for anchoring the FRP through mechanical means.

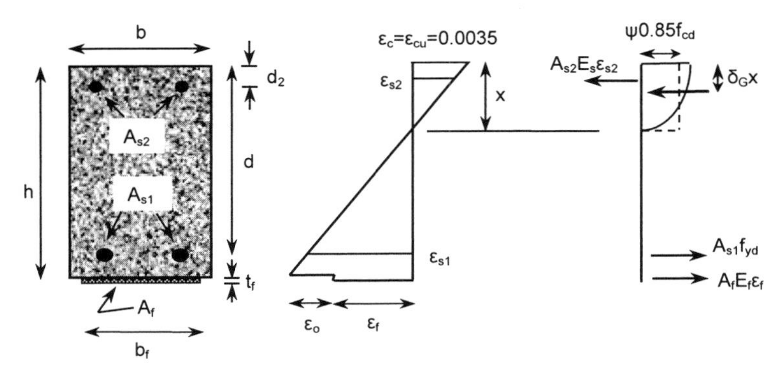

Figure 1. Cross section, strain and stress distribution for the analysis of the ultimate limit state in bending.

In addition, **bond failure modes** should be verified, which forms a rather complex aspect of the calculation. Depending on the starting point of the debonding process, the following failure modes can be identified (Figure 2).

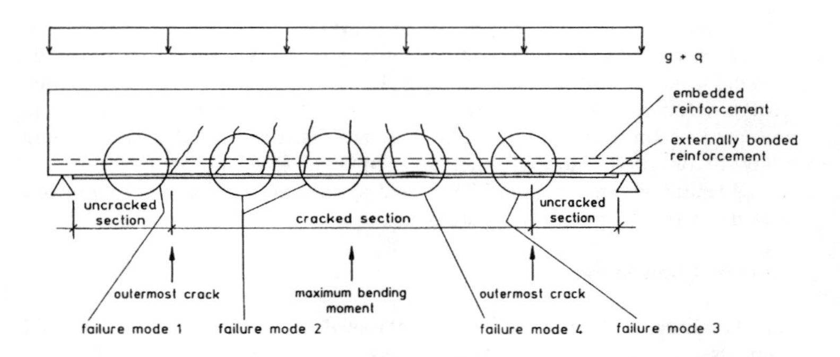

Figure 2. Bond failure modes of a concrete member with EBR (Blaschko et al. 1998).

- Mode 1: Peeling-off in an uncracked anchorage zone as a result of shear fracture through the concrete.
- Mode 2: Peeling-off caused at flexural cracks, which may propagate along the concrete-FRP interface.
- Mode 3: Peeling-off caused at shear cracks, due to the combined horizontal and vertical crack opening. However, note that in elements with sufficient internal (and external) shear reinforcement (as well as in slabs) the effect of vertical crack opening on peeling-off is negligible.
- Mode 4: Peeling-off caused by the unevenness of the concrete surface. This is avoided by adopting certain practical execution rules and limitations on concrete surface roughness.

Peeling-off caused at shear cracks (Mode 3) has not been quantified in proper detail by the research community yet (an appropriate bond model is yet to be developed). The model of Deuring (1993) is probably the most comprehensive one as of to date, but is rather complicated to apply. In a relatively recent study Blaschko (1997) proposed that peeling-off at shear cracks may be prevented by limiting the acting shear force to the shear resistance V_{Rd1} of RC members without shear reinforcement (Eurocode 2 approach) with a modification for the characteristic shear strength of concrete τ_{Rk} and the equivalent longitudinal reinforcement ratio $\rho_{eq} = [A_{s1}+A_f(E_f/E_s)]/bd$ (A_f = cross sectional area of FRP, A_{s1} = cross sectional area of tension steel reinforcement, E_f = elastic modulus of FRP, E_s = elastic modulus of steel). In case the design shear capacity falls below the required, an appropriate means of shear strengthening should be provided. Along these lines, Matthys (2000) has derived a shear resistance (based on test results) $V_{Rp} = \tau_{Rp}bd$, with a characteristic value for the shear strength $\tau_{Rk} = 0.38+151\rho_{eq}$ (MPa).

Treatment of peeling-off at the end anchorage (Mode 1) and at flexural cracks (Mode 2) may be done according to various approaches, which are described briefly in the following. Detailed treatment of these approaches is presented in *fib* (2001).

Approach 1 – Verification of end anchorage, strain limitation in the FRP. This approach involves two independent steps: first, the end anchorage should be verified based on the shear stress – slip constitutive law at the FRP-concrete interface. Then a strain limitation should be applied on the FRP to ensure that bond failure far from the anchorage will be prevented. Notably, this procedure has been followed in a number of draft design guidelines so far, mainly due to its simplicity. However, it represents a crude simplification of the real behaviour, as the FRP strain corresponding to bond failure is not a fixed value but it depends on a series of parameters, including the moment-shear relation, the strain in the internal steel and the distribution of cracks.

Approach 2 – Verification according to the envelope line of tensile stresses in the FRP. In this approach peeling-off is treated in a unified way both at the end anchorage and at any point along the FRP-concrete interface based on the interface shear stress – slip law and the envelope line of tensile stresses in the FRP (Niedermeier 2000). The main advantage of this approach is that peeling-off at the end and at flexural cracks is treated with the same model, whereas the main disadvantage is its complexity, which makes it difficult to apply as a practical engineering model.

Approach 3 – Verification of end anchorage and force transfer at the FRP/concrete interface. According to the third approach (Matthys 2000), two independent steps should be followed (as in the first one). In the first, the end anchorage should be verified based on the shear stress – slip constitutive law at the FRP-concrete interface. And in the second it should be verified that the shear stress along the interface, calculated based on simplified equilibrium conditions, is kept below a critical value (the shear strength of concrete). One disadvantage of this approach is the treatment of the same – in principle – phenomenon (peeling-off at the FRP end and far from it) with different models and another one is that it is based on a simplified stress distribution (for a homogeneous beam, fully cracked state). However, one major advantage is the simplicity of application in practical problems.

Finally, **FRP end shear failure** (Figure 3) should be taken into account. Jansze (1997) employed the *fictitious shear span* concept, to compute the shear resistance of plated beams along the lines of the Model Code (CEB 1993). The fictitious shear span concept provides a simplified engineering approach for the FRP end shear failure. It should be noted that other models have been developed too (e.g. Täljsten 1997, Malek et al. 1998), based on the analytical calculation of shear and normal stresses at the FRP end. However, such models are much more complicated to use.

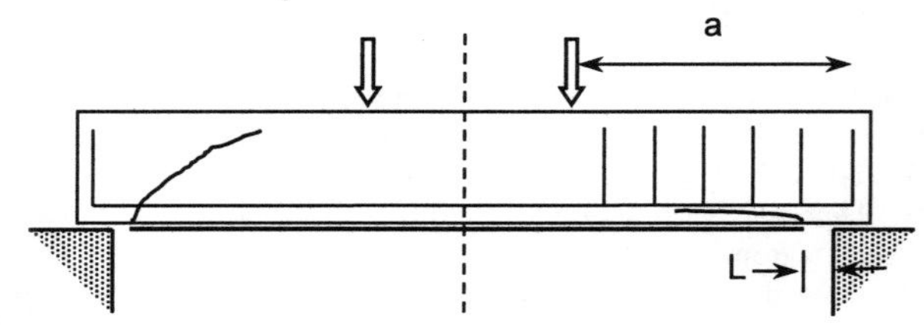

Figure 3. FRP end shear failure.

Serviceability Limit State

For the serviceability limit state (SLS), verification of **stresses** and **deflections** can be based on standard approaches (e.g. CEB Model Code 90). Compared to steel RC members the calculation becomes somewhat more complicated as the initial strain distribution during strengthening should be considered. If fracture mechanics based models are used to determine the anchorage length needed to fulfil the ULS, interface cracking in the SLS at the anchorage zone should be checked based on linear bond models. To predict the **crack width** of RC elements strengthened in flexure, the influence of both the internal and the external reinforcement, which have different bond behaviour, should be taken into account (Matthys 2000). Finally, providing proper restriction of crack widths, no **bond interface cracking** is to be expected in the SLS at flexural cracks.

Summary of Design Procedure

The procedure for dimensioning FRP-strengthened RC elements in flexure may be summarized as follows (*fib* 2001):

- For the member before strengthening: determine the resisting design moment (ULS) and check the SLS. The latter is not needed directly, but it will provide valuable information with respect to the SLS of the strengthened member (most likely to govern the design).
- From the service moment M_o during strengthening determine the initial strain ε_o at the extreme tension fibre.

- Assume full composite action and from the design moment after strengthening determine the required FRP cross section to fulfil the ULS. Verify that sufficient ductility is obtained.
- Calculate the deflections in the SLS. If the maximum allowable deflection is exceeded, determine the required FRP cross section to fulfil the deflection requirements.
- Calculate the stresses in the concrete, steel and FRP in the SLS. If allowable stresses are exceeded, determine the required FRP cross section to fulfil the stress limitation requirements.
- Verify that the provided FRP bond width is sufficient to control crack widths in the SLS. Increase the FRP width, if necessary, or, given a maximum width, increase the amount (thickness) of FRP. Bond interface cracking in the SLS is not of concern.
- Verify the resisting shear force at which bond failure due to shear cracks (vertical crack displacement) occurs (ULS). If this failure mode dominates, determine a new value of the FRP cross section.
- Verify that bond failure at the end anchorage and along the FRP (e.g. in regions where flexural cracking dominates) does not occur. If this is the case mechanical anchorage should be provided.
- Verify that FRP end shear failure is avoided. Provide shear strengthening at the ends if required.
- Verify the accidental situation. In this case, accidental loss of the EBR (e.g. due to impact, fire, etc.) is assumed, corresponding to an unstrengthened beam subjected to the loads of a strengthened beam.
- Verify the shear design resistance of the strengthened member. If needed shear strengthening should be provided.

Special Cases

Pre-tensioned or Post-tensioned Concrete Elements. In designing an FRP-strengthening of a prestressed member, the implications due to the presence of long-term phenomena should be clearly understood, as opposed to the case of conventional reinforced concrete, where the effects of shrinkage and creep are easily dealt with. Strengthening interventions usually take place when all long-term phenomena (creep, shrinkage, relaxation) have fully developed. Though this apparently favourable situation may seem to simplify the design procedure, it actually complicates the preliminary assessment phase of the existing conditions: the current state depends on all previous states, which therefore must be properly reconstructed. Thus, special care should be devoted to: construction sequence, with due consideration to all prestressing phases, correct description of long-term phenomena along with their superposition and mutual interaction, and evaluation of damage effects (due to impact, etc.) on the section stress pattern. Assessment of prestressed structures should be carried out in accordance to appropriate national standards. Alternative to this, a simplified approach can be adopted, in which all time-dependent effects are lumped into a single reduction coefficient, applied to the tendon stress, from which the pre-strengthening stress/deformation state is computed. Detailed considerations on the safety verifications as well as on modelling issues are given in *fib* (2001).

Externally Bonded Reinforcement in Compression. The elastic modulus of FRP in compression is, in general, lower than that in tension. Moreover, typical EBR configurations have very low flexural rigidity, so that local buckling may occur at relatively low stress levels. It is generally felt that FRP should not be used as compression reinforcement. However, at certain instances FRP may be subjected to compressive stresses which may be of secondary importance but not negligible. If this is the case, local buckling of the FRP should be considered. The analysis of local buckling would involve idealising the FRP as an elastic thin strip supported over an elastic medium of high rigidity. Local buckling may be avoided by placing compressive stress limitations in the FRP (yet to be established), which are expected to be satisfied in many cases, as permanent compressive stresses in concrete should be kept low too, in order to prevent excessive creep deformations. Otherwise, FRP should either not be glued in compression zones or special devices (e.g. external clamps) should be provided to fix the reinforcement against buckling.

Strengthening with Prestressed FRP. Conventional reinforced concrete theory may be applied to determine accurately the cracking and yield loads of beams with prestressed strips in flexure provided that the initial stress in the strip is included in the calculations. However, premature failure by other failure modes, as described above, must be examined. As the ultimate load in flexure is approached, cracking of the concrete will inevitably occur and the section will revert to normal reinforced concrete behaviour. In this case the ultimate shear strength of a beam strengthened with a stressed strip will be the same as that of the original beam.

The ultimate flexural strength of a beam with a stressed strip will not markedly differ from that of a beam with an unstressed strip. However, the initial strain in the strip will be added to that induced by bending so that strip failure is more likely and the failure mode of "steel yielding followed by FRP fracture" may be activated.

In calculating the shear resistance, the contribution of the strip to dowel action must be ignored unlike the main tensile steel reinforcement that can be included. The reason for this is that any vertical movement may lead to peeling-off failure resulting in debonding of the strip from the concrete. To be effective the strip would need to be enclosed by shear links.

In terms of designing end anchorages, it should be remembered that only a low fraction (in the order of 5-20%) of the ultimate strength of a prestressed strip can be transferred into the concrete by the adhesive alone (e.g. Triantafillou et al. 1992, Deuring 1993). Prestressing forces greater than this require an adequate anchor system to transfer the stressing force into the member to avoid peeling-off at the end of the strip.

Conclusions

Many aspects of flexural strengthening of RC members with externally bonded FRP reinforcement have been studied in considerable detail. Those include the identification of possible failure modes, the analysis for each mode and the establishment of a proper design method. Other aspects have received less attention and deserve further study. For instance, a simple, yet reliable and based on a firm scientific ground, model that will account for bond failures in a unified way has not been developed yet. Moreover, several gaps exist with respect to developing simple and reliable analytical models and design tools thereof for the cases of: (a) strengthening pre- or post-tensioned elements or (b) FRP in compression.

The aspects related to flexural strengthening of RC elements are directly applicable also to masonry walls subjected to out-of-plane bending. Key difference here may be the absence of steel reinforcement and the lower strength of masonry.

Acknowledgements

The authors wish to thank those members of the *fib* group EBR who contributed to the development of guidelines for the design of RC elements strengthened in flexure.

References

Blaschko, M. (1997). *Strengthening with CFRP.* Münchner Massivbau Seminar, TU München (In German).

Blaschko M., Nierdermeir R. and Zilch, K. (1998). "Bond failure modes of flexural members strengthened with FRP". In Proc. of *Second International Conference on Composites in Infrastructures*, Saadatmanesh, H. and Ehsani, M. R., eds., Tucson, Arizona, 315-327.

CEB (1993). *CEB-FIP Model Code 1990, Design Code.* Comité Euro-International du Béton, Lausanne, Switzerland, Thomas Telford.

Deuring, M. (1993). *Strengthening of RC with prestressed fiber reinforced plastic sheets.* EMPA Research Report 224, Dübendorf, Switzerland (in German).

fib (2001). *Design and Use of Externally Bonded FRP Reinforcement (FRP EBR) for Reinforced Concrete Structures.* Final Draft, Progress Report of *fib* EBR group, International Concrete Federation.

Jansze, W. (1997). *Strengthening of reinforced concrete members in bending by externally bonded steel plates.* PhD dissertation, TU Delft, The Netherlands.

Malek, A. M., Saadatmanesh, H. and Krishnamoorthy, M. R. (1998). "Prediction of failure load of R/C beams strengthened with FRP plate due to stress concentration at the plate end." *ACI Structural Journal*, 95(2), 142-152.

Matthys, S. (2000), *Structural behaviour and design of concrete members strengthened with externally bonded FRP reinforcement.* Doctoral thesis, Ghent University.

Niedermeier, R. (2000). *Envelope line of tensile forces while using externally bonded reinforcement.* Doctoral Dissertation, TU München, (In German).

Täljsten, B. (1997). "Strengthening of beams by plate bonding." *Journal of Materials in Civil Engineering, ASCE*, 9(4), 206-212.

Triantafillou, T. C., Deskovic, N. and Deuring, M. (1992). "Strengthening of concrete structures with prestressed fiber reinforced plastic sheets." *ACI Structural Journal*, 89(3), 235-244.

Shear and Torsion Strengthening with Externally Bonded FRP Reinforcement

Stijn Matthys[1] and Thanasis Triantafillou[2]

Abstract

Nowadays, there is a considerable interest in FRP (fibre reinforced polymer) reinforcement for building construction. Especially, the use of FRP for externally bonded reinforcement (EBR) to repair and strengthen existing structures is becoming more and more established world-wide. Since first applications in the late 1980s, the commercial use of externally bonded FRP reinforcement (FRP EBR) is currently increasing exponentially in many countries, its use can be regarded well documented and strengthening with FRP EBR is becoming a standard technique in a fast way.

Although a majority of the applications deal with flexural strengthening, also a deficient shear capacity often occurs. Due to its flexibility, FRP EBR is very attractive for shear strengthening. As demonstrated by several researchers, the use of FRP EBR for shear and torsion strengthening is feasible and efficient. Different calculation models have been proposed, which are still subject to further verification.

Introduction

Although applications and demonstration projects have been taken place already in the late 1980's and early 1990's, commercial used of FRP EBR started mainly in Switzerland around 1993 and soon followed in other European countries. Whereas in the beginning it was expected that the use of FRP EBR would mainly apply to strengthening of bridges, it appears that a lot of the strengthening work with this technique is done in the building sector, among

[1] Post-Doc. Res. Assoc., Magnel Laboratory for Concrete Research, Department of Stuctural Engineering, Ghent University, Belgium; Stijn.Matthys@rug.ac.be

[2] Assoc. Prof., Dept. of Civil Engrg., Univ. of Patras, Patras 26500, Greece; ttriant@upatras.gr

which the restoration of old buildings. In several of the practical cases, not only flexural but also shear (and rarely torsion) strengthening is needed, as e.g. reported by Alexander and Cheng (1996).

Whereas about 5 years ago, little research has been reported concerning the behaviour of reinforced concrete elements strengthened in shear (a lot of research on flexural strengthening being available), more recently the importance of shear strengthening has been recognized by several researchers. Indeed, existing structures with a deficient shear capacity have been reported, or a lack of shear capacity may be obtained after flexural strengthening (if the unstrengthened member has sufficient shear capacity, this may no longer be the case for the load capacity of the beam strengthened in flexure).

In the following the use of FRP reinforcement for shear and torsion strengthening is discussed, mainly with respect to a practical calculation model.

Strengthening Configuration and Efficiency

Strengthening in shear. By means of externally bonded FRP reinforcement, a wide range of strengthening configurations is possible. Some examples are shown in Fig. 1. To allow moisture exchange of the concrete and as it is normally most practical to attach the external FRP reinforcement with the principal fibre direction perpendicular to the member longitudinal axis, the 90° strip configuration (Fig. 1) is often preferred. If the FRP is aligned according to the 45° direction, more efficiency will be obtained as the fibres are provided more or less according to the principal tensile stress and as the anchorage length is longer. Sometimes it is suggested that fabrics with fibres in the 0° - 90° direction are preferable (assuming a vertical and horizontal component of the principal tensile stress, effectively carried in both fibre directions). Fig. 2 shows the dependence of the modulus of elasticity of FRP in function of the fibre orientation. From this figure it is clear that the effectiveness of the shear strengthening strongly relates to the fibre orientation (among other aspects).

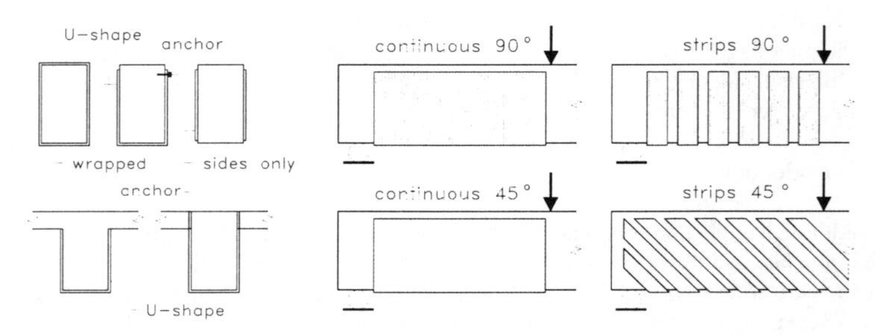

Figure 1. Shear strengthening of beams

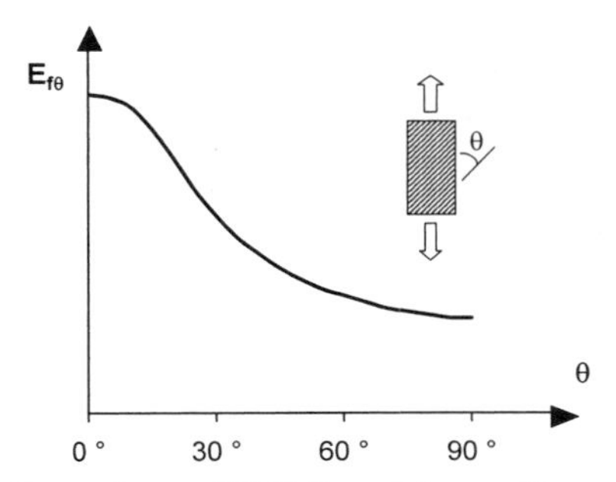

Figure 2. Dependence of FRP E-modulus on fibre orientation

To avoid (or delay) bond failure and given the limited anchorage length available, the FRP is preferably wrapped or anchored (Fig. 1). Especially, for T-shaped sections this aspect is of importance as often the compression zone, where the FRP is preferably anchored, is located in the flange. The more complex the beam shape, e.g. I-shaped sections, the more difficult (or the more mechanical fixings are needed) to avoid premature anchorage failure of the externally bonded FRP shear reinforcement.

The feasibility and effectiveness of FRP shear strengthening has been demonstrated by several researchers, e.g. JCI (1997), ACI (1999).

The shear failure of beams strengthened with FRP EBR mostly happens, in a similar way as for unstrengthened members, by diagonal tension (failure at an inclined shear crack). This failure may be initiated prematurely as a result of FRP bond failure or may correspond to fracture of the FRP. Due to strain variations along the shear crack, local debonding at both sides of the shear crack, possible bond failure, fibre orientation influence, etc. it appears that the contribution of the FRP is limited to an effective tensile strain ε_{fue}, which is generally lower than the ultimate FRP strain ε_{fu} (Täljsten 1997, Triantafillou 1998, Matthys 2000).

Strengthening in torsion. Strengthening for increased torsional capacity may be required sometimes in conventional beams and columns, as well as in bridge box girders. The principles applied to strengthening in shear are also valid in the case of torsion. The main difference between shear cracking and torsional cracking lies in the crack pattern. Torsional cracks are inclined, just like shear cracks, but they have different directions on opposite sides, following a spiral-like pattern (Fig. 3). Hence, an FRP material placed with the fibres forming an angle α with respect to the member axis may be quite effective in arresting diagonal cracking on one side but ineffective on the other side. Externally bonded FRP reinforcement will provide contribution to torsional capacity only if full wrapping around the element's cross section is applied, so that the tensile forces carried by the FRP on each side of the cross section may form a continuous loop.

Testing of concrete elements strengthened in torsion with FRP EBR has been very limited. Some tests have been conducted by Täljsten (1998), who demonstrated that torsional strengthening of beams with rectangular cross sections is feasible with CFRP wrapping.

Back
Top
α Front

Back
Top
α Front

Figure 3. Torsional (left) and shear (right) cracking

Modelling Approach

Detailed investigations on shear (torsion) strengthening of concrete members, with respect to the failure mechanisms, have been relatively limited and, to a certain degree, controversial. Most researchers have idealised the FRP materials in analogy with internal steel stirrups, assuming that the contribution of FRP to shear capacity emanates from the capacity of fibres to carry tensile stresses at a more or less constant strain ε_{fu} or ε_{fue}. Hence, it is assumed that the superposition principle can be applied as well as the truss analogy (Matthys 2000, *fib* 2001):

$$V_R = V_c + V_{ws} + V_{wf} \tag{1}$$

where, V_c, V_{ws} and V_{wf} are the shear capacity of the concrete, the steel stirrups and the external FRP reinforcement, respectively. Based on the truss analogy, the contribution of the shear links is given as:

$$V_{ws} = \frac{A_{ws}}{s_s} 0.9df_{wy}(\cot\theta + \cot\alpha_s)\sin\alpha_s \tag{2}$$

$$V_{wf} = \frac{A_{wf}}{s_f} 0.9dE_f\varepsilon_{f,eff}(\cot\theta + \cot\alpha_f)\sin\alpha_f \tag{3}$$

with, A_w the cross-sectional area of the shear links (ws: steel stirrups, wf: FRP links), s_s and s_f the stirrup spacing of the steel and FRP respectively, d the effective depth, f_{wy} the yield strength of the stirrups, $\varepsilon_{f,eff}$ the effective FRP strain, θ the angle between the diagonal shear crack and the member longitudinal axis (generally assumed as 45°) and α_s or α_f the angle of the steel or FRP stirrups with respect to the member longitudinal axis.

Using the superposition principle, it is assumed that the contribution of the concrete to the shear capacity does not significantly change compared to unstrengthened members. Where some did not found a significant change in the concrete contribution, others suggest that $\varepsilon_{f,eff}$ should be limited so to prevent a reduction in the concrete contribution at ultimate. Clearly, aspects related to the concrete shear contribution of strengthened concrete members, such as a reduced aggregate interlock contribution near ultimate load and the influence of confining action by the FRP shear reinforcement, may be subject to further study.

From the proposed modelling it follows that the prediction of the shear contribution of the external FRP reinforcement basically depends on the determination of the effective FRP strain $\varepsilon_{f,eff}$. The modelling of this strain depends on several aspects (see section 'Strengthening in shear'). As these aspects and their interaction are very difficult to model and would need extensive and specific research data (which are currently not available), a more deterministic approach as first suggested by Triantafillou (1997, 1998) can be proposed (see next section).

Effective Ultimate FRP Strain

In the deterministic approach, based on an analysis (curve fitting) of a large data base of experimental results, equations for the effective strain are proposed as a function of a correlation parameter. Triantafillou (1997, 1998) argues that $\varepsilon_{f,eff}$ will mainly depend on the available development length of the FRP (force transfer zone at both sides of the shear crack) which is a function of the bond conditions and the FRP axial rigidity (area times elastic modulus). By means of an experimental data fitting, it is proposed to calibrate $\varepsilon_{f,eff}$ as a function of $E_f\rho_{wf}$, with $\rho_{wf} = A_{wf}/(s_f b_w)$. Herewith, the FRP shear contribution $V_{f,exp}$ is taken as the difference between the experimental failure load of the strengthened and unstrengthened beams. Given $V_{f,exp}$ and assuming diagonal shear cracks with $\theta = 45°$, the corresponding strain $\varepsilon_{f,eff}$ is obtained from Eq. (3).

According to this approach a relationship for $\varepsilon_{f,eff}$ is suggested by Triantafillou (1997, 1998), as shown in Fig. 4. Later on, this was updated (Triantafillou and Antonopoulos, 2000), as shown in Figs 5 and 6. A similar approach by Khalifa et al (1998) is shown in Fig. 7. In Figs 4 and 7 the curves are plotted against test results of 70 experiments. The latter data base, collected by Matthys (2000), comprises the original data base of Triantafillou and Khalifa et al, test data by Matthys and additional data found in the literature. Based on this data base Matthys (2000) proposed relationships as shown in Fig. 8, with $\varepsilon_{f,eff}$ as a function of:

$$\Gamma_f = \frac{E_f \rho_{wf}}{f_{cm}^{2/3}(a/d)} \tag{4}$$

with, (a/d) the shear span to effective depth ratio. In Figs. 5 and 6, the curves are plotted against the data base of Triantafillou and Antonopoulos (2000).

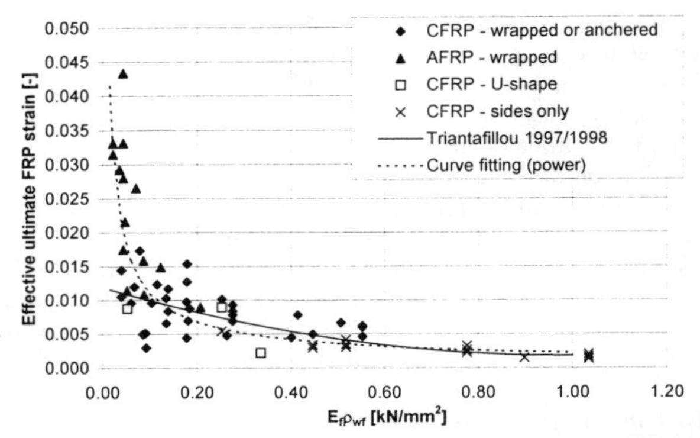

Figure 4. Effective FRP strain according to Triantafillou (1997,1998)

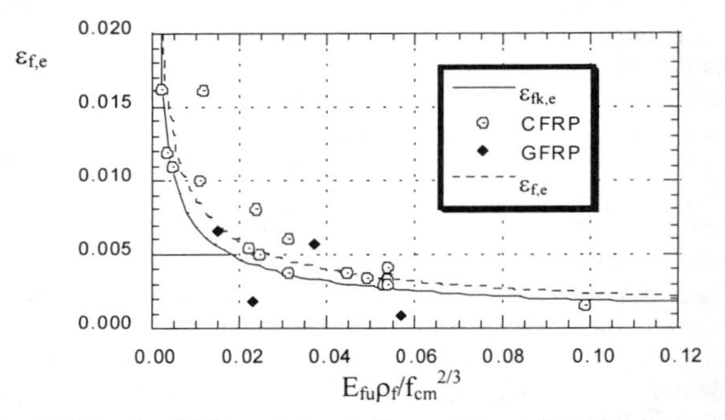

Figure 5. Effective FRP strain according to Triantafillou and Antonopoulos (2000), shear failure combined with FRP debonding

Although, the data follow a certain trend, large scatter is obtained in Figs. 4 till 8 and more experimental data may be required to validate these equations. The relatively large scatter also indicates that more research is needed to obtain a more fundamental understanding in the interaction of the phenomena influencing the effective FRP strain.

Whereas the deterministic approach presented here has the advantage of offering a practical engineering model, yet more general applicable than the single value reduction factors for the effective strain suggested by some researchers, it has the disadvantage of being fairly inaccurate. As a result large safety factors are needed for its application in design equations (*fib* 2001), which may limit the economical feasibility.

Proposals for more fundamental models have been made as well (e.g. JCI, 1997; Khalifa et al 1998, ACI 1999). Yet these models appear to be rather complex or take into account only few of the phenomena influencing the effective FRP strain, and still are subject to further verification.

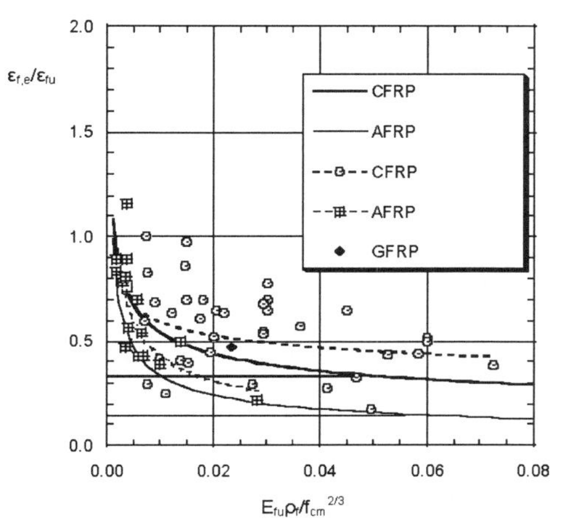

Figure 6. Effective FRP strain according to Triantafillou and Antonopoulos (2000), shear failure combined with or followed by FRP fracture

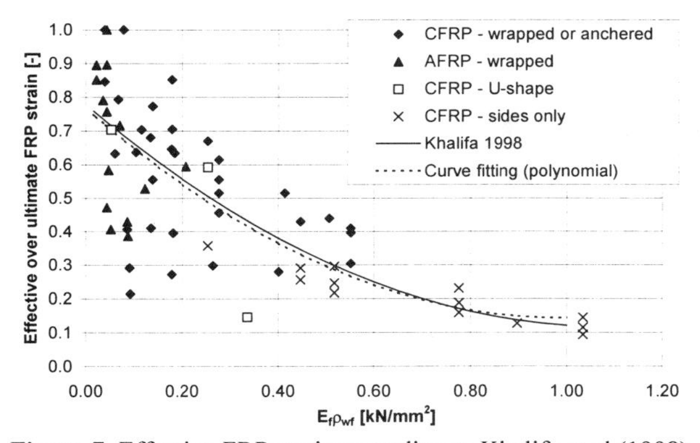

Figure 7. Effective FRP strain according to Khalifa et al (1998)

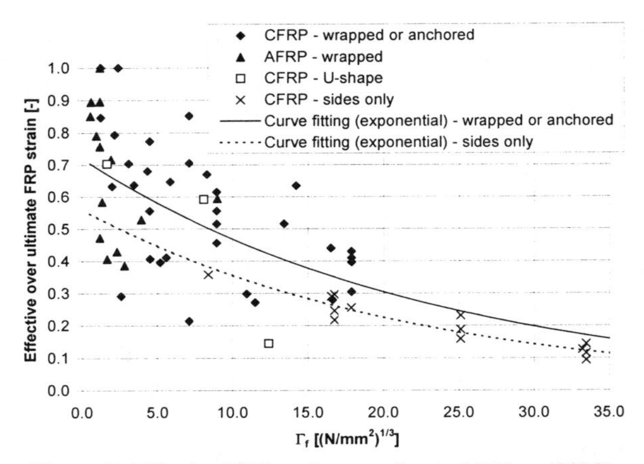

Figure 8. Effective FRP strain according to Matthys (2000)

Conclusions

By means of externally bonded FRP reinforcement, the shear and torsion strength of concrete members can be increased considerably. If sufficient FRP EBR is provided, a shear failure can be prevented, so that a flexural failure is obtained. The strengthening configuration may considerably influence the effectiveness of the shear strengthening.

The contribution of the FRP to the shear (torsion) capacity is related to an effective FRP strain which is generally lower than the ultimate FRP strain. This effective strain reflects aspects such as crack opening along the shear crack (determining the FRP strain variation along the shear crack), local debonding of the FRP when bridging the shear crack, available anchorage capacity and fibre orientation.

Increased efficiency is obtained if FRP bond failure is avoided or delayed. This may be done by providing U-shape wrapping, closed shape wrapping or mechanical fixings.

It is often assumed that the calculation of the shear (torsion) capacity can be performed according to a truss analogy, whereas the FRP contribution is obtained in a similar way as for the steel stirrups.

A fundamental modelling of the effective ultimate FRP strain is complex. Therefore, a practical engineering model based on experimental data fitting is proposed for the design by *fib* Task Group 9.3 (2001). Based on this model reasonably accurate predictions are obtained.

References

ACI (1999). *Chapter 14 - Shear strengthening with fibre reinforced polymers*. Proc. of the 4th Int. Symp. on FRP Reinforcement for Reinforced Concrete Structures, SP-188, Eds. C.W. Dolan, S.H. Rizkalla, A. Nanni, American Concrete Institute, USA, 933-1008.

Alexander, J.G.S. and Cheng, J.J.R. (1996). "Field application and studies of using CFRP sheets to strengthen concrete bridge girders". Proceedings *2nd. Int. Conf. On Advanced Composite Materials in Bridges and Structures*, Ed. M. El-Badry, The Canadian Society for Civil Engineering, Montréal, Québec, Canada, 465-472.

fib Task Group 9.3 (2001). *Design and Use of Externally Bonded FRP Reinforcement (FRP EBR) for Reinforced Concrete Structures*. Final Draft, Progress Report of *fib* EBR group, International Federation for Structural Concrete, Lauzanne.

JCI (1997). *Repair, Strengthening (Shear 1 and 2)*. Proc. of the 3rd Int. Symp. on Non-Metallic (FRP) Reinforcement for Concrete Structures, Japan Concrete Institute, Tokyo, Japan, 459-576.

Khalifa A., Gold W.J., Nanni A. and Aziz A.M.I. (1998), "Contribution of externally bonded FRP to shear capacity of RC flexural members", *ASCE Journal of Composites for Construction*, 2(4), 195-202.

Matthys, S. (2000). *Structural behaviour and design of concrete members strengthened with externally bonded FRP reinforcement*. Doctoral thesis, Department of Structural Engineering, Ghent University.

Täljsten, B. (1997). "Strengthening of concrete structures for shear with bonded CFRP-fabrics". *Recent advances in bridge engineering, Advanced rehabilitation, durable materials, non-destructive evaluation and management*, Eds. U. Meier and R. Betti, Dübendorf, 57-64.

Täljsten B. (1998), *Förstärkning av betongkonstruktioner med stålplåt och avancerade kompositmaterial utsatta för vridning*, Forskningsrapport, Luleå tekniska universitet, Avdelningen för konstruktionsteknik, Institutionen för Väg- och vattenbyggnad, 1998:01, ISSN 1402-1528 (In Swedish).

Triantafillou, T.C. (1997). "Shear strengthening of concrete members using composites". Proceedings *3rd. Int. Symp. on Non-Metallic (FRP)*

Reinforcement for Concrete Structures, Japan Concrete Institute, Tokyo, Japan, Vol. 1, 523-530.

Triantafillou, T.C. (1998). "Shear strengthening of reinforced concrete beams using epoxy-bonded FRP composites". *ACI Structural Journal*, 95(2), March-April, 107-115.

Triantafillou, T.C. and Antonopoulos, C.P. (2000), "Design of concrete flexural members strengthened in shear with FRP". *ASCE Journal of Composites for Construction*, 4(4), 198-205.

Confining Reinforced Concrete with FRP: Behavior and Modeling

Giorgio Monti [1]

Abstract

Relevant issues regarding the behavior of concrete, either plain or reinforced, confined with FRP are presented. The paper describes the state of development of three research areas: experimental, analytical (modeling) and design-oriented. The achievements attained so far are recognized and the needs for further research and future advancements are indicated.

Introduction

The behavior of confined concrete has been the subject of countless studies in the last three decades, many of which are so well known and established (and their results so widely accepted) in the scientific community that it seems superfluous to recall them. The common denominator of all of those studies was that they considered as confining device a material, steel, that, after yielding, exerts a constant confining pressure. This very fact allowed all researchers to assume that the confining pressure is a property of the confining device, thus avoiding to tackle the complex problem of the dilating behavior of concrete and of its interaction with the confining device itself, and expressing the response of confined concrete as if under hydrostatic pressure with the material parameters given as function of the steel yield strength.

The introduction of FRP as innovative confining device for concrete has changed the point of view over this, otherwise fully understood, phenomenon: FRP is an elastic material, and as such it does not yield; as a consequence, it exerts a continuously increasing confinement action on concrete. The response of concrete turns out to be completely different from what known so far, and this opened the way to a remarkable research effort that in the last few years has produced a number of valuable studies of different nature (experimental, analytical and design-oriented) with the common aim of clarifying all new aspects in this phenomenon.

[1] Professor, Dip. di Ingegneria Strutturale e Geotecnica, Università La Sapienza di Roma, Via A. Gramsci, 53 – 00197 Roma, Italy; giorgio.monti@uniroma1.it

Methods for confining concrete with FRP

Three techniques can be identified as applied either in experimental studies (reviewed in the following section) or in field applications: a) prefabricated shell jacketing, b) hand lay-up of sheets, and c) automated wrapping. Technique a) consists in the application of prefabricated shells around the (generally circular) column, either in half circles (*e.g.*, Nanni and Norris 1995), half squares (*e.g.*, Ohno et al. 1997) or circles with a slit or in continuous rolls (*e.g.*, Xiao and Ma 1997), which can be opened and placed around the column. It is worth mentioning that prefabricated jackets can be used as formwork in modifying column shapes, usually from rectangular to elliptical, in a strengthening measure. In technique b) the column can be wrapped using either FRP sheets in single or multiple layers, or FRP straps in a continuous spiral, or in discrete rings. Field applications of this technique are widely reported for both bridge piers and building columns (for a review see, for ex., ACI 1996, Neale and Labossiere 1997, Tan 1997). Finally, technique c) is based on a computer-controlled winding machine, the first of which was built in Japan in the 80s (ACI 1996), that automatically impregnates the fibers in a resin bath before winding them around the column, thus realizing an FRP jacket of controlled thickness, volume fraction and fiber orientation.

Experimental Studies and Observed Behavior

After the pioneering works by Fardis and Khalili (1981), and by Ahmad and Shah (1982), in the last ten years several tests have been conducted with the aim of understanding the peculiar aspects of FRP-confined concrete. Studies in this area are oriented towards understanding either the compressive strength enhancement or the seismic performance of FRP-confined concrete columns. Under the first group fall the tests on axially loaded columns (*e.g.*, Cousson and Paultre 1995, Nanni and Bradford 1995, Picher et al. 1996, Karbhari and Gao 1997, Demers and Neale 1999, Miyauchi et al. 1999, Purba and Mufti 1999, Saafi et al. 1999, Toutanji 1999, Xiao and Wu 2000, Zhang et al. 2000), where some of them address specific issues, such as: wrap properties (*e.g.*, Ahmad et al. 1991 for spiral confinement, Howie and Karbhari 1994 for wrap architecture, Harmon et al. 1995 for FRP stiffness), or column geometry (*e.g.*, Demers and Neale 1994, Mirmiran et al. 1998, Rochette and Labossiere 2000 for square or rectangular specimens), with an eye to durability (*e.g.*, Toutanji and El-Korchi 1999). Under the second group are tests on specimens under simulated seismic action with particular interest in the enhancement of ductility (*e.g.*, Mirmiran et al. 1996, Mirmiran and Shahawy 1997, Mirmiran et al. 1998 (specimen geometry), Saadatmanesh et al. 1994, Saadatmanesh et al. 1997a, 1997b, 1997c, Seible et al. 1997, Xiao and Ma 1997, Fam and Rizkalla 2001a).

The prevalent mode of failure observed in the above tests occurs in the confining FRP, although premature failure due to separation of FRP at the lap joint, has also been reported (*e.g.*, Demers and Neale 1994, Nanni and Bradford 1995).

All of these authors conclude that FRP-confinement is an effective way of enhancing both the load-carrying capacity and the ductility of seismically loaded

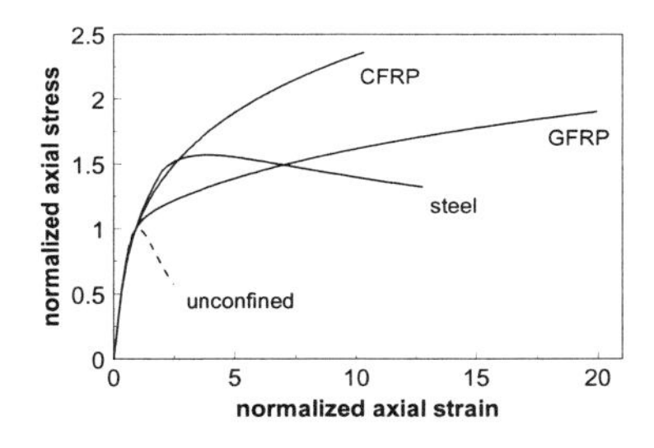

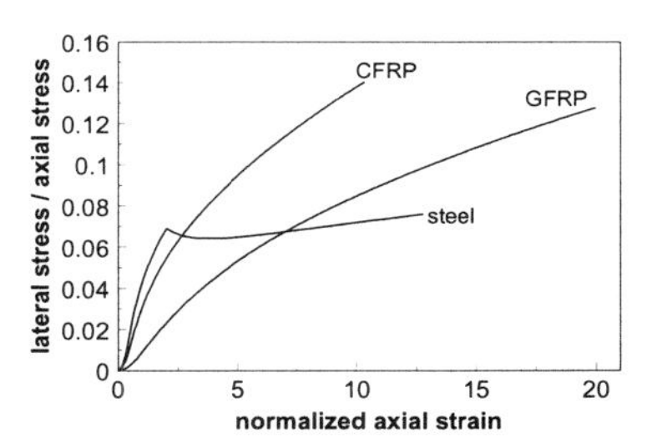

Figure 1. Behavior of concrete confined with steel, CFRP and GFRP: axial stress vs. axial strain (left), confinement effectiveness (right).

concrete columns. The amount of these increases and how they relate to the wrap properties is the object of other studies that are reported in the following section.

The most relevant findings of the above-listed experimental studies are condensed in Figure 1, where the (normalized) behavior of (both Glass and Carbon) FRP-confined cylindrical concrete specimens are compared to the more familiar steel-confined concrete. In the stress-strain relation, left, it is seen that the (G-C)FRP-confined concretes show an ever-increasing branch, as opposed to the steel-confined one, which, after reaching the peak strength, decays on a softening branch. The increasing confinement action of the elastic FRP limits the lateral strain thus delaying degradation. As regards the ultimate strain, and implicitly ductility, it should be noted that, notwithstanding the low ultimate strain of the FRP-jackets, the concrete ultimate strain is comparable (CFRP) or even greater (GFRP) than that obtained with steel. In the right figure, the jacket effectiveness, expressed in terms of ratio of the lateral stress to the *current* axial stress, is compared. The steel jacket effectiveness after yield is only due to the softening behavior of concrete, whereas in the other two cases it is the elastic behavior of the FRP jackets that increases the ratio. Here it should be evident how the two FRP materials reach almost the same level of effectiveness, but at different axial strain levels, which renders more attractive the use of GFRP jackets that also exploit ductility while maintaining the same effectiveness of CFRP jackets.

Modeling Studies

Parallel to the experimental studies cited in the previous section, studies have been carried out in developing stress-strain models to predict the response of FRP-confined concrete until failure. Most studies refer to the material behavior (*e.g.*, Cusson and Paultre 1995, Kawashima et al. 1997, Mirmiran and Shahawy 1997, Samaan et al. 1998, Spoelstra and Monti 1999, Toutanji 1999, Manfredi and Realfonzo 2001, Fam and Rizkalla 2001b); only a few to the flexural response of FRP-confined RC sections (*e.g.*, Monti and Spoelstra 1997).

Among the studies conducted so far, two major lines of thought are found: one makes use of a predefined explicit equation for the stress-strain law of concrete whose parameters are related to the confining FRP properties through regressions over experimental results, and another that implicitly obtains the concrete response by stepwise solving a set of equations that model the underlying physical phenomena. Among the models belonging to the first group, two are worth describing briefly: Samaan et al. 1998, and Toutanji 1999, while among those of the second group two are described: Spoelstra and Monti 1999, and Fam and Rizkalla 2001b.

The model by Samaan et al. (1998) is a non-normalized version of the well-known Menegotto-Pinto equation, used for steel, that provides a continuous curve whose shape is determined by four regression-based parameters that depend on the FRP properties. The authors observe that the axial stress-lateral strain curve has the same shape and that the transition zone occurs at same axial stress level, and thus conclude that the model can be extended to the lateral direction by using the same expression with different regression-based parameters.

The model by Toutanji (1999) provides a stress-strain curve characterized by two different predefined expressions. In the first expression the behavior of confined concrete is practically the same as that of unconfined concrete, because the lateral dilatation is negligible, while the second expression is influenced by the stiffness of the confining FRP system. Similarly to the previous one, this model has demonstrated its capability in modeling experimental tests under monotonic loading, though some doubts remain as for its extension to deal with cyclic behavior.

In the model by Spoelstra and Monti (1999), belonging to the second group, the concrete response, under a varying confining pressure, is thought of as a curve that crosses a family of curves describing the response under different constant pressures. This approach has demonstrated to be effective in its simplicity, though the formulation is necessarily iterative, due to the nonlinear nature of the three interacting phenomena. The advantage of such approach is that the use is avoided of non-physical parameters, as typical of many regression-based models. Extension of the model to deal with the cyclic response should be straightforward as soon as experimental tests become available.

Fam and Rizkalla (2001b) propose a model that predicts the behavior of axially loaded short concrete columns confined by FRP circular tubes, which also account for partially filled tubes with central holes. It uses a bi-axial strength failure criteria for FRP to account for possible bi-axial stress states. The model is based on equilibrium and radial displacement compatibility under variable confinement, using a step-by-step strain increment technique similar to that of the previous model.

Design-Oriented Studies

Some authors have focused on the development of predictive equations for the ultimate strength and strain of FRP-confined concrete, to be used in design. The motivation for these studies stems from the awareness that existing strength models for the axial compressive strength of steel-confined concrete are unconservative if applied to FRP-confined concrete (Mirmiran and Shahawy 1997, Samaan et al. 1998, Saafi et al. 1999). In the following, some exemplificative studies are reported that produced such equations (Saadatmanesh et al. 1994, Seible et al. 1995, Karbhari and Gao 1997, Mirmiran et al. 1998, Samaan et al. 1998, Miyauchi et al. 1999, Saafi et al. 1999, Spoelstra and Monti 1999, Toutanji 1999), after which the results of studies by three different authors (Seible et al. 1995, Mutsuyoshi et al. 1999, Monti et al. 2001) are briefly reported that make use of such predictive equations to establish design procedures for the design of strengthening measures of RC members.

Predictive Equations of the Ultimate Strength

The majority of the studies cited above developed strength equations for FRP-confined circular concrete columns in the form: $f'_{cc}/f'_{co} = 1 + k_1 f_l/f'_{co}$, where f'_{cc} and f'_{co} are the confined and unconfined concrete strength, respectively, $f_l = 2 f_f t_f/d$ is the maximum confining pressure at the ultimate strength f_f of the FRP jacket of thickness t_f and diameter *d*. Each author gives a different expression for the coefficient k_1 that measures the confinement effectiveness.

Predictive equations of different form have also been proposed, such as, for example, Karbhari and Gao (1997): $f'_{cc}/f'_{co} = 1 + 3.1\, \nu_c E_f f_l / E_c f_f + f_l / f'_{co}$, also dependent on the Poisson's ratio ν_c and modulus of elasticity E_c of concrete, and that E_f of FRP, or that by Spoelstra and Monti (1999): $f'_{cc}/f'_{co} = 0.2 + 3\sqrt{f_l/f'_{co}}$, while according to Saadatmanesh et al. (1994), Purba and Mufti (1999), the well-known equation proposed by Mander et al. (1988) can be used.

Predictive Equations of the Ultimate Strain

The ultimate strain of concrete as increased by the presence of FRP-wrapping is a fundamental quantity to know when designing strengthening measures for ductility enhancement of non-seismically detailed columns.

Only two expressions are briefly recalled here, because they appear to be more widely used. According to Seible et al. (1995), the concrete ultimate strain ε_{cu} can be computed through a formula experimentally derived for steel confinement and adapted to the case of FRP: $\varepsilon_{cu} = 0.004 + 2.5 \rho_f f_f \varepsilon_f / f'_{cc}$, where $\rho_j = 4 t_f / d$ is the jacket volumetric ratio; and ε_f is the FRP effective ultimate strain. The confined concrete peak strength f'_{cc} is computed according to Mander et al. (1998). An alternative expression, obtained through regression analyses on a previously described stress-strain model, is proposed by Spoelstra and Monti (1999): $\varepsilon_{cu} = \varepsilon_{co}\left(2 + 1.25\, E_c / f'_{co}\, \varepsilon_f \sqrt{f_l / f'_{co}}\right)$.

Scale effects, rectangular sections and discontinuous confinement

The scale effect of length-to-diameter ratio (L/d) of a column is taken into account through the coefficient (Mirmiran et al. 1998): $1.418 - 0.263(L/d) + 0.0288(L/d)^2$, that gives the confined concrete strength ratio to that of a column having $L/d = 2$.

In the case of square or rectangular columns of longer side length *D*, whose corners must be rounded with radius R_c (usually, from 5 mm to 40 mm), the effectiveness of confinement is sensibly reduced. It has been proposed (Mirmiran et al. 1998) to reduce the confinement pressure through a shape factor: $k_s = 2R_c/D$.

When the concrete cover is not sufficiently thick, it is impossible to round the corners with high values of R_c, thus resulting in low confinement effectiveness. In these cases it is expedient to reshape the column to either circular or elliptical as originally proposed by Priestley et al. (1994) for steel jacketing of rectangular columns. For elliptical sections the confinement pressure is then evaluated based on the lengths a and b of the semi-axes: $f_l = t_f\, f_f \left[1.5(a+b) - \sqrt{ab}\right] / 2ab$.

Partially wrapped columns with discontinuous confinement (either continuous spiral or discrete rings with gaps in between), show the so-called arching-effect. It has been proposed (Saadatmanesh et al. 1994) that the confinement pressure be reduced by: $k_g = (1 - s_f/2d)^2/(1-\rho_{sc})$, where s_f is the gap between straps or rings and ρ_{sc} is the ratio of steel reinforcement area to confined concrete core.

Strengthening of members

Three methods are here briefly recalled that aim at determining the amount (thickness) of FRP for strengthening an under-designed RC section.

The procedure proposed by Seible et al. (1995) of selecting the jacket thickness for a target displacement ductility factor μ_Δ for a given column starts from the calculation of the plastic hinge length L_p. From L_p and μ_Δ, the curvature ductility factor $\mu_\Phi = \Phi_u/\Phi_y$ is determined. Section analysis gives both yield and ultimate curvature, the latter being determined in terms of the ultimate concrete strain. Inserting ε_{cu} into one of the ultimate strain predictive equations described above, the required FRP thickness is determined.

For Mutsuyoshi et al. (1999) the ductility factor may be related to the shear capacity V_u, and to the moment capacity M_u of the member after retrofit, according to an empirical equation: $\mu_\Delta = \alpha + \beta(V_u a/M_u) \le 10$, where a is the shear span and the constants α, β depend on the FRP type.

An alternative design equation (Monti et al. 2001) for the ductility upgrade of circular columns having diameter D stems from the definition of a section upgrading index $I_{sec} = \delta_{sec}^{tar}/\delta_{sec}^{ava}$, representing the ratio of the target sectional ductility (to be obtained through upgrading) and the initially available sectional ductility (evaluated through assessing the existing section). The design equation yields the jacket thickness: $t_j = 0.2\, D\, I_{sec}^2 \cdot f_{cc}'^{\,ava}\, {\varepsilon_{cu}^{ava}}^2\, {\varepsilon_f}^{-1.5} / f_f$, where $f_{cc}'^{\,ava}$ and ε_{cu}^{ava} are the available strength and ultimate strain of concrete determined through preliminary assessment of the concrete structure. It should be noted that this design equation is only applicable to the case of circular sections.

Research Needs

In the three research fields examined above, *i.e.* experimental, analytical, and design-oriented, the works developed so far has helped to clarify most aspects of the behavior of concrete elements confined with FRP that were initially misinterpreted.

Yet, there are some relevant issues that need to be explored in order to arrive at a full understanding of the phenomenon and to an extension of the knowledge accumulated until now to more complex situations, such as, for example, the combined action of axial load and bending moment and the important case of cyclic loading, so far inexplicably neglected.

Most of all, a reliable predictive equation for the ultimate strain is needed.

The needs are of course of different nature for the two fields of analytical modeling and design. Experimental research should support both fields with tests purposely designed to clarify those issues in need of a deeper insight.

For example, it would be of much use to perform more experimental tests on both circular and non-circular specimens under combined axial load and bending. This would serve to shed light on the shape of the confining stress field over a cross-section subjected to a strain gradient. A common assumption that states that concrete portions at the same depth undergo the same level of confinement, still needs to be validated.

Another aspect of some interest should be explored: the interaction between the additional confining FRP and the existing one, usually provided by steel hoops, to understand how the FRP effectiveness is affected by the presence of the hoops.

It goes without saying that a prevalent goal for both analytical and design-oriented studies should be of formulating an improved stress-strain law in a single closed-form equation, without the need of stepwise following the interaction between axial and lateral strain, which would be more practical for design purposes, to describe the stress-block acting on an FRP-confined RC section. This curve should be parabolic, with parameters depending on the FRP properties, and should be used to determine the ultimate strength and ductility of RC sections under combined axial load and bending.

Additional developments are expected for the extension of the FRP-confined concrete models to the cyclic case, which would be of much use for the analysis of FRP-strengthened structural systems subjected to earthquake action. Here, it is fundamental to understand how the unloading branch slope is affected by the underlying mechanism. Expected findings should help at arriving at a constitutive law to be used in post-strengthening assessment studies of RC members subjected to cyclic action. A practical application could be, for example, the possibility of analyzing models of RC structures after the application of FRP for strengthening, in order to assess that the applied measures, designed for monotonic load, are also effective under seismic conditions. From the author's standpoint, regression-based models don't lend themselves easily to this task, because the number of parameters increases dramatically in the cyclic case and their physical meaning becomes even less clear. Models that give a physical interpretation of the essential phenomena, propose themselves as natural candidates for possible developments in that direction.

References

ACI 440R-96 (1996). State-of-the-Art Report on Fiber Reinforced Plastic (FRP) Reinforcement for concrete structures. *American Concrete Institute (ACI)*, Committee 440, Michigan, USA.

Cusson, D. and Paultre, P. (1995). Stress-strain model for confined high-strength concrete. *J. of Structural Engineering*, ASCE, 121(3), 468-477.

Demers, M. and Neale, K.W. (1994). Strengthening of concrete columns with unidirectional composite sheets. *Development in Short and Medium Span Bridge Engineering*, Canadian Society for Civil Engineering, Montreal, Canada, 895-905.

Demers, M. and Neale, K.W. (1999). Confinement of reinforced concrete columns with fibre-reinforced composite sheets – an experimental study, *Canadian J. of Civil Engineering*, 26, 226-241.

Fam, A.Z. and Rizkalla, S.H. (2001a). Concrete-filled FRP tubes for flexural and axial compression members, *ACI Structural J., in press.*

Fam, A.Z. and Rizkalla, S.H. (2001b). Confinement model for axially loaded concrete confined by circular FRP tubes, *ACI Structural J., in press.*

Fardis, M.N. and Khalili, H. (1981). Concrete encased in GFRP, *ACI J.*, 78, 440-445.

Harmon, T., Slattery, K. and Ramakrishnan, S. (1995). The effect of confinement stiffness on confined concrete, *Non-Metallic (FRP) Reinforcement for Concrete Structures, Proc. of the 2nd International RILEM Symposium*, RILEM, Ghent, Belgium, August, 585-600.

Howie, I. and Karbhari, V.M. (1994). Effect of tow sheet composite wrap architecture on strengthening of concrete due to confinement: I – Experimental studies, *J. of Reinforced Plastic and Composites*, 14(9), 1008-1030.

Karbhari, V.M. and Gao, Y. (1997). Composite jacketed concrete under uniaxial compression – verification of simple design equations. *J. of Materials in Civil Engineering*, ASCE, 9(4), 185-193.

Kawashima, K., Hosotani, M. and Hoshikuma, J. (1997). A model for confinement effect for concrete cylinders confined by carbon fiber sheets, *NCEER-INCEDE Workshop on Earthquake Engineering Frontiers of Transportation Facilities*, March, NCEER SUNY, Buffalo, USA.

Mander, J.B., Priestley, M.J.N. and Park, R. (1988). Theoretical stress-strain model for confined concrete, *J. of Structural Engineering,* ASCE, 114(8), 1804-1826.

Manfredi, G. and Realfonzo, R. (2001). Models for concrete confined by fiber composites, *in press.*

Mirmiran, A. and Shahawy, M. (1997). Behavior of concrete columns confined by fiber composites, *J. of Structural Engineering*, ASCE, 123(5), 583-590.

Mirmiran, A., Kargahi, M., Samaan, M. and Shahawy, M. (1996). Composite FRP-concrete column with bi-directional external reinforcement, *Proc. 1st International Conference on Composites in Infrastructure*, Tucson, Arizona, USA, 888-902.

Mirmiran, A., Shahawy, M., Samaan, M., El Echary, H., Mastrapa, J.C. and Pico, O. (1998). Effect of column parameters on FRP-confined concrete, *J. Composites for Construction*, ASCE, 2(4), 175-185.

Monti, G., and Spoelstra M.R. (1997). Analysis of RC bridge piers externally reinforced with fiber-reinforced plastics. *2nd Italy-Japan Workshop on Seismic Design and Retrofit of Bridges,* Roma, Italy.

Monti, G., Nisticò, N. and Santini S. (2001). Design of FRP jackets for upgrade of circular bridge piers, *J. of Composites for Constructions*, ASCE, 5(2), 94-101.

Mutsuyoshi, H., Ishibashi, T., Okano, M. and Katsuki, F. (1999), New design method for seismic retrofit of bridge columns with continuous fiber sheet – Performance-based design. *Fiber Reinforced Polymer Reinforcement for Reinforced Concrete Structures*, Eds. C. W. Dolan, S. H. Rizkalla and A. Nanni, ACI Report SP-188. Detroit, Michigan, 229-241.

Neale, K.W. and Labossiere, P. (1997). State-of-the-art report on retrofitting and strengthening by continuous fibre in Canada. *Non-Metallic (FRP) Reinforcement for Concrete Structures, Proc. 3rd Int. Symposium*, Sapporo, Japan, 25-39.

Nanni, A. and Bradford, N.M. (1995). FRP jacketed concrete under uniaxial compression, *Construction and Building Materials*, 9(2), 115-124.

Nanni, A. and Norris, M.S. (1995). FRP jacketed concrete under flexure and combined flexure-compression, *Construction and Building Materials*, 9(5), 273-281.

Ohno, S., Miyauchi, Y., Kei, T. and Higashibata, Y. (1997). Bond properties of CFRP plate joint. *Non-Metallic (FRP) Reinforcement for Concrete Structures, Proc. 3rd Int. Symposium*, Sapporo, Japan, 241-248.

Picher, F., Rochette, P. and Labossière, P. (1996). Confinement of concrete cylinders with CFRP, *Proc. First International Conference on Composite Infrastructures*, Tucson, Arizona, USA, 829-841.

Priestley, M.J.N., Seible, F., Xiao, Y. and Verma, R. (1994). Steel jacket retrofitting of reinforced concrete bridge columns for enhanced shear strength. *ACI Structural J.*, 91(4), 394-405.

Purba, B.K. and Mufti, A.A. (1999). Investigation on the behavior of circular concrete columns reinforced with CFRP jackets, *Canadian J. of Civil Engineering*, 26, 590-596.

Rochette, P. and Labossiere, P. (2000). Axial testing of rectangular column models confined with composites, *J. of Composites for Construction*, ASCE, 4(3), 129-136.

Saadatmanesh, H., Ehsani, M.R. and Li, M.W. (1994), Strength and ductility of concrete columns externally reinforced with fiber composite straps, *ACI Structural J.*, 91(4), 434-447.

Saadatmanesh, H., Ehsani, M.R., and Jin, L. (1997a). Seismic strengthening of circular bridge piers with fiber composites, *ACI Structural J.*, 93(6), 639-647.

Saadatmanesh, H., Ehsani, M.R., and Jin, L. (1997b). Seismic retrofitting of rectangular bridge columns with composite straps, *Earthquake Spectra*, 13(2), 281-304.

Saadatmanesh, H., Ehsani, M.R., and Jin, L. (1997c). Repair of earthquake-damaged RC columns with FRP wraps, *ACI Structural J.*, 94(2), 206-215.

Saafi, M., Toutanji, H.A. and Li, Z. (1999), Behavior of concrete columns confined with fiber reinforced polymer tubes, *ACI Materials J.*, 96(4), 500-509.

Samaan, M., Mirmiran, A. and Shahawy, M. (1998), Model of concrete confined by fiber composites, *J. of Structural Engineering*, ASCE, 124(9), 1025-1031.

Seible, F., Priestley, M.J.N. and Innamorato, D. (1995). Earthquake retrofit of bridge columns with continuous fiber jackets, Volume II, Design guidelines, *Advanced Composite Technology Transfer Consortium, Report No. ACTT-95/08*, University of California, San Diego, USA.

Seible, F., Priestley, M.J.N., Hegemier, G.A. and Innamorato, D. (1997). Seismic retrofit of RC columns with continuous carbon fiber jackets, *J. of Composites for Construction*, ASCE, 1(2), 52-62.

Spoelstra, M.R. and Monti, G. (1999). FRP-confined concrete model. *J. of Composites in Construction*, ASCE, 3(3), 143-150.

Tan, K.H. (1997). State-of-the-art report on retrofitting and strengthening by continuous fibers, Southeast Asian perspective – Status, prospects and research needs. *Non-Metallic (FRP) Reinforcement for Concrete Structures, Proc. 3rd Int. Symposium*, Sapporo, Japan, 13-23.

Toutanji, H.A. (1999). Stress-strain characteristics of concrete columns externally confined with advanced fiber composite sheets, *ACI Materials J.*, 96(3), 397-404.

Toutanji, H.A. and El-Korchi, T. (1999). Tensile durability of cement-based FRP composite wrapped specimens, *J. Composites for Construction*, ASCE, 3(1), 38-45.

Xiao, Y. and Ma, R. (1997). Seismic retrofit of RC circular columns using prefabricated composite jacketing, *J. of Structural Engineering*, ASCE, 123(10), 1357-1364.

Xiao, Y. and Wu, H. (2000). Compressive behavior of concrete confined by carbon fiber composite jackets, *J. of Materials in Civil Engineering*, ASCE, 12(2), 139-146.

Zhang, S., Ye, L. and Mai, Y.W. (2000). A study on polymer composites strengthening systems for concrete columns, *Applied Composite Materials*, 7, 125-138.

Strengthening Historical Monuments with FRP: a Design Criteria Review

Angelo Di Tommaso[1] and Francesco Focacci [1].

Abstract

The aim of this paper is to discuss certain aspects concerning applications of FRP materials in strengthening historical masonry buildings. The discussion covers the objectives of the reinforcement and the design criteria. As FRP materials can be used in a number of ways for the structural strengthening of masonry buildings, it is impossible to consider all possible cases here; this paper discusses a few recurrent applications such as for strengthening masonry arches or barrel and cross vaults, the flexural and shear strengthening of panels, and the belting of buildings.

Introduction

In recent years, fiber-reinforced polymer (FRP) materials have been widely used for strengthening historical masonry buildings. In Italy, especially after the earthquake of 1997, hundreds of applications of FRP for repairing damage and improving the seismic response of historical (and even non-historical) buildings have been implemented. Strips and pultruded plates were used in the main, while the use of bars and cables is limited nowadays.

The main advantages of using composite materials in these applications were found to be quick and easy installation, light weight and flexibility. As mentioned before, bars and cables are rarely adopted, though they offer advantages related to the opportunity to impress a moderate post-tensioning force on the masonry, due to the low Young's modulus, which might contain the losses due to creep, whereas post-tensioning with strips and pultruded plates poses numerous practical difficulties.

----

[1] Dipartimento di Costruzione dell'Architettura, Istituto Universitario di Architettura di Venezia, Campo Tolentini, 191, 30135 Venice Italy; adt@brezza.iuav.unive.it, foca@dada.it.

Despite the number of practical applications, standard design procedures and criteria for the FRP reinforcement of masonry structures are still not available; at present, most research work concerns experimental analyses and some analytical models (especially regarding arches and vaults), but a few design equations have been developed (Triantafillou, 1998), (Faccio et al. 2000, a, b).

Strengthening historical monuments with FRP

The strengthening of historical masonry structures becomes necessary mainly in the event of i) damage due to exceptional events (such as earthquake or fire), ii) damage occurring with time in the life of the building (such as differential soil settlement), or iii) in the event of an increase in live loads (upgrading of the structure).

In the first case, FRP materials enable the structure to be repaired and its strength restored at least to the level existing before it was damaged; in most cases, the vulnerability of the structure can be reduced and applying FRP materials ensures the structural conservation of the monument. It is worth noting that in this case the FRP application is performed when at least the dead loads are brought to bear, so the building preserves the structural scheme of the original design and the new materials only cooperate in the event of successive exceptional events.

In the second case, the cause of the damage has first to be overcome. Simply repairing the visible damage only leads to the onset of cracks in a part of the structure without any reinforcement. Once the cause has been eliminated (or much reduced), FRP materials can be used to repair and support the structure in the event of further small relative displacements taking place, distributing the corresponding force on a wider area of masonry. In this case, tensile forces act on the composite material and the creep of the resin must be taken into account.

In the third case, it has to be remembered that the reinforcement material only cooperates in sustaining the live loads (present and future), unless the structure is supported during the application of the reinforcing material.

Design criteria and material selection

In most cases, a historical masonry structure can be strengthened with a global application and one or more local applications. In general, local applications can be said to be more suitable, avoiding having to completely cover a large area of the structure and thus allowing for the flow of moisture. This becomes compulsory if there are frescoes on the other side. From the design point of view, the application of local strengthening implies the need to have a very clear understanding of the structure's behavior.

For the analysis and design of strengthening arrangement for a masonry structure, three general methods can be outlined.

The first is based on an elastic analysis of the structure; the FRP reinforcement is placed where there are tensile forces that cannot be absorbed by the masonry. The main disadvantage of elastic analysis concerns the fact that it does not allow for the fact that masonry is a no-tension material; safe equilibrium configurations can sometimes be found even without the presence of tensile stresses. In such cases, elastic analysis leads to the application of FRP strips where they are not actually needed. Moreover, in most cases, the composite only adds its contribution when the structure is in the collapse phase. However, elastic analysis provides general indications on the behavior of the structure and enables a preliminary dimensioning of the reinforcement.

The second method is based on determining the collapse mechanism of the structure under a given load system. A structure fails when a sufficient number of disconnections develop within it (Figure 1). FRP elements (strips, bars or cables) have to be placed so as to avoid the formation of said disconnections. This method implies a more refined structural analysis than for the *elastic analysis* method, but generally enables a better dimensioning of the reinforcement. Examples of its application to arches and barrel vaults are given in (Faccio, et al. 1999) and (Faccio et al. 2000).

The third method consists in identifying a resisting structure in which the compressive forces are supported by the masonry and the tensile forces are supported by the composite material.

Choosing between a unidirectional and a bidirectional FRP depends on the direction of the stresses to deal with. When the direction of the principal tensile stress consequent to a certain load condition is well known, a unidirectional material is preferable in order to maximize the resisting effect of the fibers; otherwise, or when a distribution of the load is required, a bidirectional material might be chosen.

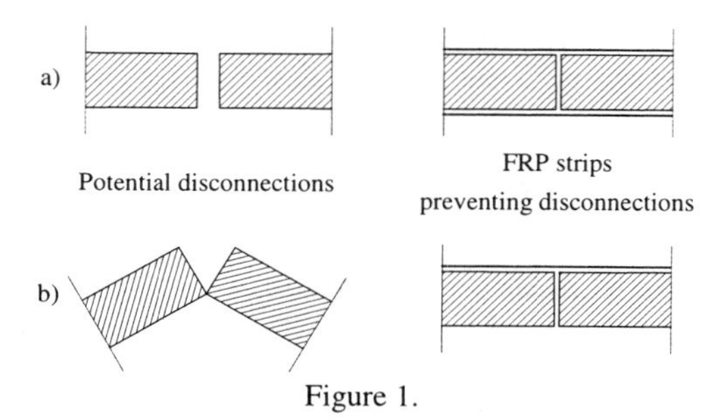

Figure 1.

Typical applications

FRP materials can be used in a number of ways for the structural strengthening of masonry buildings: only some applications referring to recurrent macro-elements in historical buildings are discussed in the following paragraphs.

Reinforcing arches and barrel vaults. Provided that the abutments are capable of sufficient lateral thrusts, the most common collapse mechanism for an arch with an unsymmetrical load corresponds to the formation of four cylindrical hinges (Fig. 1b) and a consequent displacement mechanism (Fig. 2), (Faccio, et al. 1999). FRP strips bonded to the intrados and/or to the extrados locally prevent the formation of the hinges and consequently increase the collapse load.

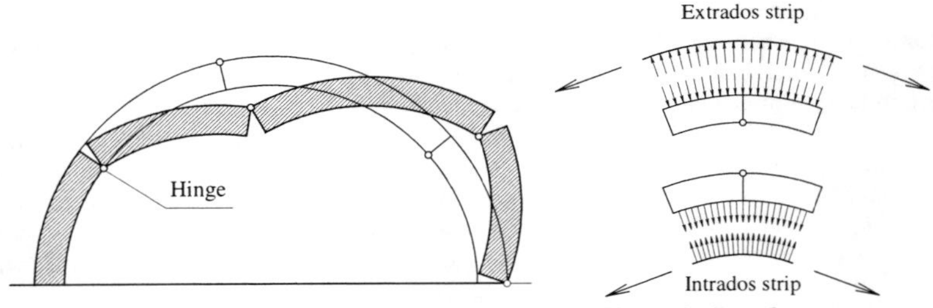

Figure 2. Collapse mechanism of an arch with unsymmetrical load.

Figure 3. Interface stress.

Depending on the arrangement of the strips, the formation of hinges can be totally avoided (in which case, the collapse mechanism will correspond to crushing of masonry), or it can be modified with respect to the structure without reinforcement. In both cases, the collapse load will be increased.

It has to be noted that either intrados or extrados strips are effective in preventing the formation of hinges. In the case of extrados strips, however, compression stresses act at the interface and any debonding of the reinforcement is consequently avoided, whereas tensile stresses act at the interface in the case of intrados strips (Fig. 3). The problem of the bond of intrados strips is analyzed experimentally and theoretically in (Aiello et al. 2000).

Unbonded FRP extrados cable anchored to the supports, with or without prestressing, could also be used to strengthen arches and barrel vaults (the case of using steel cables was experimentally investigated in (Jurina, 2000)). Using unprestressed cables, the opening of the cracks is restricted by the increase in the tensile force in the cable, corresponding to the structural displacements (a minimum stiffness of the cable is required to increase the collapse load); with prestressed cables, the radial compression at the interface also increases the load which causes the formation of hinges.

Reinforcing cross vaults. The structural behavior of full-scale cross vaults with fixed abutments is experimentally and theoretically described in (Faccio et al. 2000 b). Said paper describe the collapse mechanism of cross vaults with and without FRP reinforcement, with a localized load applied on a web. The collapse mechanism of the specimens without reinforcement (sketched in Fig. 4a), involving a lateral thrust brought to bear by the loaded web on the two adjacent webs and consequent tensile cracks in them suggested the reinforcing configuration shown in Fig. 4b. The collapse load of the reinforced specimens was 2.3 and 4.3 times the collapse load of the unstrengthened specimens.

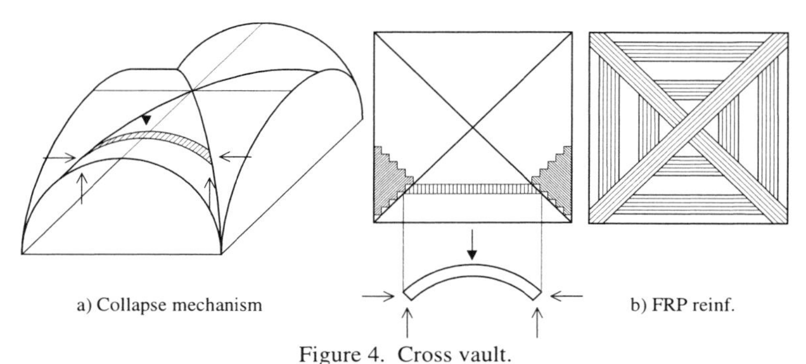

Figure 4. Cross vault.

Reinforcing domes. Two FRP reinforcement configurations for a hemispherical dome are represented in Fig. 5, the first (Fig. 5a) being based on elastic analysis of the structure (membrane theory) and the second (Fig. 5b) on collapse analysis (once cracks form along the meridians, the dome can be analyzed as a number of adjacent arches). A detailed comparison between the two approaches is reported in (Nart, 2000 a, b). Experimental investigations on domes loaded at the top with different reinforcement configurations are reported in (Faccio and Foraboschi, 2000 b).

Belting of buildings. The belting of buildings is often done with FRP strips (Fig. 6a), especially for buildings damaged by earthquakes, to improve the box effect and resist the lateral force corresponding to thrusting elements (Fig. 6b). The effectiveness of composite strips for belting masonry boxes has been experimentally observed in (Avorio et al. 1999).

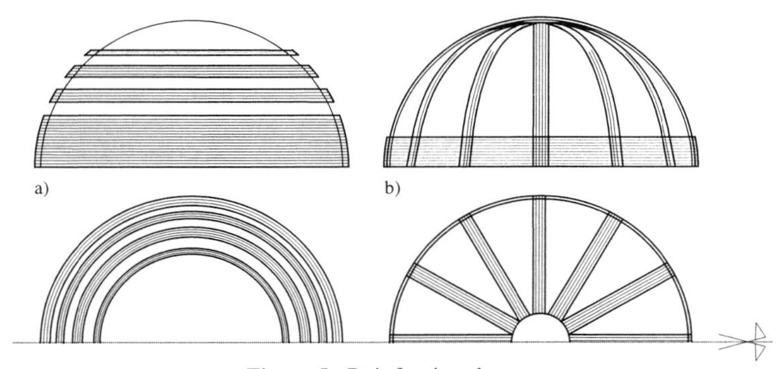

Figure 5. Reinforcing domes.

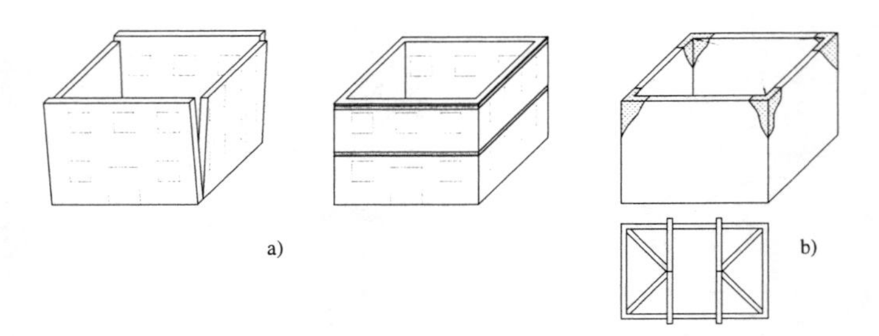

Figure 6. Belting buildings.

Shear reinforcement for tower walls. The in-plane shear resistance of walls can be significantly improved by adopting a strip configurations as shown in Fig. 7 (Schwelger, 1996), (Di Tommaso et al. 1999), achieving a resistant mechanism involving compression forces in the masonry and tensile forces in the composite, (Stratford et al. 2000), (Valluzzi et al. 2000). One-sided reinforcement of a single panel does not significantly improves the collapse load (Valluzzi et al. 2000), whereas a two-sided reinforcement greatly improves it. This finding needs further study, especially since it is unfeasible to apply FRP strips to the outer surfaces of historical towers and taking into account for the global structural behavior of the towers.

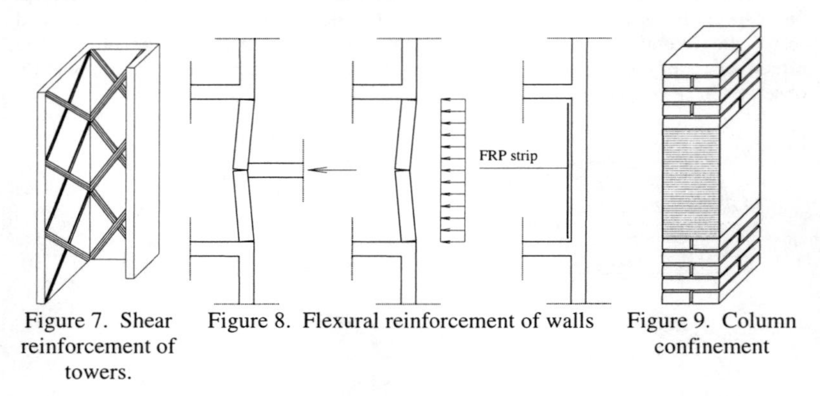

Figure 7. Shear reinforcement of towers.

Figure 8. Flexural reinforcement of walls

Figure 9. Column confinement

Flexural reinforcement for masonry walls. An axially-loaded masonry panel also subjected to out-of-plane bending due to transverse loads or the presence of horizontal structures presents the collapse mechanism illustrated in Fig. 8. The introduction of an FRP element in the tensile area strongly increases the collapse transverse load. It has to be said that FRP reinforcement only helps to counter the transverse load, since it is applied when the vertical load is already coming to bear.

Experimental investigations on masonry panels subjected to an axial (compressive) force and bending moment can be found in (Barbieri, 2000) and (Avorio, Borri, Celestini, Corradi, 1999). Design equations can be found in (Triantafillou,1998). The flexural reinforcement of masonry panels can also be achieved with FRP rods installed near the masonry surface (De Lorenzis, Tinazzi, Nanni, 2000).

Confinement of masonry columns. FRP confinement (Fig. 9) is an effective technique for improving the compressive strength and the ductility of a concrete column. Experimental and theoretical analyses have been carried out on the behavior of FRP-confined circular and rectangular concrete columns. This method can also be used to advantage for reinforcing (and increasing the ductility) of masonry columns of historical buildings, if a plaster cover is provided. FRP wrapping increases the compressive strength of short columns up to 3 times (La Tegola et al. 2000), (Avorio et al. 1999). Rounding off the corners (radius 20 ÷ 30 mm) to avoid local rupture of the fibers should be included in any good code of praxis.

Conclusions

1. FRP strengthening of historical monuments presents remarkable advantages with respect to other traditional techniques for the structural restoration and conservation.
2. The problem to develop design equations related to the aim of the application (sometimes not univocally defined) is nowadays open.
3. The "design criteria" are nowadays satisfactorily defined.

Acknowledgements

The authors are grateful for the financial support from the Ministry for the University and Research (year 2000) for the IUAV Research Group.

References

Aiello, M.A., Galati, N., La Tegola, A. (2000). "Carico di collasso di archi in muratura rinforzati con nastri in FRP." *Proceedings of the conference "Mechanics of masonry structures strengthened with FRP-materials"*, Venice, Italy (in Italian).

Avorio, A., Borri, A., Celestini, G., Corradi, M. (1999). "Sperimentazione e analisi sull'utilizzo dei materiali compositi nelle costruzioni in muratura." *L'Edilizia*, n. 9/10, 1999 (in Italian).

Barbieri, A. (2000). "Interventi strutturali su edifici storici in muratura: elementi pressoinflessi rinforzati con materiale composito." *PhD Thesis*, University of Lecce.

De Lorenzis, L., Tinazzi, D., Nanni, A. (2000). "Near Surface Mounted FRP Rods for Masonry Strengthening: Bond and Flexural Testing." *Proceedings of the conference "Mechanics of masonry structures strengthened with FRP-materials"*, Venice, Italy.

Di Tommaso, A., Barbieri, A. (1999). "Evoluzione delle tecniche per il miglioramento sismico di torri e campanili: impiego degli FRP-materials".

Proceedings of Int. Workshop on Seismic Performance of Built Heritage in Small Historic Centres, Assisi, Italy, 1999 (in Italian).

Faccio, P., Foraboschi, P., Siviero, E. (1999). "Volte in muratura con rinforzi in FRP." *L'Edilizia* n.9/10, 1999 (in Italian).

Faccio, P., Foraboschi, P. (2000), a. "Analisi limite ultima di strutture in muratura con materiali compositi incollati al contorno." *Proceedings of the conference "Mechanics of masonry structures strengthened with FRP-materials"*, Venice, Italy (in Italian).

Faccio, P., Foraboschi, P. (2000), b. "Experimental and Theoretical Analysis of Masonry Vaults with FRP Reinforcements." *Proceedings of 3rd Int. Conf. Advanced Composite Materials in Bridges and Structures.* August 2000. Ottawa, Canada, pp. 629-636.

Faccio, P., Foraboschi, P., Siviero, E. (2000). "Collapse of Masonry Arch Bridges Strengthened Using FRP Laminates." *Proceedings of 3rd Int. Conf. Advanced Composite Materials in Bridges and Structures.* August 2000. Ottawa, Canada, pp. 505-512.

Jurina, L. (2000). "L'arco armato nel consolidamento di archi e volte in muratura." *Recupero e conservazione* n.33, April/May 2000, pp. 54-61 (in Italian).

La Tegola, A., De Lorenzis, L., Micelli, F. (2000). "Confinamento di pilastri in muratura mediante barre e nastri in FRP." *Proceedings of the conference "Mechanics of masonry structures strengthened with FRP-materials"*, Venice, Italy (in Italian).

Nart, M. (2001), a. "Cupole in muratura con rinforzi in materiale composito." *Costruire in laterizio*, in press, (in Italian).

Nart, M. (2001), b. "Il rinforzo in FRP per il consolidamento delle cupole in muratura." *L'edilizia*, in press, (in Italian).

Stratford, T., Pascale, G., Manfroni, O., Bonfiglioli, B. (2000). "Shear Strengthening of Masonry Panels with GFRP: Preliminary Experimental Results.". *Proceedings of the conference "Mechanics of masonry structures strengthened with FRP-materials"*, Venice, Italy.

Schwegler, G. (1996). "Verstärkung von Mauerwerkbauten mit CFK-Lamellen." *Sonderdruck ans Schweizer Ingenieur und Architekt*, n. 44.

Triantafillou, T. C. (1998). "Strengthening of Masonry Structures Using Epoxy-Bonded FRP Laminates." *J. of Composite for Construction* ASCE Vol. 2, No. 2, May 1998, pp. 96-104.

Valluzzi, M. R., Tinazzi, D., Modena, C. (2000). "Prove a taglio su elementi murari rinforzati con tessuti in FRP." *Proceedings of the conference "Mechanics of masonry structures strengthened with FRP-materials"*, Venice, Italy (in Italian).

Seismic Strengthening with FRP: Opportunities and Limitations

Gaetano Manfredi[1] and Andrea Prota[2]

Abstract

This paper deals with seismic strengthening with composites. A general overview on the basic concepts that drive the upgrade of structures located in seismic regions is provided; methods for improving their seismic performance or reducing the seismic demand are outlined. The discussion focuses on opportunities and reasons for using Fiber Reinforced Polymer (FRP) materials in the strengthening of both reinforced concrete (RC) and masonry structures. Possible local (i.e., columns, piers, beam-column joints) and global (i.e., frames and walls) applications are presented along with the most relevant conducted studies. Advantages of techniques utilizing FRP composites for the seismic upgrade of structures are summarized; limitations and needs to be solved and analyzed by future research are indicated.

Strengthening Philosophy in Seismic Areas

Seismic upgrade of structures located in seismic areas is generally aimed either at improving their seismic performance or at limiting the seismic demand within acceptable levels. Depending on the selected strengthening technique, the behavior of the structure can be improved by means of increasing its strength and/or stiffness, boosting its deformation capacity, or allowing a more effective energy dissipation. On the other hand, a reduction of the seismic risk can be obtained by lowering the seismic demand during an earthquake (ATC 1996, BIA 1996).

Techniques such as shear walls, braced frames or buttresses have been used for the upgrade of RC frames, increasing their lateral strength and, consequently, their stiffness; in these cases, new portions (i.e., internal or external) are added to the original structure.

---

[1] Professor of Structural Engineering, Department of Structural Analysis and Design, University of Naples "Federico II", via Claudio 21, 80125 Naples, Italy; phone 081-768-3488; gamanfre@unina.it

[2] PhD Candidate, Department of Structural Analysis and Design, University of Naples "Federico II", via Claudio 21, 80125 Naples, Italy; phone 081-768-3534; aprota@unina.it

Local upgrade of single elements can allow to improve the structural deformation capacity by providing them with a higher confinement level or modifying the strength hierarchy. Columns or piers in under-designed structures (Cosenza et al. 1996) are often characterized by minimum cross section size and lack of steel reinforcement. This results in a weak column – strong beam construction, in which case, a seismic event is likely to determine the formation of plastic hinges in the columns and lead to a soft story collapse mechanism. This failure mode represents the lower bound in the strength hierarchy of the structure and it is both brittle and catastrophic. The upgrade should tend to improve its response during an earthquake in order to cause the formation of plastic hinges to occur in the beams (i.e., beam mechanism). This failure mode is preferred because it allows a more effective energy dissipation, and the undamaged columns are still able to carry vertical loads providing a larger safety margin against the collapse. Steel plates, section enlargements and shotcrete have been used for strengthening columns and beam-column joints. Nowadays, FRP materials show great potential for obtaining the same objective with low installation costs, ease of construction, no durability concerns and not significant influence on the global mass.

Another strengthening technique is based on the reduction of seismic demand, modifying the structural parameters in order to decrease the demand of forces and displacements under seismic actions. This can be achieved by base isolation or energy dissipation systems, by active control or by reducing the global mass.

Composites for Seismic Applications: Where and Why

FRP laminates and near surface mounted (NSM) rods can represent an effective tool when the strengthening philosophy is based on the upgrade of the structural elements as previous discussed. Laminates are installed by manual lay-up and impregnated in-situ; in the case of wrapping, prefabricated systems can be also adopted. Carbon (C) FRP laminates have been widely utilized, even though glass (G) and aramid (A) FRP laminates have been also applied for confining RC elements or for strengthening masonry structures. Due to their mechanical properties, CFRP NSM rods are generally selected for RC, while GFPR NSM rods are preferred for masonry. Seismic applications of FRP materials to columns and bridge piers, beam-column joints and walls are summarized in the following sections.

Columns and Piers

Shear Strengthening. The lack of transverse reinforcement can cause a brittle shear failure (Seible et al. 1997), with formation of inclined cracks due to diagonal tension, cover concrete spalling and rupture of the transverse reinforcement. Externally bonded FRP laminates with fibers in the hoop direction increase the shear capacity of deficient columns or piers. The effectiveness of this strengthening technique has been validated by many experimental studies. It has been demonstrated that the shear strength of the upgraded column or pier increases proportionally with the thickness of CFRP and AFRP jacket (Fujisaki et al. 1997, Masukawa et al. 1997, Matsuzaki et al. 2000); in particular, Priestley and Seible (1995) showed that FRP shear retrofit allows to achieve a stable hysteretic behavior with high ductility levels.

Confinement. The lack of appropriate size and spacing of ties in a column or pier can induce the collapse to occur at its end, resulting in crushing of the not confined concrete, instability of the steel reinforcing bars in compression and pull out of those in tension. Experimental tests conducted by Kobatake et al. (1993), Saadatmanesh et al. (1996), Masukawa et al. (1997), Seible et al. (1997), Fukuyama et al. (1999), Matsuzaki et al. (1999), Seible et al. (1999), Pantelides et al. (2000a), Pantelides et al. (2000b), Prota et al. (2001), Saiidi et al. (2000), Sheikh et al. (2000), confirmed the effectiveness of FRP wrapping for its function of confining the plastic hinge regions and allowing a significant enhancement of column or pier ductility capacity.

In order to carry out a proper design of FRP jacket, a model for FRP-confined concrete is needed. Mirmiran and Shahawy (1997) approached the problem by looking at the volumetric strain of concrete; Priestley and Seible (1995), Karbhari and Gao (1997), Samaan et al. (1998), Wang and Restrepo (2001) also proposed other models on the base of experimental results. Within this series of studies (Manfredi and Realfonzo 2001), models presented by Mirmiran and Shahawy (1997) and Spoelstra and Monti (1999) represent an important step towards simple and accurate criteria for the design of FRP jacket. However, the future research needs to clarify issues such as the effectiveness of confinement for rectangular sections and the reduction of FRP ultimate strain due to the corner effects, where the concentration of transverse stresses lowers the actual performance of the jacket.

The implementation of the FRP-confined concrete model in the calculation of the moment-curvature relationship for strengthened elements (Wang et al. 2000) requires considerations on both the interaction stirrups/FRP confinement and the combination of axial load and bending.

Lap splices. This is a typical deficiency concerning the lower end of columns or piers. As vertical cracks initiate in the cover, concrete dilatation occurs and eventually cover spalling is generated. Test on columns and piers (Ma and Xiao 1997, Saadatmanesh et al. 1997, Seible et al. 1997, Osada et al. 1999, Pantelides et al. 2000b) evidenced that slippage of reinforcement bars start for dilation strains ranging between 0.001 and 0.002. FRP lap confinement can limit such strain and then prevent the rapid flexural strength degradation of the vertical member. Seible et al. (1997) found the FRP wrapping thickness needed to clamp the lap splice region to be directly proportional to the effective column diameter and inversely proportional to the modulus of elasticity of the laminate. They suggest to design it based on the lower bound strain of 0.001. Experimental tests conducted by Prota et al. (2001) showed that CFRP NSM rods could also be a solution as additional reinforcement fully effective in the lap splice region.

Beam-Column Joints

The upgrade of beam-column joints represents a key issue for both buildings and bridges. Even though different types of connections exist, the concept of strength hierarchy represents the controlling concept of their seismic retrofit. The local upgrade of members should have the objective of obtaining a ductile global behavior and achieving global mechanisms.

The strengthening of columns by providing them with higher confinement level and/or with more flexural reinforcement could cause the failure to occur in the nodal zone. The increase of column strength and/or ductility improves the structural performance in terms of global behavior. In order to move up along the strength hierarchy, the panel should be also strengthened. The upgrade of both column and panel could allow to move from the previous intermediate level of the strength hierarchy (i.e., shear failure of the panel) to its upper bound (i.e., crisis of the beams). Inducing such failure mode would be the best result of a seismic repair/strengthening. Formation of plastic beam hinges would mean that a ductile and very effectively energy dissipating mechanism be achieved.

Experimental tests demonstrated that FRP composites could be a valid alternative to conventional RC or steel jackets. Pantelides et al. (2000a), Pantelides et al. (2000b), Gergely et al. (2000), Ghobarah and Said (2001) strengthened successfully with FRP laminates exterior beam-column joints with deficiency in shear strength. Antonopoulos and Triantafillou (2001) investigated 2/3-scale exterior joints looking at different design parameters. Prota et al. (2001) tested RC interior connections proposing an innovative technique based on the combined use of CFRP laminates and NSM bars (Figure 1-a). Experimental outcomes showed that a selective upgrade could be achieved choosing different combinations and location of sheets and bars in order to obtain different structural performances of joints (Figure 2).

Frames

Full-scale tests on RC frames demonstrated the potential of composites for seismic repair and strengthening. Castellani et al. (1999) used CFRP laminates for the retrofit of a RC frame designed according to Eurocode8 provisions, while a dual-system frame (Tsionis et al. 2001) designed using displacement-based criteria was first damaged and then tested after that columns and joints were retrofitted by uni-axial, bi-axial and quadri-axial CFRP fabrics (Figure 3).

In both cases pseudo-dynamic tests were conducted. Experimental results represent a first confirmation of the effectiveness of such upgrade techniques even under cyclic actions.

Walls

Experiments on RC shear walls with FRP laminates as flexural strengthened have been carried out by Lombard et al. (2000), while Masuo (1999) and Iso et al. (2000) tested RC columns with wing walls strengthened in shear. Important studies on the shear behavior of walls strengthened with FRP have been carried out by Ehsani et. al. (1997) and Triantafillou (1998). Tinazzi et al. (2000), Tumialan et al. (2000) and Tumialan et al. (2001) studied the in-plane and out-of-plane behavior of masonry walls strengthened with both laminates and NSM rods. Preliminary outcomes highlighted the potential of FRP composites for seismic strengthening of wall systems. At this time, limited data are available in literature; next research developments will need to confirm those promising results.

Composites for Seismic Applications: Advantages, Limitations and Research Needs

Advantages. The opportunity of choosing the type of fibers, their orientation, their thickness and the number of plies results in a great flexibility in selecting the appropriate retrofit scheme that allows to target the strength hierarchy at both local (i.e., upgrade of single elements) and global (i.e., localized strengthening in order to prevent local mechanisms) level. FRP materials allow to achieve these objectives with a negligible change of global mass. When they are subject to uniform tension (i.e., confinement of circular sections), composites perform very well. Their weight and installation procedure result in significant advantages from a constructability stand-point; in many cases, their application does not require to interrupt the use of the structure.

Limitations. The lack of standards in quality control of manufacturing and in-situ application and the absence of well-stated design criteria represent a barrier for the full development of the seismic strengthening with FRP. Detailing the structural behavior, composites are very sensitive to transverse actions (i.e., corner or discontinuity effects) and unable to transfer local shear (i.e., interfacial failure) (Figure 1-b). In cases of alternated loads, their contribution appears to be very low when in compression.

Research Needs. The use of GFRP fabrics requires more interest, mainly when high stiffness and strength in strengthening could reduce displacement and dissipative capacity (i.e. masonry). Effects of the cumulative damage generated by cyclic loads should be clarified for strenghtened elements. Structural behavior and design criteria should be more analyzed and defined mainly for: a) hollow columns; b) columns with high ratio between sides; c) joints, taking into account failure mechanism related to local shear transfer (Figure 1-b); d) shear strengthening of RC shear walls. Local effects on the bond mechanism under alternated load should be also studied.

References

Antonopoulos, C., and Triantafillou, T.C. (2001). "Experimental investigation of FRP-strengthened RC beam-column joints." *ASCE-Journal of Composites for Construction*, submitted.

ATC. (1996). "Seismic evaluation and retrofit of concrete buildings." Applied Technology Council (Rep.No.ATC – 40), Redwood City, Ca.

B.I.A. "The Assessment and Improvement of the Structural Performance of Earthquake Risk Buildings." New Zealand National Society for Earthquake Engineering.

Castellani, A., Negro, P., Colombo, A., Grandi, A., Ghisalberti, G., and Castellani, M. (1999). "Carbon fiber reinforced polymers (CFRP) for strengthening and repairing under seismic actions." European Laboratory for Structural Assessment, Joint Research Centre Research Report I.99.41, Ispra, Italy.

Cosenza, E., and Manfredi, G. (1996). "Some Remarks on the Evaluation and Strengthening of Underdesigned r.c. Frame Buildings." Proceedings of the Workshop US-Italy on Seismic Evaluation and Retrofit, Abrams, D., and Calvi, G.M. (eds), Columbia University, New York, 157-175.

Ehsani, M.R., Saadatmanesh, H., and Al-Saidy, A. (1997). "Shear Behavior of URM Retrofitted with FRP Overlays." *ASCE-Journal of Composites for Construction*, 1(1), 17-25.

Fujisaki, T., Hosotani, M., Ohno, S., and Mutsuyoshi, H. (1997). "JCI state-of-the-art on retrofitting." Proceedings of the Third Conference on Non-Metallic (FRP) Reinforcement for Concrete Structures (Vol. 1), Japan Concrete Institute, 613-620.

Fukuyama, H., Suzuki, H., and Nakamura, H. (1999). "Seismic retrofit of reinforced concrete columns by fiber sheet wrapping without removal of finishing mortar and side wall concrete." Fiber Reinforced Polymer Reinforcement for Reinforced Concrete Structures. Dolan, C.W., Rizkalla, S.H. and Nanni, A. (eds), ACI Report SP-188. Detroit, Michigan. 205-216.

Gergely, J., Pantelides, C.P., and Reaveley, L.D. (2000). "Shear strengthening of RC T-Joints using CFRP composites." *ASCE-Journal of Composites for Construction*, 4(2), 56-64.

Ghobarah, A., and Said, A. (2001). "Seismic Rehabilitation of Beam-Column Joints using FRP Laminates." *Journal of Earthquake Engineering*, 5(1), 113-129.

Karbhari, V.M., and Gao, Y. (1997). "Composite jacketed concrete under uniaxial compression – verification of simple design equations." *ASCE-Journal of Materials in Civil Engineering,* 9(4), 185-193.

Kobatake, Y., Kimura, K., and Katsumata, H. (1993). "A retrofitting method for reinforced concrete structures using carbon fiber." Fiber-Reinforced-Plastic (FRP) Reinforcement for Concrete Structures: Properties and Applications, Nanni, A. (ed.), Elsevier Science Publishers. 435-450.

Ma, R., and Xiao, Y. (1997). "Seismic retrofit and repair of circular bridge columns with advanced composite materials." *Earthquake Spectra,* 15(4), 747-764.

Manfredi, G., and Realfonzo, R. (2001). "Models of Concrete Confined by Fiber Composites." Proceedings of the Fifth Conference on Non-Metallic Reinforcement fo Concrete Structures, Cambridge, UK, paper no. 154.

Masukawa, J., Akiyama, H., and Saito, H. (1997). "Retrofit of existing reinforced concrete piers by using carbon fiber sheet and aramid fiber sheet." Proceedings of the 3rd Conference on Non-Metallic (FRP) Reinforcement for Concrete Structures (Vol. 1). Japan Concrete Institute, 411-418.

Matsuzaki, Y., Nakano, K., Fujii, S., and Fukuyama, H. (2000). "Seismic retrofit using continuous fiber sheets." Proceedings of the 12th World Conference on Earthquake Engineering, New Zealand. Paper no. 2524.

Mirmiran, A., and Shahawy, M. (1997). "Behavior of concrete columns confined by fiber composites." *ASCE-Journal of Structural Engineering*, 123(5), 583-590.

Osada, K., Yamaguchi, T., and Ikeda, S. (1999). "Seismic performance and the retrofit of hollow circular reinforced concrete piers having reinforcement cut-off planes and variable wall thickness." Transactions of the Japan Concrete Institute, 21, 263-274.

Pantelides, C.P., Clyde, C., and Reaveley, L.D. (2000a). "Rehabilitation of R/C Buildings Joints with FRP Composites." Proceedings of the 12th World Conference on Earthquake Engineering, New Zealand, Paper no. 2306.

Pantelides, C.P., Gergely, J., Reaveley, L.D., and Volnyy, V.A. (2000b). "Seismic Strengthening of Reinforced Concrete Bridge Pier with FRP Composites" Proceedings of the 12th World Conference on Earthquake Engineering, New Zealand, paper no. 127.

Priestley, M.J.N., and Seible, F. (1995). "Design of seismic retrofit measures for concrete and masonry structures." *Construction and Building Materials*, 9(6), 365-377.

Prota, A., Nanni, A., Manfredi, G., and Cosenza, E. (2001). "Selective Upgrade of Beam-Column Joints with Composites", Proceedings of the International Conference on FRP Composites in Civil Engineering CICE 2001, Hong Kong, China, in press.

Saadatmanesh, H., Ehsani, M.R., and Jin, L. (1996). "Seismic strengthening of circular bridge pier models with fiber composites." *ACI Structural Journal*, 93(6), 639-647.

Saadatmanesh, H., Ehsani, M.R., and Jin, L. (1997). "Repair of earthquake-damaged RC columns with FRP wraps." *ACI Structural Journal*, 94(2), 206-215.

Saiidi, S., Sanders, D.H., Gordaninejad, F., Martinovic, F.M. and McElhaney, B.A. (2000). "Seismic retrofit of non-prismatic RC bridge columns with fibrous composites." Proceedings of the 12th World Conference on Earthquake Engineering, New Zealand, Paper no. 123.

Samaan, M., Mirmiran, A., and Shahawy, M. (1998). "Model of concrete confined by fiber composites." *ASCE-Journal of Structural Engineering*, 124(9), 1025-1031.

Seible, F., Priestley, M.J.N., Hegemier, G.A., and Innamorato, D. (1997). "Seismic Retrofit of RC Columns with Continuous Carbon Fiber Jackets." *ASCE-Journal of Composites for Construction*, 1(2), 52-62.

Seible, F., Innamorato, D., Baumgartner, J., Karbhari, V., and Sheng, L.H. (1999). "Seismic retrofit of flexural bridge spandrel columns using fiber reinforced polymer composite jackets." Fiber Reinforced Polymer Reinforcement for Reinforced Concrete Structures. Dolan, C.W., Rizkalla, S.H. and Nanni, A. (eds), ACI Report SP-188. Detroit, Michigan. 919-931.

Sheikh, S.A., Iacobucci, R., and Bayrak, O. (2000). "Seismic upgrade of concrete columns with fibre reinforced polymers." Advanced Composite Materials in Bridges and Structures. Humar, J., and Razaqpur, A.G. (eds), Montreal: The Canadian Society for Civil Engineering. 267-274.

Spoelstra, M.R., and Monti, G. (1999). "FRP-confined concrete model." *ASCE-Journal of Composites for Construction*, 3(3), 143-150.

Tinazzi, D., Modena, D., and Nanni, A. (2000). "Strengthening of Masonry Assemblages with FRP Rods and Laminates," Proceedings of International Meeting on Composite Materials, PLAST 2000, Crivelli-Visconti, I. (ed), Milan, Italy, 411-418.

Triantafillou, T. C. (1998). "Strengthening of Masonry Structures using Epoxi-Bonded FRP Laminates." *ASCE-Journal of Composites for Construction,* 2(2), 96-104.

Tsionis, G., Negro P., Molina J., and Colombo A. (2001). "Pseudodinamic Tests on a 4-Storey RC Dual Frame Building," Report EUR 19902, Ispra, Italy.

Tumialan, G., Micelli, F., and Nanni, A. (2001). "Strengthening of Masonry Structures with FRP Composites." Structures 2001, Washington DC, CD-ROM , paper 40558-052-005, 8 pp.

Tumialan, G., Myers, J.J., and Nanni, A. (2000). "An In-Situ Evaluation of FRP Strengthened Unreinforced Masonry Walls Subjected to Out of Plane Loading," ASCE Structures Congress, Philadelphia, PA M. Elgaaly, Ed., CD version, #40492-046-004, 8 pp.

Wang, Y.C., Restrepo, J.I., and Park, R. (2000). "Retrofit of Reinforced Concrete Memebrs using Advanced Composite Materials." Research Report 2000-3, Department of Civil Engineering, University of Cantebury Christchurch, New Zealand, 375 pp.

Wang, Y.C., and Restrepo, J.I. (2001). "Investigation of Concentrically Loaded Reinforced Concrete Columns Confined with Glass Fiber-Reinforced Polymer Jackets." *ACI Structural Journal*, 98(3), 377-385.

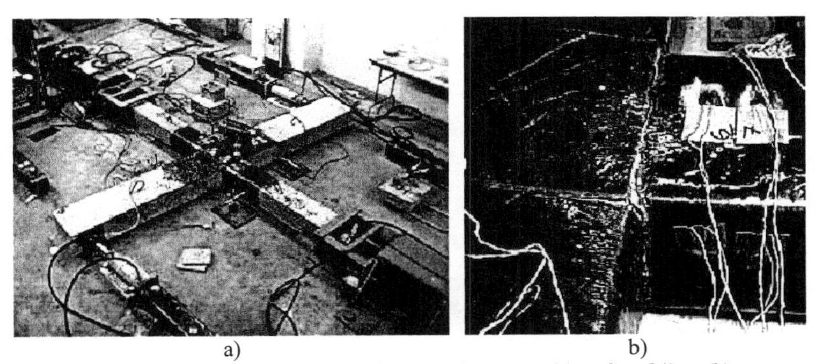

a) b)

Figure 1. Test on RC interior joint (a), column-panel interface failure (b) (Prota et al. 2001)

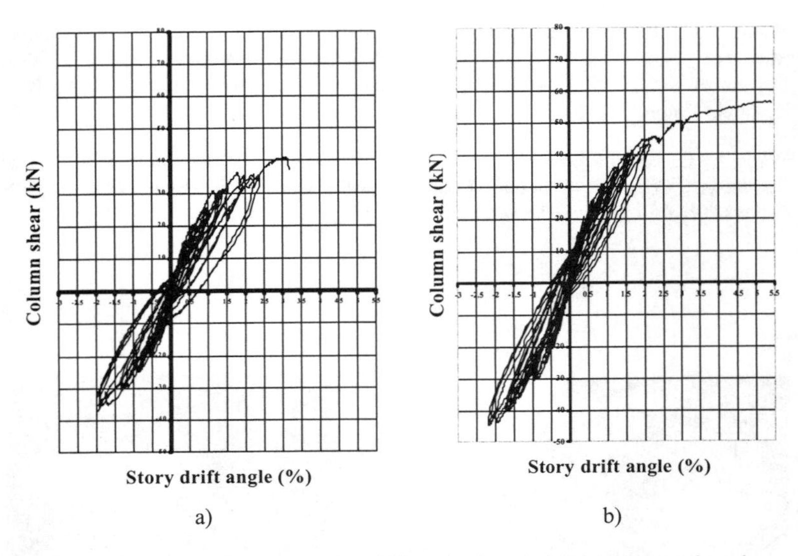

Figure 2. Column shear versus story drift angle: a) as-built joint; b) strengthened joint (Prota et al. 2001)

Figure 3. Tested RC frame (a), repaired joint (b) (Tsionis et al. 2001)

Structural behaviour of FRP profiles

Marisa Pecce[1]

Abstract

The present paper summarizes the meaningful subjects of the structural behaviour of Fiber Reinforced Polymer (FRP) profiles. Material advantages are defined, underlining that it can be fruitfully used in infrastructure where durability is a controlling factor; many bridges have also been constructed designing specific shapes and systems. Three problems are analysed focusing the main points of each one: mechanical characterization, buckling and failure mode. References of experimental tests, modeling and design issues are discussed. The state-of-the-art shows many research needs especially due to the variety of products available in the market and the lack of general provisions for experimental tests and design.

Introduction

Fiber Reinforced Polymer (FRP) profiles offer some benefits compared to traditional materials of civil engineering such as concrete and steel. A direct comparison can be done in terms of strength/weight ratio and durability, which are both higher for FRP materials.

Another peculiarity is the possibility of changing the material properties designing the volume ratio of the fibers to resin and choosing the type and orientation of the fibers. Using specific resins, the protection of the fibers against fire, thermal and environmental effects can be improved.

The meaningful reasons of using FRP profiles are the faster installation time due to the low weight and the low maintenance costs due to durability. Therefore, their great potential is for bridges exposed to aggressive environments and subjected to an intense traffic volume; in these cases, management costs related to the service interruption play a crucial role in the selection of the construction system.

Only a design that considers maintenance and life duration can make FRP profiles competitive with other materials surely characterized by lower initial construction costs.

Many bridges have been realized in the last ten years. The most famous and long were realized in Aberfeldy (Scotland) in 1992 and in Kolding (Denmark) in

---

[1] Full Professor of Structural Engineering, School of Engineering, University of Sannio, Piazza Roma, Benevento, Italy; phone (39) 081 7683488; pecce@unisannio.it

1997; in both cases FRP profiles were chosen to allow a fast and simple construction and to rapidly restore the service. Many other short span bridges have been constructed realizing specific FRP shapes and carrying out a design by testing. A typical example is the rehabilitation of the Tom's Creek Bridge in Blacksburg (Hayes et al. 2000). The old steel beams of the deck were deteriorated and the primary requirement of the design was durability; FRP beams were chosen and the developed system was tested in laboratory in real scale. The FRP bridge described in Foster et al. (2000) weighs about one-fifth as much as conventional reinforced concrete structures, and was constructed in six weeks instead of ten weeks, needed for traditional materials.

Another issue related to FRP structures is the need of designing new solutions for the section shapes and the systems of assembling; the traditional systems used for steel are not reliable since the mechanical characteristics of FRP are completely different from those of steel. New systems have been studied especially for bridge decks. The ACCS system was designed by Maunsell International Ltd. for the Aberfeldy bridge, and was assembled on the site in only ten weeks obtaining a 120m-long fully bonded element; more recently, a modular causeway system has been designed (Bank et al. 2000) and named PCC (Pultruded Composite Causeway).

A lot of tests and numerical analyses have been focused on characterizing FRP materials and studying the structural behaviour of FRP profiles; however, variability in constituent materials, fiber architecture and manufacturing processes makes difficult any generalization. Design provisions have been developed by several producers, but they are referred to specific series of profiles. Furthermore, problems such as long term behaviour have not been completely understood and model development has been carried out on the base of few experimental tests, while new products and systems are continuously proposed on the market.

The high ratio strength/elastic modulus, and the orthotropic behaviour with high ratios longitudinal/transversal Young's modulus and longitudinal/shear modulus, can be highlighted as the main aspects that influence the structural behaviour of FRP profiles, define their potential use and limit their application. It is clear that stiffness and buckling issues govern FRP constructions, resulting in stress values at service that are many times lower than those of steel structures.

The following sections deal with mechanical characterization, buckling and failure modes of FRP shapes, providing a brief background for each of these topics. The discussion is aimed at evidencing differences with traditional materials, the key issues of the design approach and the research needs.

Characterization of material

FRP profiles are constructed by pultrusion using E-glass fibers in most cases; in the second generation of shapes, a particular architecture of carbon and glass fibers is realized (hybrid elements). The pultruded shapes are obtained coupling rovings and mats of fibers, so that the final product is formed by layers (Satiropoulos et al. 1994); the fiber volume fraction varies between 30-60%. Generally thermoset resins such as polyester and vinylester are used. The cross section can be I, channel, angle and box, but also custom shapes, sometimes designed by the client, are available.

FRP pultruded profiles are orthotropic materials, and their mechanical properties depend on the fibers orientation. The behaviour is linear elastic up to failure, and five elastic constants are necessary to define the material:

- Young's modulus and Poisson ratio parallel to the fibers: E_x and ν_{xy};
- Young's modulus and Poisson ratio perpendicular to the fibers: E_y and ν_{yx};
- Shear modulus: G.

Material strength is usually evaluated in the direction of fibers, because the strength in the orthogonal direction is governed by resin. Tensile tests are difficult; the entire element cannot be gripped and for coupon specimens the grips pressure is too high for the transversal strength. Compression tests can be carried out on entire profile, but in most cases buckling or local failure occurs before material crisis.

The values of elastic constants and strength are strongly depending on the types of fibers, even though for the same type of fibers available products are different. In Table 1 the longitudinal modulus, the shear modulus and the tensile strength of shapes produced by the most important producers (Strongwell and Creative Pultrusion for USA and Fiberline for Europe) are reported; the values are referred to material properties probably estimated with coupon tests. It is important to observe that various performances are offered by each producer, in addition to those reported in Table 1.

For what concerns the other elastic constants, the typical values for the ratio E_y/E_x range from 0.3 to 0.6, while for ν_{yx} they range between 0.1 and 0.2. Moreover, the shear modulus is less variable because, for FRP with unidirectional fibers, it depends only on the resin properties (Sonti et al. 1994).

Along with the discussed typical shapes, hybrid shapes with glass and carbon fibers can be also available. They are characterized by carbon fibers concentrated in the flange to improve bending strength and they have been used for the rehabilitation design described in (Hayes et al. 2000). For hybrid elements it is essential to test the entire profile rather than coupon specimens.

Experimental tests carried out by academics for defining material properties have shown that elastic moduli and strength can be described by a Weibull distribution. The coefficients of variation (COV) observed in wide experimental analysis are in the range of 2% to 11% (Zureick and Steffen 2000) for various profiles of the same series; for the same profile, COV reached 15% (Zureick and Scott 1997) and average values for flanges and web differ of about 11%.

The evaluation of elastic constants on the entire element is difficult, while few data available show that the extension of the coupon results is not completely reliable. The pultrusion technique used for profiles gives laminate materials with various layers, therefore mechanical properties depend on the thickness and experimental results on coupon specimens can not generalized to material when the shape and the thickness change (Cosenza et. al. 1996, Nagaraj and GangaRao 1997). Characterization and quality control need to be evaluated on each series by coupon and structural tests.

The complexity of the problem and the lack of provisions about the procedure of experimental tests suggest to use high safety factors for strength; Zureick and Steffen (2000) suggest higher than four. Surely it is convenient to introduce suitable safety factors also for elastic constants in stiffness and buckling verifications.

However, high safety factors increase the construction cost, highlighting the need of a fast improvement of knowledge and quality control.

Buckling

Many types of structures have been constructed with FRP profiles, and in most cases buckling phenomena remain a fundamental key issue of the design.

Buckling is global if deformation regards the entire element and is local if only parts of the elements are involved.

When FRP is used bucking is different from steel profiles; the phenomenon is always elastic but it can be influenced by orthotropy and high shear strain of the material. Furthermore, there are effects of geometrical and mechanical imperfections, about which there is lack of information.

Variation and uncertainties of material properties, previously summarized, induced researchers to characterize the material of their specimens before analysing buckling behaviour; sometimes data about geometrical imperfections have been obtained.

Elements in compression and bending have been tested. Global buckling has been evidenced in many studies (Davalos and Pizhong 1997, Mottram 1992, Zureick and Scott 1997) for I-shaped and box profiles; recently single angle struts have been tested (Zureick and Steffen 2000), that are commonly used in trusses. Global buckling is governed by longitudinal Young's modulus, i.e. the higher one, but the results have cleared out the importance of shear strain due to the high ratio between Young's modulus and shear modulus (about 6-10). Therefore, in the formulas proposed by Zureick and Scott (1997) for evaluating the critical load, a factor depending on the section shape and the shear modulus is introduced obtaining a new expression of the global slenderness λ. The experimental data considered by the authors confirm the good reliability of this parameter. In Zureick and Steffen (2000) the influence of imperfection is evidenced.

In FRP profiles local buckling (Figure 1) is more dangerous than global one, because the sub-element (flange, web, leg) is bidimensional and its behaviour is affected also by the transversal elastic modulus, lower than the longitudinal one.

Also the pure local buckling phenomenon has been analysed in many studies (Bank et al. 1995, Cosenza et al. 1998, Yooh et al. 1996, Barbero et al. 1992, Barbero ans Raftoyianni 1993, Bank et al. 1994, Bank at al. 1994b) considering the opportunity of simplified models or finite element methods (Bank and Yin 1996, Cosenza et al. 1996, Yooh et al. 1996), for evaluating the critical load of elements in compression or in bending. The most common simplified approach considers the flange as a plate elastically restrained by the web; different numerical and analytical procedures have been proposed for the solution of the problem and the evaluation of the critical load. However, the restraining action of the section components has not been clearly defined, even though the geometrical parameters of the cross section governing the problem have been individuated (Bank and Rodhes 1983).

In (Cosenza and Pecce 2000) an analytical expression of the local buckling curve for I-shaped profiles is proposed, considering the orthotropy of the material and the restraint action of the web on the flange, that is explicitly introduced as a function of the geometrical and mechanical data of the section sub-components. The curve has been developed by a wide numerical analysis with element finite method, and its reliability as a design tool is confirmed by the comparison to the experimental results. In Figure 2 the comparison of the proposed curve to experimental results is shown; the ratio between the critical stress (σ_{cr}) and the strength (σ_u) is drawn versus the local slenderness λ, in which expression the cross-section dimensions and the elastic constant of the material are introduced. Other experimental tests need to focus on local buckling of I-shaped beams and the curve for other shapes should be defined.

Another approach for evaluating the critical load has been introduced in Barbero and De Vivo (1999); based on many experimental results and numerical simulations with a finite element model, for cases where local and/or global buckling mode occurred, a single expression for developing the buckling curve is proposed; in it, a coefficient takes into account the mode-interaction. However, the curve gives the critical load as a function of the failure load of the squat element, that for many shapes corresponds to the local buckling critical load; thus the reliability of the formulation is related to the experimental or theoretical evaluation of this failure load.

It has to be underlined that buckling curves, global or local, have always the classical expression, while slenderness or critical load definitions are different.

The examination of experimental results and the model approaches about buckling suggest that using global (even with mode-interaction) and local buckling curves is possible to optimize the length of the element for each cross section shape, and also to provide suggestions for improving the elastic constant values.

All tests show that the ratio between critical stress and strength of material depends on the type of buckling and the shape of the beam section. The lower values are reached in global buckling for slender elements and in local buckling for I-shaped beams with wide flanges. In this last case, the result does not depend on the length, and critical stress is between 30% and 50% of the strength; this points out the design importance of the phenomenon and probably the need of reshape the sections.

Failure modes

Failure modes are in most cases due to buckling, with a type of crushing (Figure 3) that depends on the element deformation in post-buckling conditions (Bank and Yin 1999).

Crushing of the material can be attained only for sections where local buckling is limited and the element is squat. However, also for these cases the element failure is not attained, whereas local crushing occurs. In compression tests the edges of the element could be damaged with a delamination of the material, therefore stiffening systems have to be adopted. In bending tests, local failure can occur where concentrated loads are applied or at supports; material strength in the direction perpendicular to the element surface is governed by resin properties and then it is very low. Moreover, local buckling of the section web probably occurs.

Particular systems in FRP such as the deck system proposed by Strongwell (Hayes et al. 2000) showed local failure due to punching around the loaded area and damage at he holes where transversal tube elements are connected with longitudinal profiles. However, the design of the deck is governed by deflection limits at a load that is 1/3 of the ultimate value. The test showed a good performance for fatigue loads with no significant loss of stiffness and strength after 3.000.000 cycles. In conclusion, the failure mode due to the material strength is practically not significant because other phenomena govern the element crushing .

Conclusive remarks

The subjects summarized have outlined research needs concerning the structural behaviour of FRP profiles. Further experimental tests need to validate the models for the evaluation of critical load in global and local buckling; design suggestions have to be developed for checking the elements with reliable safety factors. Using the models for local and global buckling the optimization of the shapes available in the market can be studied in terms of geometrical and mechanical characteristics.

Experimental procedures and guidelines for interpretation of results need mechanical characterization; more information about variability of elastic constants and strength have to be obtained identifying the parameters that govern the distribution of the values (thickness, fiber amount, type of resin, shape of the section). These data are important to use suitable safety factors in a design that does not over-dimension to cover the unknowns.

Each type of system constructed with pultruded shapes needs specific analysis and experimental tests because local failures depend on the loading pattern, the shape and the assemblage system.

References

Bank, L.C., Gentry, T.R., Nuss, S.S., Hurd, A., Lamanna, A., Duich, S., Oh, B. (2000). "Construction of a Pultruded Composite Structure: A Case of Study" J. of Composites for Construction, ASCE, 4 (3), 112-119.

Bank, L.C., Gentry, T.R., Nadipelli, M. (1994). "Local Buckling of Pultruded FRP Beams- Analysis and design" Proceedings of 49th Annual Conference, Composites Institute, The Society of Plastic Industry, Session 8-D, 1-6.

Bank, L.C., Nadipelli, M., Gentry, T.R. (1994). "Local Buckling and Failure of Pultruded FRP Beams" J. of Materials in Engineering Material and Technology, 1(116), 233-237.

Bank, L.C., Yin, J. (1996). "Buckling of Orthotropic Plates with Free and Rotationally Restrained Unloaded Edges" Thin-Walled Structures, 24, 83-96.

Bank, L.C., Yin, J. (1999). "Failure of Web-flange Junction in Postbuckling Pultruded I-Beams" J. of Composites for Construction, ASCE, 3 (4), 177-184.

Bank, L.C., Yin, J., Nadipelli, M. (1995). "Local buckling of pultruded beams – nonlinearity, anisotropy and inhomogeneity" J. of Construction and Building Materials, 9(6), 325-331.

Banks, W.M., Rhodes, J. (1983). "The Instability of Composite Channel Sections" Proceedings 3rd Int. Conference Composite Structures: Applied Science Publisher, 442-452.

Barbero, E., Fu, S.H., Raftoyiannis, I. (1993). "Local Buckling of FRP Beams and Columns" J. of Materials in Civil Engineering, 5(3), 339-355.

Barbero, E.J., DeVivo, L. (1999). "Beam-Column design equations for wide-flange pultruded structural shapes" J. of Composites for Constructions 1999; 3(4): 185-191.

Barbero, E.J., Tomblin, J., Ritchey, R.A. (1992). "Local Buckling of FRP structural Shapes" Proceeding of 47th Annual Conference, Composite Institute, The Society of Plastic Industry, Session 15-E, 1-7.

Cosenza, E., Lazzaro, F., Pecce, M. (1996). "Experimental Evaluation of Bending and Torsional Deformability of FRP Pultruded Beams" Proceedings of the 2st International Conference on Advanced Composite Materials in Bridge and Structures. Montreal, Canada: Canadian Society for Civil Engineering, 117-124.

Cosenza, E., Lazzaro, F., Pecce, M. (1998). "Local buckling of FRP profiles: experimental results and numerical analysis" Proceedings of 8° European Conference on Composite Materials, Napoli, Italy, 331-338.

Davalos, F.J., Pizhong, Q. (1997). "A analytical and experimental study of lateral and distorsional buckling of frp wide-flange beams" J. of Composites for Construction, 1(4), 150-159.

Foster, D.C., Dan Richards, P.E., and Bogner, B.R. (2000). "Design and Installation of Fiber-Reinforced Polymer Composite Bridge" J. of Composite for Construction, ASCE, 4 (1), 33-37.

Hayes, M..D., Lesko, J.J., Haramis, J., Cousin, T.E., Gomez, J., and Massarelli, P. (2000). "Laboratory Testing and Evaluation of a Composite Bridge Superstructure" J. of Composites for Construction, ASCE, 4 (3), 120-128.

Mottram, J.Y. (1992). "Lateral-torsional buckling of pultruded I-beams" Composites, 23(2), 81-92.

Nagaraj, V., GangaRao, H.V.S. (1997). "Static behaviour of pultruded GFRP beams", J. of Composites for Construction , 1(3), 120-129.

Pecce, M., Cosenza, E. (2000). "Local buckling curves for design of FRP profiles", Thin Walled Structures, 37, 207-222.

Satiropoulos, S.N., GangaRao, H.V.S., Mongi, A.N.K. (1994). "Theoretical and Experimental Evaluation of FRP Components and Systems" J. of Structural Engineering, ASCE, 120 (2), 464-485.

Sonti, S.S., Barbero, E.J., Winegardner, T. (1994). "Determination of Shear Properties for RP Pultruded Composites" 49th Annual Conference, Composites Institute, The Society of Plastic Industry,. Session 2-A, 1-5.

Yooh, S.J., Scott D.W., Zureick, A. (1996). "An experimental investigation of the behaviour of concentrated loaded pultruded columns" Proceedings of the 2st International Conference on Advanced Composite Materials in Bridge and Structures. Montreal, Canada, Canadian Society for Civil Engineering, 309-317.

Zureick, A., and Steffen, R. (2000). "Behaviour and design of Concentrically Loaded Pultruded Angle Struts" J. of Structural Engineering, ASCE, 126 (3), 406-416.

Zureick, A., Scott, D. (1997). "Short-term behaviour and design of fiber-reinforced polymeric slender members under axial compression" J. of Composites for Construction, 1(4), 140-149.

Table 1. Characteristics of FRP shapes available in the market.

	Longitudinal modulus [GPa]	Shear modulus [GPa]	Tensile strength [MPa]
Strongwell	17.2-17.9	2.9	207
Creative Pultrusion	28.6	3.4	275-316
Fiberline	14-40	2.5-4.0	200-400

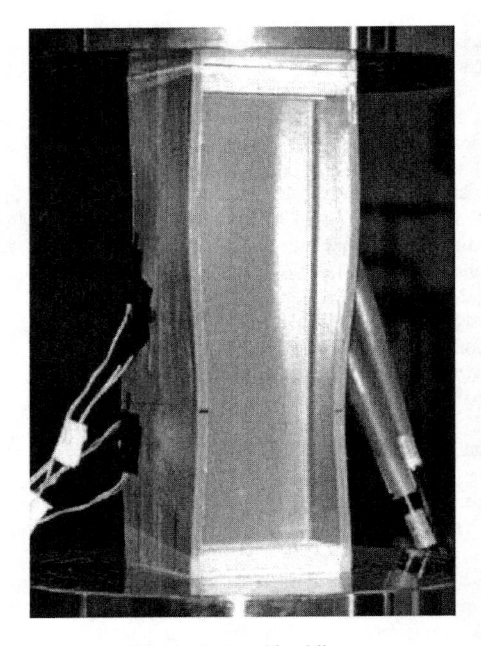

Figure 1. Local buckling.

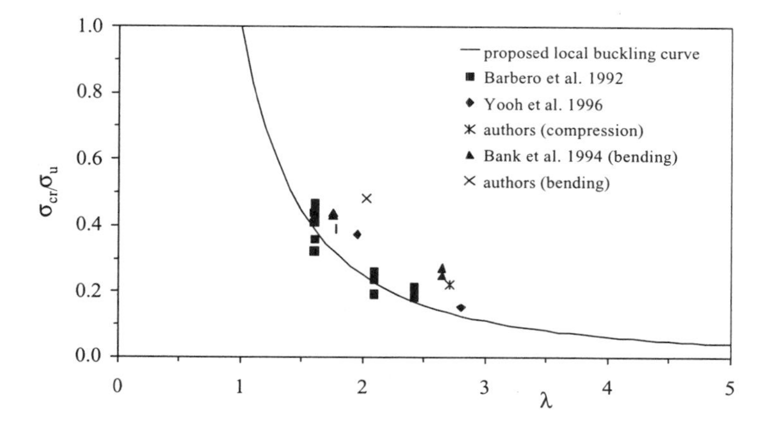

Figure 2. Local buckling curve proposed by Pecce and Cosenza, 2000.

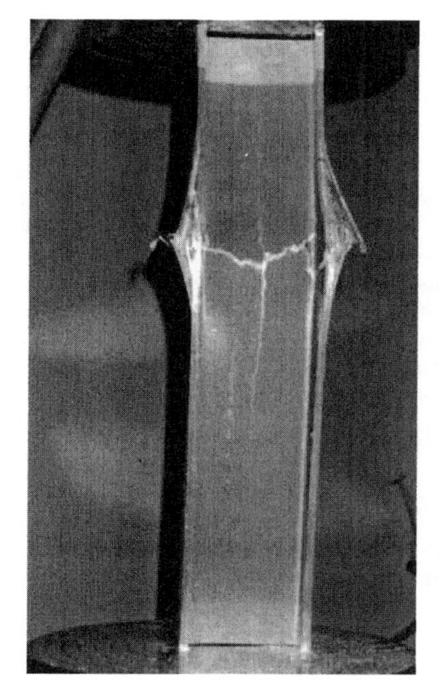

Figure 3. Example of failure post-buckling.

Analysis and Design of Connections for Pultruded FRP Structures

J. Toby Mottram[1]

Abstract

Presented is an international perspective on analysis and design of bolted connections (no adhesive bonding) for structures of Pultruded Fiber Reinforced Plastic (PFRP) shapes and systems. It is observed that guidance in the manufacturers' design manuals has been written using in-house experience and technology transfer. The manuals are found to be limited and to give different guidance. Results of physical tests on single-bolted joints by academics are found to support parts of the pultruder's guidance. Advanced Ultimate Limit State (ULS) design (EUROCOMP Design Code (Clarke 1996)) relies on finite element stress analysis and failure analysis for one of the three distinct modes of net-tension, shear-out and bearing. Physical tests have shown that other failure modes (or combinations) exist and that, to design for Serviceability Limit State (SLS), there is 'first' failure (initial material damage) at a load below maximum. Analysis by way of the finite element method is helping to determine how joint forces are distributed amongst a group of bolts and is starting to expose the local stress fields causing material failure and damage progression. Some observations are given on aspects of analysis and design that are priorities for future R&D.

Introduction

Pultruded FRP (PFRP) shapes and systems consist of pultruded thin-walled composite profiles with overall dimensions up to 1000mm (typically 300mm or less.) They have prismatic section and first generation structural shapes are I, angle, channel and box (Anon 1989, 1995, 1999). Reinforcement is E-glass fiber in the two forms of continuous rovings and continuous filament (or strand) mats. The matrix is based on a thermoset resin such as polyester or vinylester, and often contains filler and other additives. They are used in primary load-bearing structures (Turvey 2000)

---

[1]BSc, PhD. Reader, School of Engineering, University of Warwick, Coventry, CV4 7AL. United Kingdom; phone (+44) 024 765 22528; jtm@eng.warwick.ac.uk

with mechanical fastening (steel bolts in bearing) being the preferred method of connection (see Mottram and Turvey 1998). Primary joints are expected to provide strength and stiffness to a structure throughout its life (Clarke 1996). Failure of such joints would constitute major structural damage and be hazardous to life. The safe and reliable design of bolted joints is therefore clearly a priority.

This paper considers design guidance and numerical analysis. This is not only relevant to joining the current range of pultruded profiles, but also provides the underpinning knowledge for its extension to bolted joints with second generation profiles, which are reaching the marketplace as the pultrusion industry matures.

Design Guidance and Physical Testing

To help construction professionals cope with the growing use of pultrudates several manufacturers have written design manuals (Anon 1989, 1995, and 1999), based on in-house and national levels of knowledge and understanding. Each provides recommendations for bolted web-cleated ('clip') beam-to-column connections and other simple frame joints. Their details often mimic those found in steel construction (Turvey 2000) and their resistance is not governed by the recommended bolt configurations (Mottram and Turvey 1998).

Here we shall focus our attention on the bolted connection itself (the target in Sections 5.2 of Clarke (1996)) to see what design guidance has to offer. We shall consider joint configurations, such as shown in Figure 1. These consist of plate-to-plate connections with eccentric or concentric loading in the plane. Such joints often have a group of bolts as shown in Figure 1(a). To establish overall dimensions (e.g. plate thickness (t), hole (or bolt) diameter (D), width (W) and edge distance (E)) we use physical test data from single-bolted joints (Figure 1(b)). Often the concentric loading is tensile. Table 1 presents suggested and experimentally determined joint minimum parameters for the strongest single-bolted joints. The American design manuals of Strongwell (Anon 1989) and Creative Pultrusions Inc. (Anon 1999) have taken their joint parameters from the ASCE Manual No. 63 (Anon 1984). The writers of manual No. 63 had lifted the parameters from the 1960 marine design manual for GRPs by Gibbs and Cox Inc. Their recommended minimum distances are for hand lay-up GRP materials (isotropic in the plane) of unknown thicknesses and unknown bolt and joint details. We can expect the parameters to change from those in Table 1 when joints are of pultruded material (orthotropic in the plane). Fiberline Composites parameters are from Danish Standard DS456 (for steel). Company and external laboratory tests have confirmed these; no results have been disclosed.

The parameters in Table 1 from Rosner and Rizkalla (1995) and Cooper and Turvey (1995) are supported by single-bolted joint test data, where the E/D and W/D ratios are varied. Both test series used EXTREN Series 500 flat sheet materials with a polyester matrix and they were short-term with monotonic tensile loading. To explain why the two studies propose different E/D and W/D minimum parameters we need look no further than the differences in the testing. Rosner and Rizkalla used 9.53mm (3/8 inch), 12.7mm (1/2 inch) and 19.05mm (3/4 inch) thick flat sheets with a hole 1.6mm (1/16 inch) larger than the 19.05mm diameter of the high-strength plain shank steel bolt. Standard steel washers of outer diameter $2D$ were used. The double lap-shear tests had outer PFRP plates and an inner steel plate. To comply further with guidance by Strongwell (Anon 1989) the bolt torque was 32.5 N.m (24 ft-lbs) (as specified for Strongwell's proprietary Fiberbolt studs and nuts). They tested at 2 mm/min, loading at 0° (longitudinal), 45° and 90° (transverse) to the direction of pultrusion. Cooper and Turvey in contrast used 6.35mm (1/4 inch) thick flat sheet with a hole slightly larger (0.1 to 0.3mm) than the plain shank M10 steel bolt (grade 8.8). No washers were used. The double lap-shear tests had outer steel plates and the inner plate was the PFRP specimen. They choose three different torque levels 0 N.m (pin bearing), 3 N.m (lightly clamped) and 30 N.m (fully clamped) in their 0° tests, with loading at 10 kN/min. The positive effect of lateral constraint on increasing the torque from 3 to 30 N.m increased the mean failure load by 50%. Recommending assembly with fully clamped bolts, Cooper and Turvey's parameters were based on the resistance data at the lower torque of 3 N.m.

Other single-bolted joints test series using flat sheet materials are available (Abd-El-Naby and Hollaway 1993, Erki 1995, Yuan et al. 1996, Turvey 1998, Yuan and Liu 2000, Turvey and Wang 2001). These papers provide data required to develop generalized design guidance. Comparing these six tests series it is found that each set out with their own specific goals. Abd-El-Naby and Hollaway, for example, considered the effect of friction and dimension parameters, while Erki tested joints under compression and compared Strongwell's Fiberbolt studs and nuts with steel bolts. Yuan et al. (1996) investigated how strength changed with hole clearance (0 to 6.35mm), and Turvey and Wang considered temperature (RT to 80°C).

The purpose of specifying the minimum joint dimensions in Table 1 is to have the 'strongest' bolted joints whose mode of failure is by bearing (this mode is perceived to provide a benign (gradual) failure mechanism (Chapt. 2 in Mottram and Turvey 1998)). The other principal modes (see Clarke 1996 for illustrations) of net-tension, cleavage and shear-out are deemed unacceptable (if avoidable) because their failure mechanisms are sudden and catastrophic. A joint's mode of failure and its load-displacement characteristic are dependent on its detailing parameters. All the test series cited give a discussion on the visual modes of failure (post-test). On many occasions failure was not simply one of the three distinct modes used in design calculations, it was another or a combination of net-tension, bearing, and shear-out. Shear-out and cleavage modes can be considered as special cases of bearing failure.

Several papers (Rosner and Rizkalla 1995, Cooper and Turvey 1995, Erki 1995) present typical joint load-displacement plots. Vangrimde (2000) has suggested that the usefulness of the joint displacement is dependent on what it measures. He states that designers are not interested in hole elongation, but rather in the local bearing deformation. He advocates that the design parameter should be the bearing displacement and not the elongation of the hole as prescribed in ASTM D5961. In order to determine a 'bearing' displacement the second measurement point must be a distance from the hole where the stress concentrations have disappeared.

Vangrimde's observation is one reason why the plots cannot readily be compared. When the mode of failure is bearing, the load-displacement characteristics for 0° joints are seen to have the same form (Rosner and Rizkalla 1995, Cooper and Turvey 1995). The initial part shows virtually linear elastic behavior to the maximum load (a change in slope above 80% of maximum load indicates 'first' failure). It is assumed that material damage is the reason for the stiffness change, and since this initial failure state has been designed the 'first' failure concept is required. Following development of the ultimate (full bearing) failure the load reduces to between 70 and 80% of maximum and remains constant (or increases if failing material is laterally constrained) while displacement increases several fold. Under conditions where the load is able to follow a joint's post-failure stiffness the response provides evidence for pseudo-ductility (Rosner and Rizkalla 1995). Joint collapse in real structures is going to be dynamic. The stresses causing failure will be from a loading that will, for a period, remain constant post-failure. Under real conditions we find that bearing failure does not always guarantee joint 'ductility'. 'Joint ductility' will only be realized if the maximum load is for a joint displacement several times higher than the displacement at the joint's 'elastic' limit. This requires lateral constraint via adequate bolt torque and contacting surface area (Cooper and Turvey 1995).

When the parameters (*E*/*D*, *W*/*D*) are too low, or when the orientation of the material is >30°, the mode of failure is not bearing and appears distinctly brittle. Now the reported load-displacement characteristics have a very rapid falling branch after maximum load ('no' ductility) (Rosner and Rizkalla 1995, Erki 1995, Turvey 1998). With the roving reinforcement perpendicular to the load, the failure in all 90° joint tests is brittle and net-tension. Based on quoted material longitudinal (0°) and transverse (90°) strengths (Anon 1989), Erki (1995) and Turvey (1998) found joint efficiency was > 1.5 higher when the PFRP material was at 90°. This does not mean the 90° joint is the better configuration. The 0° joint has the higher ultimate load (1.3 times) because its material strength is twice the 90° material strength.

Turvey and Wang (2001) conducted single-bolted joint tests at elevated temperature. Their pioneering research shows that when the temperature is 65°C (the maximum specified in the design manuals) the mode and resistance can dramatically change from that at 25°C. This means that the room temperature parameters in Table 1 might not be appropriate at the highest operating temperatures.

Turvey (Chapt. 2 in Mottram and Turvey 1998) gives a summary on the three test series on multi-bolted concentrically loaded double lap-shear joints.

Manufacturers Design Manuals

It is instructive to summarize and remark on other differences in the limited guidance in the manufacturers design manuals. Strongwell (Anon 1989) include bolt bearing and shear allowable load tables. Fiberbolt loads are obtained with four as the factor of safety. The shear loads for stainless steel threaded bolts are of unknown origin. Creative Pultrusions Inc. (Anon 1999) give additional guidance for 'clip' frame connections. Complying with steel construction they recommend high strength (minimum 700 N/mm^2) steel bolts to ASTM A325 with grade five coarse threads. For a 12.7mm (1/2 inch) bolt diameter, the listed low torque is 39 N.m (37.5% of bolt proof load) and the high torque is 77 N.m (75%). This increases to 77 N.m and 113 N.m, respectively, for a 16.3mm (5/8 inch) bolt diameter. To assist designers there are ULS bearing strengths given at 4% hole elongation (ASTM D5691). Such hole deformation strengths are typically 36% of the material bearing strength. There is a necessity to obtain new bearing strengths for SLS design (i.e. at 'first' failure).

It is recognized that too high a bolt torque (will depend on bolt diameter, etc.) is to be avoided to prevent crushing of the PFRP material (Clarke 1996). Stress relaxation from creep is going to reduce the compressive pre-load such that the bolt torque (axial compressive force) will reduce. It is clear that a plain bolt shank in the contact zone will increase the loads at 'first' and ultimate failure (e.g., testing by Rosner and Rizkalla (1995) and Cooper and Turvey (1995)). These factors are important to a joint's long-term durability and structural integrity.

Fiberline Composites in Denmark have been the most pro-active in preparing guidance for general bolted joint design. It is based on the joint parameters in Table 1. Their manual (Anon 1995) covers flat plates (thicknesses from 3 to 20mm) fastened by A4 stainless steel bolts (M6 to M48) in a lap-joint configuration. Simple design equations, based on either bearing, or net-tension or shear-out failure, are given to determine lap-joint strength. Tables give design loads at the ULS in the longitudinal and transverse material directions using a factor of safety of three. No bolt torque is assumed in design (on site Fiberline use a torque of > 100N.m with plain M16 bolts). These tables enable any connection configuration to be designed.

Analysis and the EUROCOMP Design Code

One major disadvantage of bolting (without adhesive bonding) is that it is very difficult to have a connection strength that is 50% of the material strength of the plates being joined. When combined with other uncertainties, the current factors of safety (Anon 1995, Clarke 1996) lead to PFRP bolted lap-joint strengths that are <15% of material strength available. Physical testing is obviously important in obtaining data that can be used to determine 'optimum' dimension parameters, such as given in Table 1, or to proof test a new multi-bolted joint design. Such testing is expensive and will not provide all the information the designer needs.

To provide new information numerical analysis by the finite element method is starting to be used. The PFRP material is assumed to have linear elastic properties. There are two main types of model. One is used to determine how the external loading is distributed to the bolts in a mutli-bolted joint (a source analysis) (Clarke 1996). It uses a coarse mesh and models the bolts by elastic springs. The second type of model (can use the source analysis results) has a refined mesh for the local region around the bolt hole. It can be used to predict the local stress field, for example, due to the bearing bolt. Maximum stresses are used in a failure analysis (assuming the mode is net-tension, shear-out or bearing) to determine joint strength. When a bolt is present the model involves the complication of the sliding contact phenomenon (Steffen et al. 1999, Mottram and Padilla-Contreras 2001).

The two types of analysis are required in the design procedures given in the EUROCOMP Design Code (EDC). EDC (Clarke 1996) is based on the best knowledge prior to 1994, though its design procedures remain rudimentary. Within its scope are bolted connections as shown in Figure 1(a). There are two design approaches known as simplified and rigorous. These are generic and in their development the code writers wanted to transfer know-how from aerospace research.

The simple approach is given in the code. It is based on a number of design requirements. For multi-row joints there are fastener load distribution factors to allow for by-pass loads. To determine joint strength, each bolt's stress field is calculated via a combination of six basic load cases. For each load case stress charts, giving stress concentrations for net-tension, bearing and shear-out failures, have been prepared using unspecified finite element analysis. There are a number of obvious problems with the charts (e.g., the laminate materials are not defined, symmetry in a stress plot is absent when it should not be), which ensures they cannot be used.

Due to lack of verification, the rigorous method, involving damage tolerance, is in the handbook section (Clarke 1996). Design by this approach requires the two types of finite element stress calculations and knowledge of characteristic distances (see Chapt. 1 in Mottram and Turvey 1998), and will lead to stronger and more efficient joint lay-outs. There are a number of other problems with the EDC 'aerospace' design approaches. Plate thicknesses are often much less than in construction and clearance is relatively small at $< 0.05D$. The rigorous method uses a Point Stress Criterion (PSC) whose suitability is now openly being questioned by its users in the aerospace sector. The rigorous design approach is however the only one able to cope with a design in which the bolt group is under general in-plane loading (as in Figure 1(a)).

Concluding Remarks

It is clear that we do not possess sufficient information to prepare consistent design guidance for all bolted joint configurations. It has been found that the manufacturers have neither published joint test data nor have they used code methodology to generate design information. The limited guidance is only for design at ultimate joint failure. Several academic test series confirm parts of the guidance and point to where there is a lack of knowledge and understanding. For example, there is a perception that bearing failure is benign and 'ductile', and it might not be. There is an absence of any guidance for design at serviceability. Understanding 'first' failure damage is essential, as its presence will reduce the long-term resistance if a joint experiences fatigue, stress corrosion, etc. Other important joint requirements, e.g. bolt torque, hole clearance, by-pass load, friction, damage tolerance, environmental degradation are not properly specified. It is concluded that the structural integrity of bolted joints designed to the current guidance is unknown. Such a situation is not a long-term option. By harnessing the results of co-ordinated physical testing and numerical analysis, and combining these with a reappraisal of what is known now we can improve on today's design guidance. It should be a goal of all interested parties to develop recognized guidance (in codes of practice) leading to greater acceptance of primary load-bearing pultruded structures.

Acknowledgement

The author expresses his appreciation to EPSRC for its support (grant GR/N11797).

References

Abd-El-Naby, S. F. M., and Hollaway, L. (1993). "The experimental behaviour of bolted joints in pultruded glass/polyester material. Part 1: Single-bolt joints," *Composites*, 24 7, 531-538.

Anon. (1984). *Structural plastics design manual*, American Society of Civil Engineers Manuals and Reports on engineering practice,' No. 63, ASCE, NewYork.

Anon. (1989). *EXTREN design manual,* Strongwell, Bristol, Va.

Anon. (1995). *Fiberline design manual for structural profiles in composite materials*, Fiberline Composites A/S, Kolding, Denmark.

Anon. (1999). *The new and improved Pultrex pultrusion design manual of pultrex standard and custom fiber reinforced polymer structural profiles*, Creative Pultrusions Inc., Alum Bank, Pa.

Clarke, J. L., (ed.) (1996). *Structural design of polymer composites - EUROCOMP Design code and handbook*, S. & F. N. Spon, London.

Gibbs and Cox, Inc. (1960). *Marine design manual for fiberglass reinforced plastics*, McGraw-Hill, New York.

Cooper, C., and Turvey, G. J. (1995). "Effects of joint geometry and bolt torque on the structural performance of single-bolt tension joints in pultruded GRP sheet material," *Composite Structures*, 32 (1-4), 217-226.

Erki, M. A. (1995). "Bolted glass-fibre-reinforced plastic joints," *Can. J. Civ. Eng.*, 22, 736-744.

Mottram, J. T., and Turvey, G. J. (eds.), (1998). *State-of-the-art review on design, testing, analysis and application of polymeric composite connections*, COST C1 Report, DG XII European Commission, Office for Official Publications of the European Communities, Brussels & Luxembourg.

Mottram, J. T., and Padilla-Contreras, E. (2001). "Pin-bearing behaviour of pultruded structural materials,' to be presented at *Inter. Conf. CCC-2001 – Composites in Construction*, 10-12 October, Porto, published by Balkema.

Rosner, C. N., and Rizkalla, S. H. (1995). "Bolted connections for fiber-reinforced composite structural members: Experimental program," *J. Mat. in Civ. Engrg.*, ASCE, 7 (4), 223-231.

Steffen, R. E., Rami H-A., and Zureick, A. H. (1999). "Analysis and behavior of pultruded fiber-reinforced polymer bolted connections," *in Proc. Fifth ASCE Materials Engineering Congress*, ASCE, Reston, 76-83.

Turvey, G. J. (1998). "Single-bolt tension on pultruded GRP plate - Effect of tension direction relative to pultrusion direction," *Composite Structures*, 42(4), 341-351.

Turvey, G. J. (2000). "Bolted connections in PFRP structures," *Prog. in Struct. Engng and Mater.*, 2, 146-156.

Turvey, G. J., and Wang P. (2001). "Effect of temperature on the structural integrity of bolted jointsd in pultrusions," to be presented at *Inter. Conf. CCC-2001 – Composites in Construction*, 10-12 October, Porto, published by Balkema.

Vangrimde, B. (2000). *Assemblages boulonné matériaux composites verre-polyester*, PhD thesis, École Polytechnique de Montréal, Canada.

Yuan, R. L., Liu, C. J., and Daley, T. (1996). "Study of mechanical connection for GFRP laminated structures," *in Proc. 2nd Inter. Conf. in Advanced Composite*

Materials in Bridges and Structures, Canadian Society of Civil Engineers, Montréal, 951-958.

Yuan, R. L., and Liu, C. J. (2000). "Experimental characterization of FRP mechanical connections," *in Proc. 3rd Inter. Conf. on Advanced Composite Materials in Bridges and Structures*, Canadian Society of Civil Engineers, Montréal, 103-110.

Table 1. Suggested and experimentally-determined bolted joint minimum parameters (at Room Temperature)

Source	Plate thickness t (mm)	Bolt diameter / Plate thickness D/t	Edge distance / Bolt diameter E/D	Side distance / Bolt diameter S/D	Width distance / Bolt diameter W/D	Clearance hole size (mm)	Washer Diameter / Bolt Diameter
Strongwell (Anon 1989)	6.35 to 19.05	1.0 to 3.0	2.0 to 4.5 (3.0)[1]	1.5 to 3.5 (2.0)[1]	4 to 5 (5)[1]	1.6	-
Fiberline (Anon 1995)	3 to 20	0.5 to 16.0	2.5 & 3.5	2.0	4	1.0	2
EUROCOMP[2] (Clarke 1996)	Unspecified	1.0 to 1.5	≥ 3	$\geq 0.5W/D$	≥ 3	$\leq 0.05D$	>2
Creative Pultrusions Inc. (Anon 1999)	6.35 to 12.7	Unspecified	2.0 to 4.5 (3.0)[1]	1.5 to 3.5 (2.0)[1]	4 to 5 (5.0)[1]	1.6	2.5
Rosner & Rizkalla (1995)	9.53 to 19.05	0.5 to 1.0	5[3]	Single bolt	5[3]	1.6	-
Cooper and Turvey[4] (1995)	6.35	1.6	3	Single-bolt	4	Close fit (0.1 to .3)	-

Notes:

1. Recommended minimum design value.
2. General glass fiber reinforced plastics (including PFRPs).
3. D is hole diameter (bolt diameter and hole clearance).
4. From joint tests with tensile load in direction of pultrusion.

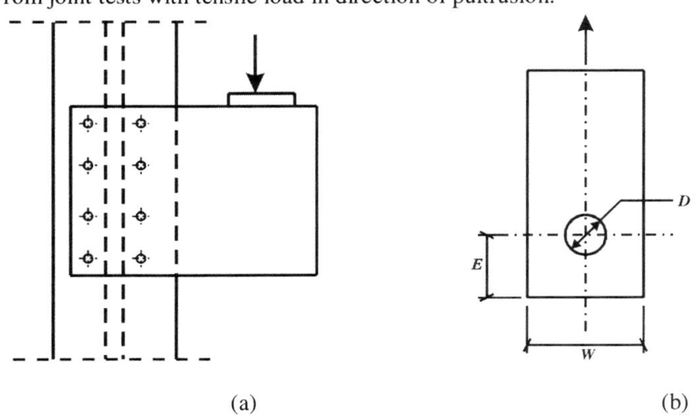

Figure 1. (a) Eccentrically loaded joint. (b) Single-bolted joint dimensions (D is hole (or bolt) diameter, E is end distance, W is width).

Pultruded GRP Frames: Simple (Conservative) Approach to Design, a Rational Alternative and Research Needs for Improved Design

Geoffrey J. Turvey[1]

Abstract

The simple approach currently used to size the beams and columns of a number of small and large pultruded GRP braced frames is outlined. It is emphasised that frame design tends to be governed by serviceability and elastic buckling criteria, because of the flexibility of the GRP profiles and the assumption that all joints behave as if pinned. Semi-rigid design analysis is advocated as a rational alternative approach to frame design. It promises significant improvements in structural performance and lower costs. Recent research, which provides limited quantification of the potential benefits offered by the semi-rigid design approach, is cited. Finally, suggestions are made with regard to the experimental, theoretical/numerical and design research required for semi-rigid frame design to be adopted with confidence.

Introduction

Pultrusion (Meyer 1985) is recognised as a rapid and economic process for the manufacture of straight prismatic FRP (Fibre Reinforced Plastic) profiles of any open or closed cross-sectional shape. Generally, pultruders manufacture two categories of profile, often referred to as standard and custom profiles. The former usually have a simple cross-sectional shape and are often available as stocked items, whereas the latter are made to order and may have complicated cross-sectional shapes and fibre architectures. Here attention is focused on *structural grade* standard profiles. Their cross-sectional shapes: open (I, channel and angle), closed (hollow circular, square and rectangular) and solid (rod, bar and flat plate) mimic those used in steelwork construction. The profiles are made of E-glass fibres (rovings and CFM) embedded in a matrix, which is a mixture of polyester or vinylester resin and low cost filler (clay or chalk). The fibre volume fraction of the profiles varies between 30 and 60% depending on the profile's cross-sectional shape and the type of action it is intended to support. Thus a solid circular tension rod has 60% of fibres by volume, whereas an I-beam only has 35 – 40%. Moreover, as the modulus of E-glass fibre is only about 30% of that of mild steel, the stiffnesses of pultruded GRP profiles may be an order of magnitude lower than similar sized mild steel profiles.

----

[1]Senior Lecturer, Engineering Department, Lancaster University, Bailrigg, Lancaster, LA2 9LF, UK, phone + 44 (0)1524 593088; g.turvey@lancaster.ac.uk

The consequence of the low stiffness of structural grade pultruded GRP profiles is that they have only been used for frame structures, when several of their advantageous properties, eg. low self-weight, high corrosion resistance and low maintenance costs, may be exploited simultaneously to outweigh the low stiffness disadvantage. Moreover, the profiles have been used mainly to fabricate secondary load-bearing frame structures, which are used to support chemical plant, pipe-work, raised platforms and walkways. It is only in the last few years that they have been used in relatively large building frames.

Examples of Pultruded GRP Frame Structures

An example of a frame structure, used to support a raised platform, is shown in Figure 1. Many simple frames of this type have been constructed over the past two decades. A recent, larger scale example is the 19.2m high Fort Story stair tower frame shown in Figure 2. These two frames are based on conventional vertical columns, which may be continuous over several storeys, and horizontal beams. In order to inhibit sway deformations, inclined bracing members are provided from the mid-span of the beam to the column feet in one or more bays of each storey. Figure 3 shows a very large rectangular box-like building frame 9 x 11m in plan and 10m high, which was recently erected in the Middle East. Equipment housed inside the building, imposed a positioning tolerance of ± 1mm, which is much less than the straightness tolerance of the GRP profiles. Construction tolerances of ± 10mm were, therefore, adopted and all equipment mountings were made adjustable to meet the positioning tolerance. Even so, it was important for the frame to be as stiff as possible and, therefore, very stiff cross-diagonal bracing and bonded and bolted plated joints were used throughout. Figure 4 shows a pitched roof portal frame building in which bonded and bolted web plate joints were used to achieve rigidity at the apex and rafter - column joints. An example of an unusual frame design is the five-storey Eyecatcher building, shown in Figure 5. The frames derive their sway resistance from the triangulated structural form of the inclined truss and the vertical column prop, which runs the height of the frame.

Simple (Conservative) Approach to Frame Design

Because of the low inherent stiffness of structural grade pultruded GRP profiles and the present limited understanding of joint stiffness and strength, frames are designed as *no-sway* or *braced*. Bracing is used in one or more bays to inhibit the *mechanism* collapse mode.

The design procedure used for the frames in Figures 1 and 2 is similar to the simple design method for braced steel frames. The ends of the beams are assumed simply supported and the connections are assumed to resist only vertical shear forces. However, the beam sizes are generally determined from stiffness considerations, i.e. the deflection serviceability criterion dominates. This is in stark contrast to the situation for steel frames, where ultimate strength is the governing limit state criterion. Lateral buckling is generally *designed out* by taking account of lateral restraint provided by flooring or intermediate beams. Local buckling is not usually critical, because the deflection serviceability criterion means that stresses in the compression flange and web are generally low. However, where checking is deemed necessary the flanges and web of the beam profile are treated as individual elastic othrotropic plates, i.e. the rotational stiffnesses at the web-flange junctions are ignored, and simple formulae are used to determine the compression and shear buckling stresses. Column sizes are determined from overall elastic buckling considerations and effective lengths take account of the end restraints due to continuity etc. Also, local buckling in column profiles may be checked, albeit conservatively, using the approach adopted for beam profiles. The axial loads in the bracing members are determined from a laterally loaded pin-jointed frame analysis. The bracing members are sized on the basis of overall elastic buckling failure, as for the columns. The end connections of the bracing members are designed to resist tension and compression based on simple bearing, net tension etc type calculations using factored material strengths given in manufacturer's design handbooks (see Anon. 1989).

Because of the simplifying assumptions made with regard to member interaction, frame design amounts to a set of simple procedures based on elastic analysis for sizing individual members (beams and columns) and based on material strength analysis for sizing connections. It is only for the design of the sway bracing that resort is made to elastic pin-jointed frame analysis. Clearly, elastic analysis plays a much more dominant role in GRP frame design than it does in the design of frames made of ductile materials such as steel or aluminium. Details of the simple procedures for the design of pultruded GRP members are set out in the design handbooks (Anon. 1989 and Anon. 1995). Moreover, two simple worked examples explaining their application to the design of secondary load bearing braced frames, used to support raised platforms, have also been published (Anon. 1989 and Hollaway 1994).

A more rigorous approach was adopted for the design of the frame shown in Figure 3, because of the stringent deformation criteria that had to be satisfied. A rigid-jointed frame analysis was used to size the beams, columns and the bracing members. The effects of shear deformation were also considered in the design of the beam members and, in addition to testing, nonlinear Finite Element analysis was used to check the performance of the stiff bonded and bolted plated joints.

For the portal frames, shown in Figure 4, the design procedure was similar to that used in the elastic design of steel frames, i.e. an elastic rigid-jointed frame analysis was carried out, but the members were sized according to the deflection serviceability criteria rather than their elastic moment capacity. It was also assumed that the bonded and bolted plated web-joints were rigid.

Again, for the frames, shown in Figure 5, the members were sized on the basis of upper and lower bound elastic frame analyses, i.e. rigid and pin-jointed analyses. Vertical and horizontal deflections were limited to 1/200 and 1/350 respectively. Material property and other data were taken from the manufacturer's design manual (Fiberline A/S 1995). The beams in the frames are box-sections fabricated by bonding plates externally to the flanges of channel sections and internally to their webs. The design performance of these members was established by testing. The structural performance of the joints was not established.

What has not yet been mentioned in the foregoing rather brief outline of the simple procedures, used in the design of pultruded GRP frames, is the use of very large load and strength factors. They are necessary because of the uncertainties associated with material properties (creep and fire performance), aspects of member behaviour (localised loads) and the stiffness, strength and reliability of the connections, particularly where adhesives are used. Unfortunately, conservative design procedures, large load factors, large strength reduction factors, a limited range of profiles and large profile size increments result in stocky frames with high initial costs and poor aesthetics, all of which reduce their competitiveness vis-à-vis conventional materials.

A Rational Alternative: Semi-Rigid Design

The present lack of knowledge of the structural performance of pultruded GRP frames means that designers must err on the side of caution. Conservative design procedures will continue to be used until performance data and validated design guidance become available. Apart from one simple example, there is very little information *per se* on the design of frames in the manufacturer's design guides (Anon. 1989 and Anon. 1995). The bulk of the design information comprises: material property data, procedures for sizing beams and columns and limited guidance on the design of connections. Even the EUROCOMP design guide (Clarke 1996), perhaps the most code-like document for composites design currently in print, does not offer any guidance on frame design.

In the short term, the key to more economical frame design, using *off-the-shelf* structural grade profiles, lies in exploiting the inherent rotational stiffnesses of beam to column and column to base connections, which are currently discounted. For this to be possible, it is necessary to develop procedures and guidance, which will allow the designer to quantify (preferably by analysis) the rotational stiffness and rotation capacity at first failure of the connections. This, in turn, will allow beams to support higher loads at the deflection limit, and will also raise the buckling loads of columns. Moreover, exploitation of joint rotational stiffness will also allow the size of the bracing profiles to be reduced, bringing further cost benefits.

Recently, it has been shown (Turvey 1997 and Turvey 2000), both experimentally and theoretically, that by taking account of the initial rotational stiffnesses of the end connections, rather than treating them as pin joints, mid-span deflections of beams reduce significantly. For pultruded GRP I-beams with bolted end connections employing GRP web and flange cleats mid-span deflections reduce by about 40% at span to depth ratios of 20. Even greater reductions are achievable with stainless steel cleats, which are often preferred by fabricators. Similar theoretical studies (Zheng and Mottram 1996 and Mottram and Zheng 1998), based on semi-rigid, shear-deformable plane frame analysis, have shown that substantial improvements in structural performance arise. These are examples of the type of research that needs to be undertaken to provide designers with the quantitative data they need to achieve more efficient frames at little or no extra cost.

Research Needs for Improved Design Analysis of Pultruded GRP Frames

The previous section has discussed a change in structural design philosophy, which has the potential to lead to more economical frame design in pultruded GRP materials. Of course, there are other measures, which could also produce structural performance benefits, eg. the development of *second generation* profiles (based on hybrid fibre architectures) and frame systems (based around *snap-fit* connection technology). An example of the potential of the former development is the 914mm deep web-stiffened box-beam (Witcher 1998). This beam uses stitched-fabric in the webs and 6% by volume of CFRP rovings in the flanges to achieve a doubling of its flexural modulus compared to a similar all-GRP profile. However, these developments are in the nature of medium term objectives, which will generate their own research requirements. More pressing research requirements relate to frames fabricated from the current range of GRP profiles.

Presently, the static serviceability and ultimate load response of pultruded GRP frame structures is poorly understood. Likewise, little is known about their dynamic response. A few small, single-bay, single-storey, rectangular, portal frames have been tested in the USA (Mosaliam 1990) and the UK (Bell 1992 and Turvey 1996) under static short and long term loading. Some recent, as yet unreported, UK experimental and analytical research on two rectangular portal frames – one with conventional bolted web and flange cleat connections and the other with bolted moulded connections - has shown that semi-rigid frame analysis is capable of providing accurate response predictions up to collapse. Figure 6 shows one of the 2m span by 1.93m high portal frames being tested under sway loading. In these frame tests, local buckling in the 102x102x6.4mm beam and column profiles did not arise and lateral buckling was prevented. Also, both frames had the same overall rectangular geometry.

It is clear that much more testing and analysis of frame behaviour is required in order to accelerate progress towards acceptance of the semi-rigid approach to design, which is already accepted in steel design codes. It is convenient to list under the following headings aspects of pultruded GRP frame behaviour where knowledge is scant or even entirely lacking and where research would bring benefits:-

Experimental Research. Frame tests should be undertaken to assess the following:-

- The effects of beam span to column height ratios for both the serviceability and ultimate limit state conditions.
- The use of different sizes and shapes of beam and column profile. (Most of the frames tested have used the same size of wide flange profile in the beam and columns. However, many frames use channel profiles.)
- The effects of various conventional bolted and/or bonded joint configurations, possibly using aluminium or stainless steel angles with/without side gussets, and plated joints.
- The effects of various bracing arrangements for minimising sway deformations.

- The effects of long-term loading on the creep behaviour.
- The effects of hot-wet and fire conditions.

Theoretical/Numerical Research. This should include the following:-

- Develop semi-rigid frame analysis to enable collapse response to be predicted.
- Develop the Component Method for predicting the rotational stiffness and first damage rotational capacity of bolted and/or bonded connections.
- Develop analysis tools for predicting the effects of introducing localised loads into parts (webs/flanges) of profiles.

Design Research. This research should be aimed at:-

- Assessing the test/analysis research to define serviceability limit states.
- Assessing the test/analysis research to reduce load and resistance factors.
- Assessing the test/analysis research to provide guidance on frame stability.

Concluding Remarks

A description has been given of the salient features of the simple procedures used in the design of a number of pultruded GRP frame structures. The similarity of these procedures to those used in the simple design method for steel frames has been noted. However, it has been emphasised that stiffness and elastic stability are the dominant criteria for GRP frame design. Moreover, it appears that there is no consistency of approach in the application of stiffness criteria amongst those engaged in GRP frame design.

It has been suggested that semi-rigid analysis could provide a more rational approach to frame design. Some recent research, which quantifies the benefits arising from taking account of the inherent rotational stiffness of the frame connections, has been cited in support of this suggestion. Finally, a list of research tasks has been compiled, which, if successfully completed, would enable designers to adopt the semi-rigid design method with considerable confidence.

References

Anon. (1989). *EXTREN fiberglass structural shapes – design manual*, Strongwell, Bristol, Virginia.

Anon. (1995). *Fiberline design manual for structural profiles in composite materials*, Fiberline Composites A/S, Kolding, Denmark.

Bell, S. (1992). *GRP structure*, Final Year Undergraduate Project Report, Engineering Department, Lancaster University.

Clarke, J.L. (Ed.) (1996). *Structural design of polymer composites – EUROCOMP design code and handbook*, E. & F.N. Spon, London.

Hollaway, L. (1994). *Handbook of polymer composites for engineers*, Woodhead Publishing Limited, Cambridge.

Meyer, R.W. (1985). *Handbook of pultrusion technology*, Chapman and Hall, New York.
Mosallam, A.S. (1990). *Short and long-term behavior of a pultruded fiber reinforced plastic frame*, PhD Thesis, Catholic University of America, Washington, DC.
Mottram, J.T., and Zheng, Y. (1998). "Analysis of a pultruded frame with various connection properties." 2nd International Conference on Composites in Infrastructure (ICCI'98), University of Arizona, II, 261-274.
Turvey, G.J. (1996). "Testing of a pultruded GRP pinned base rectangular portal frame." *Structural design of polymer composites: EUROCOMP design code and handbook*, E. & F. Spon, 775-783.
Turvey, G.J. (1997). "Analysis of pultruded GRP beams with semi-rigid end connections", *Composite Structures*, 38(1-4), 3-16.
Turvey, G.J. (2000). "Flexure of pultruded GRP beams with semi-rigid end connections", *Composite Structures*, 47(1-4), 391-403.
Witcher, D.A. (1998). "Development and production of heavy structural FRP composites for use as primary structural components." 2nd International Conference on Composites in Infrastructure (ICCI'98), University of Arizona, II, 327-340.
Zheng, Y., and Mottram, J.T. (1996). "Analysis of pultruded frames with semi-rigid connections." 2nd International Conference on Advanced Composite Materials in Bridges and Structures (ACMBS/2), Canadian Society for Civil Engineering, Montreal, 919-927.

Figure 1 Platform support frame (photo: Strongwell, Bristol, Virginia)

Figure 2 Stair tower frame at Fort Story, Virginia (photo: Strongwell, Bristol, Virginia)

Figure 3 Hunter project frame (photo: David Kendall, CETEC Consultancy Ltd)

Figure 4 Pultruded GRP portal frames (photo: Strongwell, Bristol, Virginia)

Figure 5 Eyecatcher building (photo: Fiberline Composites A/S, Kolding, Denmark)

Figure 6 Sway load test on a portal frame with bolted web and flange cleat joints

Outcomes

Composites in Construction: Present Situation and Priorities for Future Research

Antonio Nanni[1], Edoardo Cosenza[2], Gaetano Manfredi[3], and Andrea Prota[4]

Abstract

This paper was written at the conclusion of the International Workshop titled "Composites for Construction: A Reality," held in Capri, Italy, July 20-21, 2001. Its objective is to collect and summarize main ideas, opinions, evaluations, and proposals that emerged during the discussions held by the participants at break-up and general sessions. For simplicity, the reporting follows the order of such sessions and includes elements from the general sessions when applicable. The paper is intended to help professionals and scientists interested in the study and implementation of composites for the built infrastructure by providing a vivid snapshot of the state-of-the-art situation, opportunities, barriers, and needed research for the immediate future.

Introduction

The format of the two-day International Workshop titled "Composites for Construction: A Reality," consisted of two phases: technical presentations and a combination of break-up and general sessions. The purpose of the presentations phase was to set the tone for the discussion to follow. The presentations stemmed from written papers included in these proceedings and circulated among workshop participants prior to the meeting. The purpose

---

[1] V&M Jones Professor of Civil Engineering, Center for Infrastructure Engineering Studies, 224 ERL, University of Missouri – Rolla, Rolla, MO 65409, USA; phone 573-341-4553; nanni@umr.edu

[2] Full Professor, Dept. of Structural Analysis and Design, University of Naples Federico II, Via Claudio 21, 80125 Naples, Italy; phone 081-7683489; cosenza@unina.it

[3] Professor of Structural Engineering, Department of Structural Analysis and Design, University of Naples "Federico II", via Claudio 21, 80125 Naples, Italy; phone 081-768-3488; gamanfre@unina.it

[4] PhD Candidate, Department of Structural Analysis and Design, University of Naples "Federico II", via Claudio 21, 80125 Naples, Italy; phone 081-768-3534; aprota@unina.it

of the discussion sessions was to debate the issues with the intent of drawing considerations on the present situation and formulating an action plan for the future of international research.

In order to facilitate the process, presentations and discussion sessions were organized according to a matrix that had been tested in the past and found successful (Nanni 1994, Nanni et al. 1996). The matrix is shown in Table 1.

Table 1: Workshop Organization

	Reinforced & Prestressed Concrete	Structural Shapes & Systems	Repair & Strengthening
Materials, Durability & Characterization			
Analysis, Design & Codes			
Manufacturing, Construction, Economics & Marketing			

Three break-up sessions were organized according to the columns of the matrix first, followed by a general session. Then, three break-up sessions were held according to the rows of the matrix followed by a general session. This allowed maintaining specificity regarding the use of FRP composites in their major application fields (the vertical columns) without losing perspective of the general issues cutting across them (the horizontal rows).

Break-up Sessions: Outcomes of the Discussion

Six summaries of the break-up and general sessions based on the records of the meetings are given below. In a sense, these summaries can be considered as the minutes of the workshop.

Reinforced and Prestressed Concrete

The session was chaired by G. Balazs with M. Guadagnini as the recorder. The following members contributed to the discussion: E. Cosenza, S. Faza, D. Gremel, A. Katz, A. Nanni, K. Pilakoutas, L. Taerwe, R. Tepfers, T. Ueda, and P. Waldron.

Material properties of FRP rebars or tendons were first debated. Participants agreed that, at present, FRP reinforcement for concrete structures is limited to special applications. The reasons for this were recognized to be primarily cost and lack of durability information. Industry representatives underlined that the material costs are not expected to decrease substantially in the foreseeable future. This is because fibers and resins are already commodity products. On the other hand, it was pointed out that there is a mass-market opportunity for FRP reinforced concrete (RC) or prestressed concrete (PC).

The most significant need is that of test methods to characterize the properties of FRP bars and tendons for both quality control and performance during their intended uses. It was highlighted that the existing standards, such as ASTM test methods for material properties, are not specific to FRP product forms for concrete reinforcement. It was also pointed out that a unified approach to link test methods to reduction or safety coefficients used in design codes is sorely needed.

There was wide consensus on the idea that poor durability of existing constructed facilities is the argument to create a larger market share for FRP reinforcement and make it more competitive. From here, the development of reasonable and credible standard durability test methods supported by international societies such ACI, fib and JCI is a high priority for future research. Researchers should focus on durability, while manufacturers should address cost related issues. The idea of conducting round robin tests to efficiently validate new methods found the agreement of all session participants.

Another key issue was recognized to be the development of specifications for different grades of FRP rebars along with a clear definition of their bond properties. There was consensus that fire resistance is not of primary importance since FRP reinforcement is typically not adopted in ordinary buildings. Conversely, FRP behavior under high temperature should be further investigated.

The disagreement of the workshop participants on the future of glass FRP (GFRP) rebars evidenced the fact that long-term performance should be clarified in order to better understand the potential and marketability of GFRP versus that of carbon (CFRP) rebars (which are still too expensive for widespread use). A better understating of GFRP long-term performance could also allow the modification of the reduction factors presently adopted by design guidelines that may be unjustifiably conservative.

Structural Shapes and System

The session was chaired by T. Mottram with M. Polese as the recorder. The following group members contributed to the discussion: D. Halpin, J. Lesko, R. Lopez-Anido, M. Pecce and G. Turvey.

The group first discussed aspects of processing. It was observed that glass fibers are generally used for structural profiles, yet the final cost of an element is still too high for widespread market acceptance. The majority of FRP shapes and systems are manufactured by pultrusion, for which no production standards are yet available. In addition, manufacturers continuously modify their processes, causing the risk that researchers characterize obsolete elements that are not representative of field use. It was mentioned that efforts towards standardization are being made in Europe and the USA and that 'standard' pultruded products have been available for many years. It was observed that VARTM, SCRIMP and similar new composite processes can be used to manufacture large and complex FRP elements. Several pultrusion companies concentrate on pultrusion products in their manuals. When developing new construction systems the group expressed their wish that the industry consider integrating different processing methods, such as pultrusion, SCRIMP and filament winding. FRP systems need to be manufactured using the most suitable and cost-effective processing methods.

In terms of product development, hybrid glass/carbon pultruded FRP shapes have been manufactured in order to lower cost and maximize performance. Tests on hybrid beams have shown that the carbon fiber-matrix interface was a weak link and this weakness may lead to a delamination problem.

The group recognized that significant work needs to be undertaken in materials characterization in order to define standard test methods for compression, tension, shear and durability characterization. There was a need for an international consensus to be reached on the type of distribution used for analyzing test data; the US seemed to prefer a Weibull distribution, while the Europeans are inclined to the log-normal distribution because of its strong links to the Eurocode design philosophy. A matrix of credible material strength and modulus reduction factors should be determined following carefully controlled physical testing. The group saw the need for direct collaboration between industry and academia that would lead to the development and maintenance of an international database of test results.

The lack of time meant that only a few issues related to element behavior were discussed. A major barrier to innovative research in academia is the lack of access to the large processing facilities needed to make construction FRP shapes and systems. The influence of processing imperfections on structural performance of pultruded shapes needs to be better understood. When conducting tests to understand, for example, the instability of pultruded columns it is important to control and know the load eccentricity. Its un-quantified presence is one reason for a large scatter in published test results. Difficulties in determining material and element shear stiffnesses were another important aspect required to be studied further. Regarding ultimate behavior of FRP shapes, the mechanisms related to premature local failure were recognized by the group as priorities to be investigated and consistently modeled.

The lack of codes and standards was discussed from a construction standpoint. This issue is a priority for future R&D. Contractors are likely to ask for guidelines and specifications for bidding projects and carrying out a life-cycle cost analysis. The unavailability of recognized information is currently a barrier to potential job opportunities and fair pricing (i.e., contractors will add a high cost factor due to the risk of using unproven technology). The same uncertainties were evident regarding the experience of the workforce. The need for a comprehensive training program for workers is potentially another huddle to the growth of this new technology.

The following priorities emerged from the group's intensive discussion:

- Integration of different processes should be explored for reducing cost and improving the marketability of FRP shapes and systems
- Development of standards for production processes and test methods which should be the result of an active cooperation between producers and researchers
- Unified reduction (safety) factors which, for different load patterns, combine environment and natural resistance of the material
- Development of design guidelines accounting for imperfections, tolerances, and premature failures (i.e., instability, delamination)
- Development of construction specifications that would be a reference for pricing and bidding projects.

Repair and Strengthening

The session was chaired by T. Triantafillou with F. Ceroni as the recorder. The following members contributed to the discussion: A. Di Tommaso, G. Fallis, W. Gold, P. Kelley, G. Manfredi, S. Matthys, G. Monti, K. Neale, A. Prota, K. Tan, and J. Teng.

The discussion on FRP for strengthening and upgrade applications focused mainly on the use of FRP composites as externally bonded reinforcement (EBR) to concrete structures since this technology is becoming used extensively worldwide.

Types of fibers used in strengthening applications were addressed first. Nowadays, carbon fibers are generally used for their excellent durability and stiffness. In some applications (e.g., seismic upgrades), a lower-stiffness higher-deformation fiber (i.e., glass or aramid) would be preferable in order to allow higher energy dissipation capacity. However, durability concerns are limiting the number of GFRP applications. There is a need for international standard tests to truly assess fiber performance and reliability. In addition to fibers and saturant resins, such characterization would include reduction factors for long-term behavior of the other components.

Fire resistance was also discussed. It was concluded that it may not play a major role in strengthening applications aimed at correcting deficiencies due to seismic, blast or wind forces. However, it remains a fundamental roadblock in strengthening applications aimed at correcting construction/design errors or for the repair/upgrade of conventional buildings.

The modeling of members upgraded with FRP composites was also covered. There was a consensus that the continuous development of new models does not represent an effective solution towards a comprehensive analysis/prediction framework. Researchers should focus on checking and improving existing models, with an effort to introduce parameters having a clear physical meaning. Participants agreed on the idea of separating the theoretical work aimed at developing mechanical models from the development of practical algorithms to be used by practitioners for design.

In terms of serviceability behavior, members with EBR appeared to need continuous monitoring only in the case of critical structures (i.e., historical monuments, bridges, etc.). The effects of a complete FRP encasement on moisture migration mechanisms, causes of galvanic corrosion, and accelerated steel reinforcement corrosion were mentioned. These phenomena could potentially decrease the long-term performance of the structure and need clarification. For the ultimate behavior, better understanding could be attained with appropriate numerical models. For field projects, it was also stated that a post-event assessment should be conducted every time the upgraded structural system is subjected to severe load conditions. For masonry structures, the effectiveness of a one-side FRP laminate application should be further studied in comparison with the FRP reinforcement placed on both faces of a wall.

From a manufacturing standpoint, it is important to develop and test mechanical anchorages (specially for masonry systems), and, in general, to devote more attention to design/construction details. Emphasis should be placed on standard quality control procedures. Education of engineers and training of the workforce should be promoted. It is imperative to recognize the importance of concrete substrate preparation (i.e., sealing of cracks, cover replacement, surface roughening) and installation conditions (i.e., surface temperature, moisture and exposure, atmospheric conditions during application and curing) for a successful project. Engineering practitioners and contractors pointed out that a design-build approach would be a good way of marketing new technology, which can be very competitive when simple solutions are adopted. A life-cycle cost analysis could become a facilitator of acceptance given that material costs might still be high for widespread acceptance.

Materials, Durability and Characterization

The session was chaired by J. Lesko with M. Polese as the recorder. The following members contributed to the discussion: G. Balazs, A. Di Tommaso, A. Katz, R. Lopez-Anido, K. Neale, J. Teng, R. Tepfers, and P. Waldron.

Durability was the first aspect discussed during this break-up session. The debate underlined the need for a clear understanding of specific durability/performance mechanisms relevant to FRP systems such as fatigue, creep, moisture, chemistry (e.g., influence of the additives), high and low temperatures, fire, UV radiation, degradation, biological effects, and galvanic currents. In the case of strengthening or use as internal reinforcement for concrete, along with these issues, the interaction of FRP and concrete could affect the durability of each individual material and should be addressed. The analysis of such mechanisms could take advantage of standardization of test conditions and definition of accelerated test techniques with the objective of generating performance-based specifications.

In terms of models, some micro-mechanics and rate-kinetics models are available in the literature but seldom applied. There is too much emphasis on empirical curve-fitting type formulations. Calibrated models based on non-linear analysis should also be developed as they could describe the behavior of a bond glue-line or address size effects. Service limits should be defined and the role of low elastic modulus on the performance of FRP systems should be clarified. Serviceability performance should be further analyzed with particular focus on displacements and in-service residual stiffness. For the behavior at ultimate, design protocols should be based on toughness and an optimization of the reduction factors should be achieved. Recyclability and poor performance in the event of fire or high-temperature exposure (e.g., effects of low glass transition temperature) appear to be barriers to widespread implementation.

FRP materials' operational limits and the impact of design and construction detailing on durability need to be specified. From the economic standpoint, attention should be paid to a more efficient use of the FRP material itself. Finally, tools for carrying out life-cycle costing and demonstrating advantages of reduced maintenance should be developed.

Analysis, Design and Codes

The session was chaired by L. Taerwe with F. Ceroni as the recorder. The following members contributed to the discussion: E. Cosenza, G. Manfredi, S. Matthys, G. Monti, T. Mottram, M. Pecce, K. Pilakoutas, K. Tan, T. Triantafillou, J. Turvey, and T. Ueda.

The first part of the break-up session was devoted to the analysis and design of FRP reinforced or strengthened concrete members. The shear behavior of FRP RC elements is not well understood. In the case of members with EBR, computing the value of the effective strain of the laminate using the truss model analogy represents a controversial issue. For columns with EBR, the interaction between internal steel stirrups and FRP confinement, the cyclic behavior under seismic loads, the damage tolerance properties (e.g., residual state of stress), the combined bending, and the effectiveness of confinement for rectangular sections are aspects to be further studied. For RC members, serviceability aspects (i.e., crack width, crack spacing, stress limitation) should be analyzed. Generic models able to predict debonding failures should be developed. New research on FRP RC should focus on long-term behavior.

In the case of masonry structures strengthened with FRP, local and global behaviors need to be considered. Masonry applications present also bond and anchorage challenges in need of further investigation. In terms of analysis, it should be determined whether models for conventionally reinforced masonry are valid in the case of FRP reinforcement. Criteria for the selection of fiber type (i.e., glass or carbon) and laminate architecture (i.e., mono-, bi- or tri-dimensional) of FRP systems should be developed.

Geometrical and material imperfections are the main issues for the accurate analysis of structural FRP profiles. Optimization of FRP cross-sectional shapes (with a different approach in comparison to steel) and the improvement of their jointing/connection methods appear to be crucial for their development as an alternative to steel construction. A better understanding of local failures (i.e., concentrated loads, punching shear, instability) and lateral-torsional buckling problems should be achieved along with models for the connections.

For the design of structures reinforced or strengthened with FRP, the need for a unified internationally-accepted design philosophy was pointed out, even though the difficulty of overcoming national/regional preferences was recognized. A major effort is required to properly address design/construction detailing and model calibration should be based on many data sets. The safety factors for materials should be fixed, while different reduction factors should be defined depending on the applications. Participants agreed on the importance of tests after strengthening in order to check its efficiency and assess the improvement of structural performance.

The situation of different codes was finally examined; it was underlined that a harmonization of different guidelines should be pursued in order to achieve common agreement on stress limitations in concrete, steel, and FRP for strengthened elements.

Manufacturing, Construction, Economics and Marketing

The session was chaired by P. Kelley with A. Prota as the recorder. The following members contributed to the discussion: G. Fallis, S. Faza, W. Gold, A. Grandi, D. Gremel, and A. Nanni.

The first part of the break-up session was devoted to identifying the main opportunities for application of FRP for: a) strengthening of existing structures (externally bonded FRP laminates or near surface mounted (NSM) FRP rods): and b) new construction (internal FRP reinforcement and structural shapes/systems).

Many strengthening opportunities exist such as: replacement of steel reinforcement due to corrosion loss, NSM rods for crack control or correction of negative moment deficiency, pipeline upgrades, joint repair, and historical vaulted structures. Consensus was reached on the fact that strengthening of un-reinforced masonry (URM) for wind or seismic loads, upgrade of chimneys and silos, strengthening of opening in slabs, strengthening of beams/slabs for additional service loads and timber strengthening possibly represent the cases where FRP would find the best immediate use and marketability.

The potential uses of FRP rebars as internal reinforcement to concrete structures was discussed for pavement dowels, cast-in-place or precast marine structures, ornamental concrete, aggressive environments and grindable cut-off walls. Participants agreed that the most promising applications would be in bridge deck top reinforcement, soil nails, and retaining walls for marine construction. It was recognized that at this time and in the immediate future the use FRP shapes and systems presents fewer job opportunities; however, the applications with most potential are piling, bridge decking or stay-in-place formwork.

It was underlined that the marketplace is extremely receptive of simple ideas. Conversely, market opportunities and trends need to be communicated to researchers to challenge their ability and creativeness. Acceptance of FRP technology comes with increased liability for all parties involved in the construction process. Such liability should be equitably distributed among material producers, engineers, and contactors. A design-build approach may promote the alliances of manufacturers, designers, contractors and users, to demonstrate and capitalize on the benefits of new technology.

Conclusions and Recommendations

The workshop participants determined that, in addition to the legacy offered by these proceedings, there should be specific action items that the participants themselves would commit to carry out or at least initiate in the immediate future. For each action item, volunteers were identified. The list is shown below (A = Asia, E = Europe, NA = North America):

- Internet-based network: U. Naples (E) and U. Sheffield (E)
- Durability of internal reinforcement: Balazs (E), Lesko (NA), Ueda (A)
- Shear strengthening: Triantafillou (E), Nanni (NA), Fukuyama (A)
- Flexural strengthening: Matthys (E), Kelley (NA), Ueda (A)
- Bond anchorage: Monti and Pecce (E), Teng (A)
- Reliability coefficients: Monti (E), Neale (NA), Tan (A)
- Stress rupture: Taerwe (E), Kelley (NA)
- Blind prediction for experiments on strengthened slabs with openings: Kelley (NA), Nanni (NA), Neale (NA)

References

Nanni, A., "Coordinated Program for Research on Advanced Composites in Construction (RACC)," NSF, Washington, D.C., Sept. 1994, pp. 50.

Nanni, A., A. Di Tommaso, and J.J.R. Cheng, "International Research on Advanced Composites in Construction (IRACC-96)," NSF, Washington, D.C., Aug. 1996, pp. 43.

Acknowledgements

The authors' contribution is limited to a distillation of the thoughts and ideas generated during the workshop by its participants under the leadership of the session chairs assisted by their recorders. The authors would like to thank G. Balazs, P. Kelley, J. Lesko, T. Mottram, L. Taerwe and T.Triantafillou for reviewing the manuscript.

List of Participants

Dr. Gyorgy L. Balazs, Technical University of Budapest, Budapest, Hungary
balazs@vasbeton.vbt.bme.hu

Mr. Alberto Balsamo, University of Naples "Federico II", Naples, Italy
albalsam@unina.it

Mr. Luigi Coppola, MAPEI S.p.a. - Milan, Italy
seg_tec@mapei.it

Dr. Edoardo Cosenza, University of Naples Federico II, Naples, Italy
cosenza@unina.it

Dr. Angelo Di Tommaso, University of Venice, Venice, Italy
adt@iuav.it

Mr. Garth Fallis, Vector Construction Group, Winnipeg, Manitoba
garthf@vectorgroup.com

Dr. Salem Faza, Concrete Reinforcement Technologies, Jacksonville, FL, USA
salemfaza@hotmail.com

Mr. William Gold, Master Builders Technology, Cleveland, USA
will.gold@mbt.com

Mr. Alberto Grandi, SIKA ITALIA S.p.a., Milano, Italy
grandi.alberto@it.sika.com

Mr. Doug Gremel, Hughes Brothers Inc, Steward, NE, USA
doug@hughesbros.com

Mr. Fabio Guerrini, SIKA ITALIA S.p.a., Milano, Italy
guerrini.fabio@it.sika.com

Dr. Daniel W. Halpin, Purdue University, West Lafayette, IN, USA
halpin@ecn.purdue.edu

Dr. Amnon Katz, Technion – Israel Institute of Technology, Israel
akatz@tx.technion.ac.il

Mr. Paul L. Kelley, Simpson Gumpertz & Heger Inc., Arlington, MA, USA
plkelley@sgh.com

Dr. Jack John J. Lesko, Virginia Tech, Blacksburg, VA, USA

jlesko@vt.edu

Dr. Roberto Lopez-Anido, The University of Maine, Orono, ME, USA
RLA@umit.maine.edu

Dr. Gaetano Manfredi, University of Naples Federico II, Naples, Italy
gamanfre@unina.it

Dr. Stijn Matthys, Magnel Laboratory for Concrete Research, Gent, Belgium
Stijn.matthys@rug.ac.be

Dr. Giorgio Monti, University of Roma La Sapienza, Rome, Italy
giorgio.monti@uniroma1.it

Dr. J. Toby Mottram, University of Warwick, UK
jtm@eng.warwick.ac.uk

Dr. Antonio Nanni, University of Missouri-Rolla, Rolla, MO, USA
nanni@umr.edu

Dr. Kenneth W. Neale, Université de Sherbrooke, Québec, Canada
kenneth.neale@courrier.usherb.ca

Dr. Luigi Nicolais, Regional Government of Campania, Naples, Italy
ass.nicolais@regione.campania.it

Dr. Marisa Pecce, University of Sannio, Benevento, Italy
pecce@unina.it

Dr. Kypros Pilakoutas, The University of Sheffield, UK
k.pilakoutas@sheffield.ac.uk

Dr. Luc Taerwe, Magnel Laboratory for Concrete Research, Gent, Belgium
luc.taerwe@rug.ac.be

Dr. Kiang Hwee Tan, National University of Singapore, Republic of Singapore
cvetankh@nus.edu.sg

Dr. Jin-Guang Teng, The Hong Kong Polytechnic University Hong Kong
cejteng@polyu.edu.hk

Dr. Ralejs Tepfers, Chalmers University of Technology, Goteborg, Sweden
tepfers@bm.chalmers.se

Dr. Thanasis Triantafillou, University of Patras, Patras, Greece
ttriant@upatras.gr

Dr. Geoffrey J. Turvey, Lancaster University, UK
g.turvey@lancaster.ac.uk

Dr. Tamon Ueda, Hokkaido University, Sapporo, Japan
ueda@eng.hokudai.ac.jp

Mr. Pasquale Zaffaroni, MAPEI S.p.a. - Milan, Italy
seg_tec@mapei.it

Dr. Peter Waldron, The University of Sheffield, UK
p.Waldron@sheffield.ac.uk

Participants to the Workshop

List of reviewers of the papers included in the volume

Dr. Tarek Alkhrdaji, Structural Preservation Systems, talkhdaji@structural.net

Mr. Gregg Blaszak, BG International, gblaszak@bg-intl.net

Dr. Sergio F. Breña, University of Massachusetts-Amherst, brena@ecs.umass.edu

Mr. John P. Busel, Market Development Alliance, jbusel@mdacomposites.org

Dr. Francois Buyle-Bodin, University of Lille, Francois.Buyle-Bodin@eudil.fr

Dr. Omar Chaallal, Ecole de Tech Superieure, ochaallal@ctn.etsmtl.ca

Dr. Edoardo Cosenza, University of Naples, cosenza@unina.it

Dr. Saeed Daniali, University of Washington, sdaniali@u.washington.edu

Dr. Dat Duthinh, Nat. Inst. of Standards & Technology Building, dduthinh@nist.gov

Dr. Mamdouh El-Badry, The University of Calgary, melbadry@ucalgary.ca

Dr. Giovanni Fabbrocino, University of Naples, fabbroci@unina.it

Dr. Kent A. Harries, University of South Carolina, harries@engr.sc.edu

Mr. Nestore Galati, University of Missouri-Rolla, galati@umr.edu

Dr. Karl Gillette, President Edge Structural Composites, karl@edgest.com

Dr. Victor Giurgiutiu, University of South Carolina, giurgiut@engr.sc.edu

Mr. Maurizio Guadagnini, University of Sheffield, M.Guadagnini@sheffield.ac.uk

Dr. Pawan Gupta, Halsall Associates Ltd., pawanrgupta@hotmail.com

Mr. Pei Chang Huang, Co-Force Taiwan, coforce@ms57.hinet.net

Dr. Vistasp M. Karbhari, University of California, San Diego, vkarbhari@ucsd.edu

Dr. Howard Kliger, H.S. Kliger & Assoc., hskliger@voicenet.com

Dr. Jesus Larralde-Muro, California State University Fresno, jesuslm@csufresno.edu

Dr. L. Javier Malvar, NAVFAC Pavement Consultant, malvarlj@nfesc.navy.mil

Dr. Gaetano Manfredi, University of Naples, gamanfre@unina.it

Mr. O. S. Marshall, U.S. Army- CERL, Orange.S.Marshall@erdc.usace.army.mil

Mr. Francesco Micelli, University of Lecce, francesco.micelli@unile.it

Dr. Amir Mirmiran, University of Cincinnati, amir.mirmiran@uc.edu

Dr. Giorgio Monti, University of Rome La Sapienza, giorgio.monti@uniroma1.it

Dr. Sarah Mouring, United States Naval Accademy, mouring@usna.edu

Dr. Antonio Nanni, University of Missouri-Rolla, nanni@umr.edu

Mr. Sinaph Namboorimadathil, University of Missouri-Rolla, sinaph@umr.edu

Dr. Kenneth W. Neale, University of Sherbrooke, kenneth.neale@courrier.usherb.ca

Mr. Renato Parretti, University of Missouri-Rolla, parretti@umr.edu

Dr. Marisa Pecce, University of Sannio, pecce@unina.it

Dr. Michael F. Petrou, University of South Carolina, petrou@engr.sc.edu

Dr. José A. Pincheira, University of Wisconsin – Madison, jpin@engr.wisc.edu

Mr. Andrea Prota, University of Missouri-Rolla, prota@umr.edu

Dr. Guillermo Ramirez, University of Kansas, willy@ku.edu

Dr. Roberto Realfonzo, University of Naples, robrealf@unina.it

Dr. D. V. Reddy, Florida Atlantic University, dvreddy@oe.fau.edu

Dr. Bahram M. Shahrooz, University of Cincinnati, mojaddb@email.uc.edu

Ms. Danielle Stone, University of Missouri-Rolla, stone@umr.edu

Dr. Khaled Soudki, University of Waterloo, soudki@uwaterloo.ca

Dr. Kiang Hwee Tan, National University of Singapore, cvetankh@nus.edu.sg

Dr. Houssam Toutanji, University of Alabama, toutanji@ebs330.eb.uah.edu

Dr. Gustavo Tumialan, University of Missouri-Rolla, jaime@umr.edu

Ms. Stephanie L. Walkup, Wiss, Janney, Elstner Associates, Inc.,swalkup@wje.com

Mr. Xinbao Yang, University of Missouri-Rolla, xinbao@umr.edu

Subject Index

Page number refers to the first page of paper

Author Index

Page number refers to the first page of paper